MOLECULAR BIOLOGY
INTELLIGENCE
UNIT

Muscle Development in Drosophila

Helen Sink

Skirball Institute of Biomolecular Medicine
Molecular Neurobiology Program
and Department of Pharmacology
New York University School of Medicine
New York, New York, U.S.A.

LANDES BIOSCIENCE / EUREKAH.COM
GEORGETOWN, TEXAS
U.S.A.

SPRINGER SCIENCE+BUSINESS MEDIA
NEW YORK, NEW YORK
U.S.A.

MUSCLE DEVELOPMENT IN DROSOPHILA

Molecular Biology Intelligence Unit

Landes Bioscience / Eurekah.com
Springer Science+Business Media, Inc.

ISBN: 0-387-30053-8   Printed on acid-free paper.

Springer Science+Business Media, Inc., 233 Spring Street, New York, New York 10013, U.S.A.
http://www.springer.com

Please address all inquiries to the Publishers:
Landes Bioscience / Eurekah.com, 810 South Church Street, Georgetown, Texas 78626, U.S.A.
Phone: 512/ 863 7762; FAX: 512/ 863 0081
http://www.eurekah.com
http://www.landesbioscience.com

Printed in the United States of America.

9 8 7 6 5 4 3 2 1

Library of Congress Cataloging-in-Publication Data

Muscle development in drosophila / [edited by] Helen Sink.
     p. ; cm. -- (Molecular biology intelligence unit)
  Includes bibliographical references and index.
  ISBN 0-387-30053-8 (alk. paper)
   1. Drosophila--Physiology. 2. Muscles--Differentiation.   I. Sink,
Helen. II. Series: Molecular biology intelligence unit (Unnumbered)
   [DNLM: 1. Drosophila melanogaster--genetics. 2. Drosophila
melanogaster--growth & development. 3. Muscle Cells--physiology.
4. Muscle Development--physiology.   QX 505 M985 2006]
QL537.D76M87 2006
573.7'515774--dc22

                                       2005030264

CONTENTS

EDITOR

Helen Sink
Skirball Institute of Biomolecular Medicine
Molecular Neurobiology Program
and Department of Pharmacology
New York University School of Medicine
New York, New York, U.S.A.
Email: sink@saturn.med.nyu.edu
Chapter 1

CONTRIBUTORS

Susan M. Abmayr
The Stowers Institute
  for Medical Research
Kansas City, Missouri, U.S.A.
Email: sma@stowers-institute.org
Chapter 8

Mary Baylies
Program in Developmental Biology
Sloan Kettering Institute
Memorial Sloan Kettering Cancer Center
New York, New York, U.S.A.
Email: m-baylies@ski.mskcc.org
Chapter 7

Rolf Bodmer
The Burnham Institute
La Jolla, California, U.S.A.
Email: rolf@burnham.org
Chapter 4

Ana Carmena
Program in Developmental Biology
Sloan Kettering Institute
Memorial Sloan Kettering Cancer Center
New York, New York, U.S.A.
Chapter 7

Devkanya Dutta
National Centre for Biological Sciences
Tata Institute of Fundamental Research
Bangalore, India
Chapter 11

Manfred Frasch
Brookdale Department of Molecular,
  Cell and Developmental Biology
Mount Sinai School of Medicine
New York, New York, U.S.A.
Email: Manfred.Frasch@mssm.edu
Chapter 6

Eileen E.M. Furlong
Developmental Biology
  and Gene Expression Programmes
EMBL
Heidelberg, Germany
Email: Eileen.Furlong@embl.de
Chapter 13

Volker Hartenstein
Department of Molecular Cell
  and Developmental Biology
University of California Los Angeles
Los Angeles, California, U.S.A.
Email: volkerh@mcdb.ucla.edu
Chapter 2

Y. Tony Ip
Program in Molecular Medicine
University of Massachusetts
  Medical School
Worcester, Massachusetts, U.S.A.
Email: Tony.Ip@umassmed.edu
Chapter 3

Haig Keshishian
Molecular, Cellular and Developmental
  Biology Department
Yale University
New Haven, Connecticut, U.S.A.
Email: haig.keshishian@yale.edu
Chapter 10

Kiranmai S. Kocherlakota
Integrative Biosciences
  Graduate Program
The Huck Institutes of Life Sciences
The Pennsylvania State University
University Park, Pennsylvania, U.S.A.
Chapter 8

Hsiu-Hsiang Lee
Brookdale Center for Developmental
  and Molecular Biology
Mount Sinai School of Medicine
New York, New York, U.S.A.
Chapter 6

Louise Nicholson
Molecular, Cellular and Developmental
  Biology Department
Yale University
New Haven, Connecticutt, U.S.A.
Chapter 10

Michael V. Taylor
Cardiff School of Biosciences
Cardiff University
Park Place, Cardiff, U.K.
Email: TaylorMV@cf.ac.uk
Chapter 14

Mark Van Doren
Department of Biology
Johns Hopkins University
Baltimore, Maryland, U.S.A.
Email: vandoren@jhu.edu
Chapter 5

Jim O. Vigoreaux
Department of Biology
University of Vermont
Burlington, Vermont, U.S.A.
Email: jvigorea@uvm.edu
Chapter 12

K. VijayRaghavan
National Centre for Biological Sciences
Tata Institute of Fundamental Research
Bangalore, India
Email: vijay@ncbs.res.in
Chapter 11

Talila Volk
Department of Molecular Genetics
Weizmann Institute of Science
Rehovot, Israel
Email: lgvolk@wicc.weizmann.ac.il
Chapter 9

Noriko Wakabayashi-Ito
Program in Molecular Medicine
University of Massachusetts Medical
  School
Worcester, Massachusetts, U.S.A.
Chapter 3

Stephane Zaffran
Department of Developmental Biology
Institut Pasteur
URA 2578 CNRS
Paris, France
Email: zaffrans@pasteur.fr
Chapter 6

CHAPTER 1

An Introduction to Muscle Development in *Drosophila*

Helen Sink*

Abstract

Muscles have multiple roles in organisms. These roles range from facilitating conscious and unconscious movement, to maintaining posture, stabilizing joints, generating body heat, and moving substances within the body. To successfully carry out these functions, each muscle must develop as the right muscle type in the right location, achieve and maintain appropriate size, possess a correctly ordered contractile apparatus, and be capable of responding to contraction-inducing stimuli. A fundamental question is: How does a muscle develop so that it possesses the properties required for correct functioning? This book reviews how the fruit fly, *Drosophila melanogaster*, is being used to address this question.

Introduction

Researchers have been utilizing cell cultures, and invertebrate and vertebrate model organisms to determine how muscles develop. Each approach has distinct advantages and disadvantages for studying muscle development. In this book, we focus on the many discoveries coming from studies of muscle development in the fruit fly, *Drosophila melanogaster*. Given the tremendous progress that has been made through utilizing *Drosophila*, it arguably offers the greatest combination of advantages for the study of muscle development.

Drosophila has long been a favored model organism of biologists. It possesses features that are ideal for an experimental organism. It has a relatively short life span, small size, is readily available from Stock Centers and research laboratories, is relatively cheap to maintain in large numbers, and is anatomically simple while possessing an interesting range of quite complex behaviors. What has made *Drosophila* extremely appealing to researchers, however, is its amenability to genetic manipulation. *Drosophila* genetic studies traditionally involve generating DNA lesions to negate gene function, followed by analysis of the resultant phenotype (i.e., defect). Mutations in genes of interest are then maintained in vivo as stable mutant fly lines. This powerful approach enables researchers to characterize the endogenous role(s) of a protein encoded by a gene of interest. The approach has been enhanced and supplemented in recent years by new experimental approaches and resources (Chapter 13).

The Muscle Pattern of *Drosophila*

Drosophila is a holometabolous insect i.e., it undergoes metamorphosis (Fig. 1). At an ambient temperature of 25°C, the *Drosophila* life cycle takes 9 days. During the first 24 hours following fertilization, an embryo develops from a single cell. At the end of embryogenesis

*Helen Sink—Skirball Institute of Biomolecular Medicine, and Department of Pharmacology, New York University School of Medicine, 540 First Avenue, New York, New York 10016, U.S.A. Email: sink@saturn.med.nyu.edu

Muscle Development in Drosophila, edited by Helen Sink. ©2006 Eurekah.com and Springer Science+Business Media.

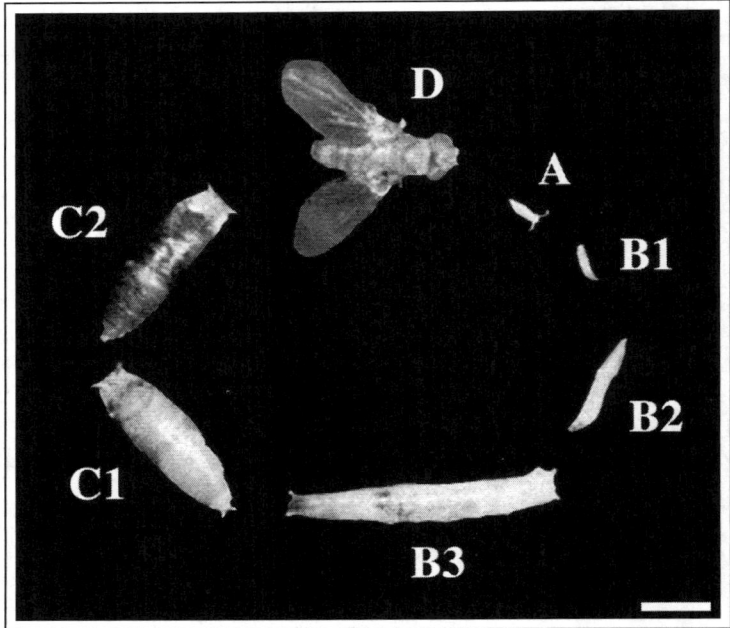

Figure 1. Photographic presentation of the life cycle of *Drosophila*. Stage in the cycle: A) embryo, B1) first instar larva; B2) second instar larva; B3) third instar larva; C1) early pupa; C2) late pupa, and D) adult. Modified from an image kindly provided by Mary Baylies.

hatching occurs, and the first instar larva is ready to live independently. The larva grows through two more larval instars over several days. At the end of the third instar, the larva pupates. In the puparium metamorphosis occurs, and much of the larval body tissue is destroyed. In its place new adult tissues form, predominantly from cells in the imaginal discs (i.e., clusters of cells set aside in the embryo and larva). At the end of metamorphosis, a new adult fly emerges (ecloses) from the puparium.

As a result of its holometabolous life cycle, *Drosophila* can virtually be considered as two distinct organisms. The larva is worm-like in appearance, and is predominantly a burrowing, crawling creature. In contrast, following metamorphosis, the adult fly coordinates three pairs of articulated legs and a pair of wings for walking and flight. Adult flies also engage in courting and mating behaviors, and in the case of the female, in egg laying. Not surprisingly the musculature of the larva and adult fly differ, reflecting the specialized needs at the two disparate life stages.

The four major musculature types in the *Drosophila* larva and adult are cardiac muscle, somatic gonad (also known as gonadal mesoderm), visceral muscle and somatic muscle. Each musculature type is associated with a specific pattern, function, and morphology in the embryo/larva and adult, and is discussed by Volker Hartenstein (Chapter 2).

Larval somatic muscles are responsible for mediating conscious and reflexive movements (e.g., crawling; recoiling from noxious stimuli) through distortion of the epidermis/cuticle and compression of the hydrostatic skeleton. The larval somatic muscles are single, syncytial (multinucleate) fibers, that possess characteristic sizes, shapes and sites of epidermal attachment. The adult fly has new and/or radically modified body parts including: a head structure with parts that differ from those of the larva, thoracic wings and articulated legs, and mature reproductive organs in the abdomen. The post-metamorphosis somatic musculature (Chapter 2) reflects the need to move and support these adult structures.

In *Drosophila* the visceral musculature is composed of two types of syncytial fibers—circular (binucleate) muscles and longitudinal (multinucleate) muscles. Visceral muscles coordinate the movement of nutrients and waste through the gastrointestinal system. They also serve to shape the gastrointestinal tract. While there has been considerable interest in the embryonic/larval visceral musculature (Chapter 6), the *Drosophila* adult visceral musculature has not received much attention. As a result, it is currently necessary to assume that its arrangement is similar to that of other Dipterans (Chapter 2). Similarly, while the embryonic somatic gonad is being actively studied (Chapter 5), less is known about the adult forms (Chapter 2).

The cardiac musculature (dorsal vessel) is responsible for aiding the movement of hemolymph—analogous to blood—in the organism (Chapter 4). Unlike the somatic and visceral musculature, but like the somatic gonad, the cardiac musculature is not syncytial. Rather, the cardiac myoblasts maintain their separate identities, existing in close apposition to form the dorsal vessel. In the embryo, the heart is basically a linear tube on the dorsal side, while in the adult it is similar with modest modifications (Chapter 2).

Mesoderm Formation in the *Drosophila* Embryo

So how do cardiac, visceral, somatic gonad and somatic musculature arise? In Chapter 3 Noriko Wakabayashi-Ito and Y. Tony Ip consider the earliest events in the formation of muscle—the progression of the *Drosophila* embryo from a single cell layer (blastula) to three cell layers (gastrula). The three cell layers of the gastrula are the external ectoderm layer and internal endoderm and mesoderm layers. It is the mesodermal cells that ultimately develops into the four types of musculature. The critical processes relating to muscle development during this period include (i) fating of blastoderm stage embryo ventral cells to be mesoderm in the gastrula and (ii) the migration of mesoderm cells into the embryo and their spread as a monolayer over the internal epidermis (Chapter 3).

Transition from a blastula to a gastrula requires the patterning of the blastula's cells to establish ventral, lateral and dorsal territories along dorsal-ventral axis. As expanded upon in Chapter 3, this dorsal-ventral axis formation arises from the sequential activation of molecular pathways in three developmental phases, commencing in the oocyte and continuing into the early embryo. The blastoderm stage embryo's ventral cells are fated to be mesoderm, and move into the interior of the embryo. This invagination and internalization of the cells during gastrulation forms the mesoderm layer in the gastrula interior.

Development of the Cardiac Musculature, Visceral Musculature, Somatic Gonad and Fat Bodies

Once the mesodermal cells have spread out over the internal epidermal surface of the gastrula, molecularly derived positional information (relative to the anterior-posterior and dorsal-ventral axes) determines their fate to a specific musculature type.

Rolf Bodmer (Chapter 4) discusses how the dorsal-most edge of the outer layer of the dorsal mesoderm becomes specified as the cardiac mesoderm, and gives rise to the cardiac musculature. The dorsal vessel consists of a linear tube composed of an inner row of myocardial cells and an outer row of pericardial cells. The anterior portion of the dorsal vessel is analogous to an aorta, while the posterior portion can be considered as the heart proper. The posterior heart has a series of inlet valves known as ostia, while a single valve separates the anterior aorta and posterior heart. Despite the seemingly simple anatomy of the cardiac musculature, there is considerable molecular complexity/diversity among the comprising cells.

Much of dorsal vessel development involves the critical transcription factor Tinman. Rolf Bodmer describes how the expression of *tinman* becomes more and more restricted over time - from the presumptive mesoderm in the blastula to the cardiac precursors in the embryo. The cardiac precursors in turn display combinatorial patterns of expression of other transcription factors, and divide to produce the cardiac cell population. The regulation of these cell divisions, some of which are symmetrical and others asymmetrical, is also considered in Chapter 4.

Finally, although the dorsal vessel is basically a linear tube, it does have morphologically distinct subdivisions i.e., the heart proper and the aorta. The role of the Hox genes in regionalizing the dorsal vessel is presented in the chapter.

Mark Van Doren (Chapter 5) provides an overview of the development of the somatic gonad (also known as gonadal mesoderm) and fat bodies. The somatic gonad cells associate intimately with the germ cells in the gonad. While their functions in the embryo and larva are unclear, in the adult they probably produce peristaltic contraction waves that aid oogenesis and spermatogenesis. The fat bodies function in the *Drosophila* larva as a site of fat storage and energy metabolism. In addition they are important for the innate immune response, steroidogenesis, and for protection in challenging environmental situations (e.g., mechanical impact, heat stress).

As discussed in Chapter 5, after gastrulation, positional information leaves cells destined to be the fat bodies and somatic gonad separated from one another across several parasegments. Consequently the fat body-fated mesodermal cells migrate, find and recognize one another to form the fat bodies during the next phases of development. Somatic gonad precursors must also find one another. In addition, they coordinate their movements and developmental program with those of the germ cells. Together they migrate along a predominantly shared route, and intermingle while remaining separate entities. At their destination they cease migratory behavior, recognize one another, and ultimately form and sustain the final compacted gonad structure. Interestingly, evidence of gonad sexual dimorphism is evident in the embryo. These differences are further compounded during metamorphosis, with the resultant gonad and associated musculature structurally correlating to the adult fly's gender.

The visceral musculature is responsible for forcing the passage of food and food-related waste through the larval gastro-intestinal tract via peristaltic contractions. The foregut and hindgut produce peristalsis with a single layer of circular muscles. In contrast midgut movements are controlled by the contractions of a circular muscle layer and an outer longitudinal muscle layer. In addition to mediating movement of the digestive system, the visceral muscles contribute to its shaping.

Hsiu-Hsiang Lee, Stephane Zaffran and Manfred Frasch (Chapter 6) discuss how the visceral mesoderm derives from the post-gastrulation dorsal mesoderm. The molecular pathways that specify the visceral mesoderm activate the expression of a crucial transcription factor, Bagpipe, in the visceral mesoderm precursors. The visceral muscle precursors are then subdivided into two different types of cells—founder cells and fusion-competent myoblasts. The process of subdividing the visceral precursors into founder cells or fusion-competent myoblasts requires communication between somatic muscle precursors in the ventro-lateral mesoderm and the adjacent visceral mesoderm precursors. The visceral founder cells and fusion-competent myoblasts then fuse with one another to form the syncytial visceral muscle fibers. Underscoring the fusion process are proteins first identified in studies of myoblast fusion in the somatic muscles (Chapters 6 and 8).

Development of the Somatic Musculature

Ana Carmena and Mary Baylies (Chapter 7) describe how positional information induces cells in the somatic muscle domain to strongly express the transcription factor Twist. Twist is the master regulator of somatic muscle fate—its absence results in a lack of somatic muscles, and its forced expression in other areas of mesoderm prompts somatic muscle-like development by those cells. Once specified as somatic mesoderm, there is a further subdivision of the cells into competence domains. Within each competence domain a subset of cells is selected as an equivalence group. Signaling among cells within an equivalence group results in one cell becoming fated as a progenitor. Each progenitor divides, giving rise to either a pair of embryonic founder cells or an embryonic founder cell and an adult muscle precursor (Chapters 7 and 11). The remaining cells in the equivalence group will be fusion-competent myoblasts. In the embryo the founder cells will serve to "seed" the muscle fibers, while the fusion-competent myoblasts contribute to generating muscle bulk by fusing with the founder cells (Chapter 8).

Chapters 2 and 7 show that the somatic musculature consists of an orderly array of muscle fibers. The morphology of the muscle fibers differ from one another in terms of size (i.e., number of nuclei in the fiber), position within muscle layers, position along the anterior-posterior and dorsal-ventral axes, and orientation and location of insertion sites. At a molecular level, a number of genes ("identity genes") have been discovered that are expressed in different muscle subsets. Such identity genes serve, in a potentially combinatorial manner, to determine the individual expression profile of a given muscle, and hence imbue it with its unique qualities. These issues are covered in Chapter 7.

The somatic muscles in *Drosophila* (and vertebrates) are unusual tissues. They are syncytia rather than a tight aggregate of individual cells as is seen in most other tissues. Susan Abmayr and Kiranmai Kocherlakota (Chapter 8) discuss how the two myoblast types in the *Drosophila* embryo—founder cells and fusion-competent myoblasts—accomplish fusion to form syncytial fibers.

Susan Abmayr and Kiranmai Kocherlakota consider how fusion occurs between the two different myoblast types rather than between a single myoblast type (Chapter 8). Briefly, the fusion-competent myoblasts respond to a founder cell-derived signal, and move towards the founder cell. The fusion-competent myoblast and founder cell then recognize each other as fusion partners. Once recognition occurs, the plasma membranes of the two cells become closely aligned at the site of contact, electron dense vesicles essential to the fusion process are recruited and retained directly beneath the plasma membranes at the contact sites, and the vesicles release their electron dense material at the plasma membranes. The plasma membranes then break down, with the excess membrane being removed. Ultimately the remaining ends of the cells' plasma membranes become one contiguous plasma membrane, and the cytoplasms unify into one cytoplasm, resulting in the syncytial muscle fiber. The molecular basis of these processes is gradually being uncovered, and is presented in the chapter.

In order to produce functionally meaningful contractions, the developing muscles must produce distortions of the epidermis and pressure on the hydrostatic skeleton. As such, muscle attachment to the epidermis is essential for somatic muscle function. In Chapter 9, Talila Volk describes how as the myotubes enlarge through fusion the fusion process, they actively seek and interact with epidermal cell partners to form extracellular matrix rich hemi-adherence junctions.

During development, the enlarging myotubes possess motile distal structures that "sample" epidermal cells—rejecting some while accepting others as their partners. The structure and exploratory behavior of the distal myotube ends is reminiscent of growth cones on growing nerves. Interestingly this extends beyond mere morphological similarities, as both types of leading edges structures use common molecular guidance mechanisms.

Explorations by the myotubes must cease when contacts with the target epidermal cells are made. Once contacted, bi-directional communication occurs. The outcome of this communication is that the epidermal attachment cells differentiate into specialized cells known as "tendon cells". The myotube ceases its exploration, and with the tendon cells form the hemi-adherence junctions. Now, when the muscle fiber responds to signals from the nerve, its contractile force can act upon the epidermis and hydrostatic skeleton and modify these body elements to produce a particular movement.

Motoneurons communicate with somatic muscles to elicit contraction. As the motoneuron's cell body is in the central nervous system, a considerable distance from the muscles, they have developed highly specialized extensions to enable communication with their distant muscle targets—the motor axons. Louise Nicholson and Haig Keshishian (Chapter 10) cover the four major phases of motor axon development: (i) directed outgrowth; (ii) muscle target selection, (iii) synapse formation and (iv) synaptic refinement. These four processes require spatially and temporally coordinated expression of molecular cues in the embryonic periphery (both nonmuscle and muscle surfaces), and appropriate temporal expression of receptors and response cascades in the motor axon. As evident in Chapter 10, major insights into the molecular mechanisms facilitating these motor axon behaviors have come from studies in *Drosophila*.

Louise Nicholson and Haig Keshishian present the molecular guidance cues known to regulate motor axon development (Chapter 10). They discuss how proteins that facilitate directed axon outgrowth are often reemployed for the matching of motor axons with their appropriate muscle targets. Once a muscle fiber is recognized as the correct target, the motor axon's growth cone ceases its motile, exploratory behaviors, and a synapse forms through bidirectional communication between the axon and muscle. The motor axon forms the presynaptic structure that will release a chemical signal to elicit muscle contraction. On the muscle side of the synapse, a post-synaptic specialization forms that is capable of receiving and responding to the motor axon's signal. Finally, contacts with nontarget muscles made during the motor axons exploratory phases are eliminated. Again, in addition to presynaptic electrical excitability, molecules used in axon guidance and target recognition are reemployed during the pruning process. Cumulatively, the sequential reemployment of some molecules, and the novel use of others, results in a highly stereotypic pattern of innervation at each muscle fiber (Chapter 10).

Metamorphosis and the Adult Musculature

Most of the body of the third instar larva, so carefully constructed during the embryonic and larval life stages, is destroyed during metamorphosis. In its place adult tissues are generated, often with new patterns and morphologies. Of the different muscle types, the changes are arguably most striking for the somatic muscles. Each adult muscle is a bundle of syncytial fibers, rather than the single syncytium seen in the embryo. Also the pattern of somatic muscle is greatly modified to meet the needs of the adult fly. This is particularly striking in the thorax where new appendages—wings and legs—demand underlying muscles to facilitate a spectrum of movement and energetic application. Devkanya Dutta and K. VijayRaghavan (Chapter 11) discuss the different cellular and molecular mechanisms that underpin these dramatic changes.

An interesting aspect in the Chapter 11 discussion comes through considering how somatic muscle generation during metamorphosis utilizes mechanisms similar and dissimilar to those employed earlier in the embryo. It is perhaps not surprising to encounter again in metamorphosis founder cells as "seeds" for some adult muscle fibers. Similarly one might have predicted that a select subset of epidermal cells would be specified as muscle attachment sites as they are in the embryo. There are, however, some surprises. Firstly, not all of the muscles are "seeded" by founder cells. Rather a muscle subset is "seeded" by larval muscles that avoid histolysis. Secondly, nerves can play a critical role in muscle formation. In the most extreme case, absence of innervation leads to failure of muscle formation. Finally, a transcription factor that serves to specify the identity of an embryonic muscle subset's founder cells is differently deployed during metamorphosis. Its function then is to determine whether fusion-competent myoblasts differentiate so that they contribute to the formation of a specific muscle subset. So while the study of muscle development in metamorphosis can replicate what is known about muscle development in the embryo, as also shown in Chapter 11, it also reveals new and unexpected twists.

The contractile apparatus in the adult fly muscles has been the focus of intense investigation; hence its molecular composition and organization are well documented. Jim Vigoreaux (Chapter 12) draws attention to the ordered interior of somatic muscle fibers. A somatic muscle is composed of myofibrils, with each myobfibril based on a repeated structural unit—the sarcomere. The sarcomere itself is composed of thin and think filaments (the myofilaments), and cross-linking molecular structures. The orderly assemblages of sarcomeres and myofibrils constitute the muscle contractile apparatus, which is responsible for translating the appropriate muscle stimulus into force via muscle contraction. This contraction occurs via a signal response cascade that culminates with thick filament myosin contacting thin filament actin with the swiveling action of the contacting myosin head pulling against the actin. The result is the sliding of the thick and thin filaments past each other, with the resultant shortening (contracting) the muscle fiber.

Yet how the underlying proteins assemble into the supramolecular sarcomeric units and the sarcomeric units into myofibrillar assemblies awaits clarification. As Jim Vigoreaux discusses, the proteins comprising the contractile apparatus may be responsible for their own assembly. He describes how a variety of genetic approaches in *Drosophila* have been brought to bear on elucidating this possibility, and points to the use of both traditional and evolving genetic and nongenetic approaches for continuing to unravel the molecular basis for sarcomere and myofibril assembly.

Approaches to Studying *Drosophila* Muscle Development

A significant part of *Drosophila's* appeal as an experimental organism has been the capacity to use genome-wide genetic approaches to identify developmentally essential genes. Traditionally this has meant employing mutagenesis screens. If a particular developmental process can be assayed for normality, a mutagenesis screen can be used to identify the genes that are important for the developmental process. Such a screen involves mutagenizing the flies, then screening their progeny for defects in the developmental process of interest. The genetic lesion that produces the phenotype is then identified. This approach requires no *a priori* assumptions or knowledge about the types of proteins that facilitate the process, and has proven an extremely powerful approach for identifying genes/proteins involved in a number of developmental processes. However other genetic approaches have been developed over the years, including gain-of-function screens, Dominant enhancer and Dominant suppressor screens, and enhancer trapping approaches.

With the sequencing of the *Drosophila* genome, and the development of new DNA, RNA and protein handling technologies, it is potentially possible to assay the developmental relevance of every gene and its protein product. In Chapter 13, Eileen Furlong presents the different "-ome" approaches that are available. They include: (i) transcriptome analysis using microarrays to analyze genome-wide changes in transcription levels, (ii) phenome mapping, involving perturbing gene function with double stranded RNA and examining resultant phenotypes; (iii) localizome mapping, which determines through in situ hybridization where the transcripts of each gene in the genome are expressed in vivo, and (iv) interactome mapping, which seeks to uncover molecular pathways by determining protein-protein interactions. Eileen Furlong covers how these approaches are being applied, and may be applied in the future, to elucidate the molecular genetic basis of muscle development.

Conclusion

Clearly much has been discovered, and will continue to be discovered, about muscle development by using *Drosophila* as a model organism. A critical question then becomes: how pertinent are the discoveries in *Drosophila* for understanding muscle development in higher organisms such as humans? In Chapter 14 Michael Taylor has undertaken a broad review of the literature associated with muscle development in both *Drosophila* and a range of vertebrate systems, and has drawn together comparisons that serve to highlight similarities and differences. His discussion provides an invaluable assessment of the immense value, and also distinct limitations, of using *Drosophila* to elucidate the cellular, molecular and genetic basis of muscle development.

Acknowledgements

Many thanks to Ron Landes for providing the opportunity to produce this book; Cynthia Conomos for dealing with the practicalities to make it a published reality, and the authors of the following chapters for giving so generously of their time, effort and knowledge. Finally, thanks to Monica Singh for insisting it was a good idea to do this book. It was.

CHAPTER 2

The Muscle Pattern of *Drosophila*

Volker Hartenstein*

Abstract

The musculature of insects is composed of an external, multilayered array of body wall muscles (somatic musculature), an internal layer of visceral muscles that surround the digestive tract and gonads, and specialized myo-epithelial tubes forming the vascular system. All three types of muscle are represented in *Drosophila* at the larval and adult stage. In this chapter, the somatic, visceral, gonadal and cardiac muscle pattern of *Drosophila* will be described. An attempt will be made to discuss possible homologies between the muscle pattern of *Drosophila*, which is a member of a highly derived clade of insects, and that of more primitive insects and other animal groups.

Somatic Musculature

Evolutionary Perspective: The Body Wall Musculature of Soft-Bodied Invertebrates

Soft-bodied invertebrates, exemplified by platyhelminths (nonsegmented worms; Fig. 1A-C) and annelids (segmented worms; Fig. 1D-F), have a body wall that includes an outer epidermal layer and three internal muscular layers.[1] Directly beneath the epidermis are the circular muscles, followed more internally by the diagonal muscles and longitudinal muscles. The somatic musculature of a juvenile flatworm (*Macrostomum sp*[2]) is shown in Figure 1C, and illustrates the orderly arrangement of the body wall musculature of a primitive invertebrate. Muscle fibers can be mononucleate (containing one nucleus) or multinucleate (containing several nuclei i.e., a muscle syncytium). As hardened skeletal elements are absent, the muscle fibers insert at the basal lamina, a dense layer of extracellular proteins secreted by epidermal and muscle cells. In nonsegmented worms, muscle contraction acts in conjunction with the noncompressible body fluid ("hydrostatic skeleton"). Contraction of longitudinal muscles shortens the body, while contraction of circular and diagonal muscle stretches the body. Coordinated contractions of a subset of muscles propel the animal forward by undulating movements of the body. Specialized patterns of movement related to flight, prey capturing, or mating can be effected by contractions of local subsets of the body wall musculature.

In annelids, the body wall musculature is organized in a metameric (that is, segmentally reiterated) pattern (Fig. 1D). Longitudinal and diagonal fibers insert at the transverse septa formed by the coelomata, which are segmentally arranged body cavities lined by mesodermally derived epithelia.[3,4] This muscle organization allows for a type of movement called peristalsis, which is well suited for burrowing and crawling through a soft substrate. Peristalsis consists of a wave of segmental contraction that passes over the body from front to end (forward locomotion) or end to front (backward locomotion). During peristalsis, the longitudinal musculature

*Volker Hartenstein—Department of Molecular Cell and Developmental Biology, University of California Los Angeles, Los Angeles, California 90095, U.S.A. Email: volkerh@mcdb.ucla.edu

Muscle Development in Drosophila, edited by Helen Sink. ©2006 Eurekah.com and Springer Science+Business Media.

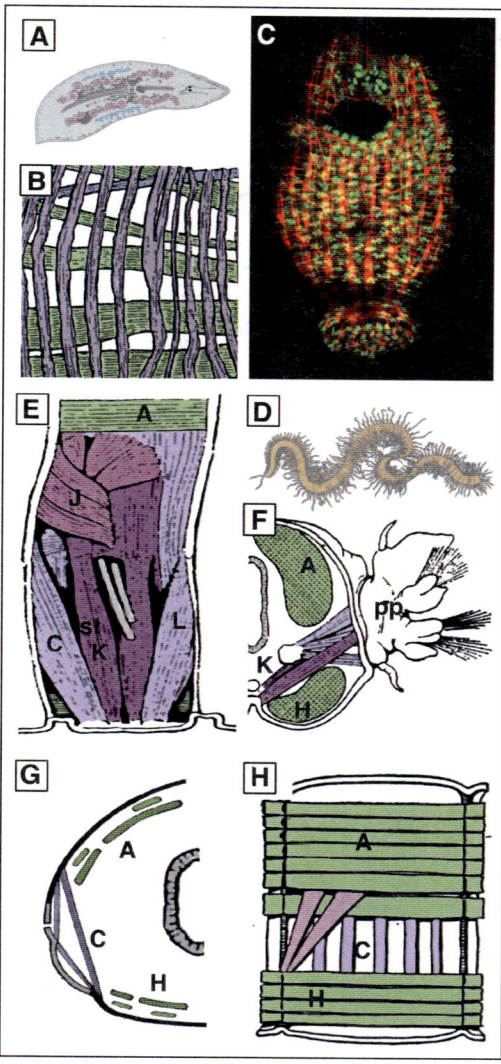

Figure 1. The muscle archetype of soft bodied invertebrates and arthropods. A) Platyhelminths (flatworms) are unsegmented worms that move by coordinated beating of epidermal cilia. B) Subepidermal plexus of somatic muscles in flatworm, comprising inner layer of longitudinal fibers (green), outer layer of circular fibers (purple), and sparse intermediate layer of oblique fibers (purple; after ref. 28). C) Orthogonal subepidermal musculature (labeled by phalloidin, red; epidermal nuclei are stained green by the nuclear marker Sytox) of the microscopic flatworm *Macrostomum sp.* (from ref. 2). D) Annelids are segmented worms bearing appendages (parapodia). E,F) Subepidermal somatic musculature of the marine annelid *Nereis* is formed by dorsal "A" and ventral "H" longitudinal muscles and lateral transverse muscles "C, J, K, L". Panel E shows a flattened hemisegment from the inside, panel F represents a cross section. Some lateral muscles insert at parapodia (pp) and setae (st; long bristles on parapodia; after ref. 4). G,H) Somatic muscle archetype in insect. G) Transverse section of abdominal hemisegment. H) View of flattened abdominal hemisegment from the inside. Similar to annelids, insects have a dorsal and ventral layer (both divided into an external and internal sublayer) of longitudinal muscles and lateral system of transverse fibers. Capital letters refer to a key of topologically and functionally defined groups of muscles created by Snodgrass[4] (see Figs. 3 and 4).

of a given segment (e.g., segment 1) contracts, thereby shortening and widening the segment, and providing a "grip" of that segment on the surrounding substrate. Subsequently, segment 2 contracts, and segment 1 extends (by relaxing the longitudinal muscles and contracting the circular muscles). At this moment, the "grip" of the body on the surrounding substrate is provided by segment 2, while segment 1 has extended forward, thereby pushing the front end of the worm ahead. Peristaltic movement plays an important role in locomotion of arthropods (see below).

The complete and symmetric layer of circular, diagonal and longitudinal muscle fibers is modified in many segmented worms. Typically, longitudinal fibers become restricted to the dorsal and ventral side of the body, whereas transverse and diagonal fibers predominate laterally (Fig. 2D,F). The same prototypical organization of body wall muscles is seen in arthropods (Fig. 1G,H; see section below). Where longitudinal and transverse fibers overlap, the ancient layering (circular=transverse superficial, longitudinal deep) is still maintained (Fig. 1G,H).

Segmental appendages, called parapodia, appear already in annelids as outgrowths of the lateral body wall (Fig. 2F). Parapodia play a role in locomotion, although the way they help propelling the body forward appears to be quite different than locomotion in arthropods.[3] This supports the idea that parapodia in annelids and "true" limbs in arthropods have evolved independently. Be that as it may, the way in which the body wall musculature becomes integrated into the appendage shows remarkable similarity between annelids and arthropods. A subset of transverse and diagonal muscles of the lateral body wall (called extrinsic limb muscles) insert at the base of the limb, thereby providing the means for forward and backward motion (promotion, remotion) and rotation of the limb. Additional lateral muscles come to lie inside the limb (intrinsic limb muscles). They function to shorten or lengthen the limbs (in soft-bodied worms) or to lift and depress the limbs (in articulated arthropod limbs; see section A3 and (Fig. 4) for further details).

The Somatic Musculature of the *Drosophila* Larva

Somatic muscles of insects show many similarities to the prototypical pattern of dorsal and ventral longitudinals, and lateral verticals, described above for annelids. This is particularly the case for the larval stages of many holometabolous insects where, as in annelids, the cuticle is soft. Dorsally and ventrally one typically finds an external and internal layer of longitudinal muscles[4] (Fig. 3G,H). Contraction of these muscles shorten the length of the segment uni- or bilaterally. A lateral array of vertical muscles is able to compress the segment in the dorsal-ventral axis. Lateral transverse muscles typically form a single, external layer; in addition, a few more internally located vertical muscles occur, usually at the segment boundary. Although strongly reduced in number, longitudinal and/or oblique lateral muscles also exist.

Pattern of Thoracic and Abdominal Larval Muscles

The above described general pattern of muscles is recognizable in the trunk of the *Drosophila* larva (Fig. 2A). To identify individual muscles, both a numbering system (1) and a lettering system reflecting topology[5,6] are in use. The dorsal muscles insert lateral of the dorsal vessel (i.e., the larval heart) and extend over approximately 30% of the circumference of the larva. They form an external layer (dorsal oblique or DO muscles, running from antero-dorsally to postero-ventrally) and an internal layer (dorsal acute or DA muscles, extending from antero-ventrally to postero-dorsally). The number of fibers varies slightly in the different segments. There are four DO fibers in each thoracic segment, and five in abdominal segments A2-A7. All DO fibers span the entire length of the segment, inserting at intersegmental apodemes. Apodemes are internally projecting ridges of the body wall, formed by specialized epidermal cells called apodeme cells or tendon cells. The main, intersegmental apodeme forms the boundary between neighboring segments. In addition, several shorter apodemes exist within segments. Like DO muscles, abdominal DA fibers (three in each segment) also insert at the intersegmental apodemes. Thoracic segments T2 and T3 possess four DA fibers; three are full length (reaching the intersegmental apodemes), and a fourth, shorter muscle (DA1) is located further dorsally.

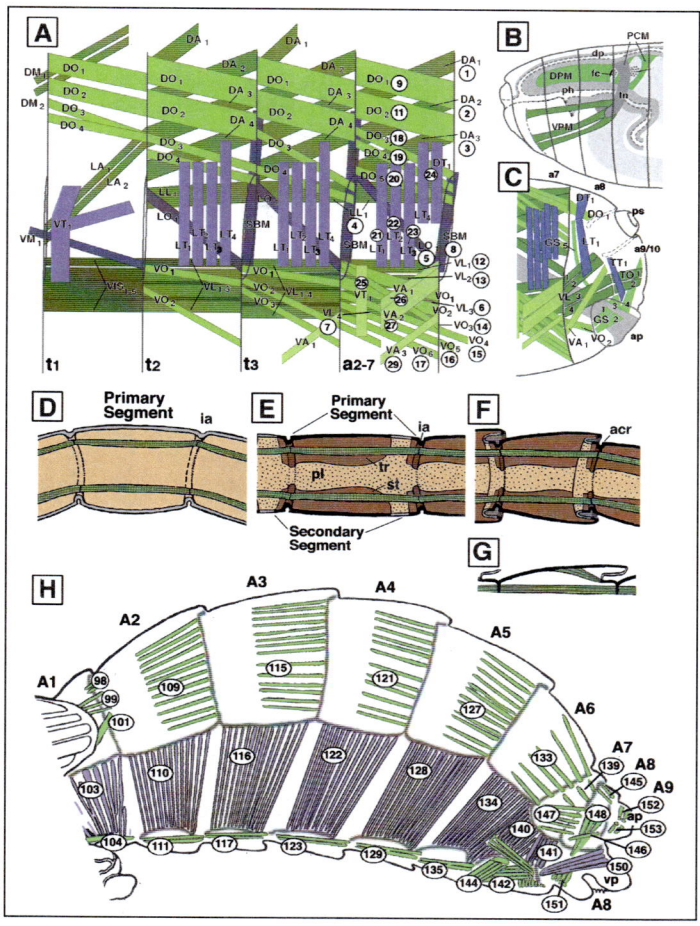

Figure 2. Somatic musculature of *Drosophila*. A-C) Larval musculature (after ref. 6). A) Schematic diagram of musculature of thoracic (t1-t3) and abdominal (a2-7) trunk segments. This external surface view of flattened segments showing ventral and dorsal longitudinals and lateral transversals. Letters reflect the muscle nomenclature used in 5 and 6 (for abbreviations of muscle groups, see text), numbers refer to nomenclature of Anderson et al.[29] Two of the abdominal muscles, referred to by #10 and #28 in Anderson et al[29] could not be identified on the map of Bate (2). B) Musculature of the larval head (derived from gnathal segments and clypeolabrum). C) Musculature of the tail segments (a8-10). D-F) Primary and secondary segmentation (after ref. 4). D) Longitudinal muscles of the soft-bodied larva insert at intersegmental apodemes (ia; primary segmentation). E,F) Cuticle of adult differentiates into sclerotized plates (dorsal tergites, tr; ventral sternites, st) surrounded by soft pleura (pl). Tergites and sternites incorporate parts of two neighboring segments (secondary segmentation). The intersegmental apodeme (ia) that forms the segment boundary in the larva becomes the antecostal ridge (acr) of the adult body wall. G) Schematic length section of abdominal body wall, depicting typical double-layered arrangement of longitudinal muscles (short externals, long internals). H) Muscle pattern of the *Drosophila* female adult abdomen (after ref. 12). Numbers refer to muscle groups defined by Miller.[12] Segments A1 – A7 each possess a group of short dorsal longitudinals (98, 99, 101, 109, 115, 121, 127, 133, 139), of ventral longitudinals (104, 111, 117, 123, 129, 135, 142), and of lateral transversals (tergosternals; 103, 110, 116, 122, 128, 134, 140). Musculature of segments A7 and A8 are reduced: dorsal muscles to anal plate (145, 146); dorsal muscle to uterus (147); dorsal muscle to gonopod (148); ventral muscle to uterus (149); lateral and ventral muscle to gonopod (150, 151); dorsal and ventral rectal muscles (152, 153). Other abbreviations: ap, anal plate; dp, dorsal pouch; DPM, dorsal pharyngeal musculature; fc, frontal commissure; ph, pharynx; ps, posterior spiracle; tn, tentorium; PCM, pericommissural muscles; vp, vaginal plate; VPM, ventral pharzngeal musculature.

DA muscles are absent in T1. Instead, one finds two long, intersegmental muscles (DM1 and DM2) that reach from the gnathocephalon to the apodeme between T1 and T2 (Fig. 2A).

The lateral musculature of the *Drosophila* larva comprises an external set of five (A1-7) or four (T2-3) transverse muscles in the center of each segment, and a single internal transverse muscle, called segment border muscle (SBM) that extends along the intersegmental apodeme (Fig. 2A). The lateral transverse musculature is modified and reduced in T1 which carries a single fiber (VT1). In addition to the transverse muscles, segments T2 to A7 each possess on longitudinal lateral (LL1) and one oblique lateral (LO1) fiber that run further internally than the transverse fibers. These muscles are absent from T1, which has instead two lateral acute muscles (LA1/2).

Ventral muscles are arranged in two layers (thorax) or three layers (abdomen), respectively. The innermost layer is formed by the ventral longitudinals (VL) which comprise four fibers in each abdominal segment and T3, and three fibers in T1 and T2. These VL muscles insert at the intersegmental apodemes. In addition, one finds two longer VL fibers in the thorax, one (VIS1) stretching from anterior T1 to posterior T2, and one (VIS5) from anterior T1 to posterior T3. Ventral oblique muscles (VO) are found in T2 to A7 where they form a layer outside the ventral longitudinals. Two VO fibers exist in T2, three in T3, and six in A1-7. VOs are absent from T1. Whereas the VOs of T2/T3 and most VOs of the abdomen insert at intersegmental apodemes, two abdominal VOs (VO4 and 5) are intersegmental and use apodemes located within the segment (anterior and posterior intrasegmental apodeme; (Fig. 2A). A third layer of ventral muscles is found in A1-7 and T3. These are the ventral acute muscles (VA) of which there exist three in the abdominal segments and one in T3. The VAs are short, spanning from the posterior intrasegmental apodeme to the intersegmental apodeme.

Pattern of Muscles in the Larval Head and Tail

The pattern of somatic muscles of the larval head and tail is highly reduced compared to the thoracic and abdominal pattern outlined above. Comparative embryological studies indicate that the insect head is formed by five or six modified segments, tipped by the acron, a nonsegmented region that lacks mesoderm. The three segments anteriorly adjacent to the thorax form the mouthparts (gnathal segments) and are most similar to the segments of the trunk. Further anterior, there are at least two additional segments (intercalary and antennal segments), as well as the labrum which is viewed differently as either a sixth segment or as an appendage of the intercalary segment (for discussion of insect head segmentation, see refs. 7-9). In primitive insects, each of these head segments produces a set of muscles, which at least in part can be recognized as serial homologs of the muscles of the trunk segments. However, *Drosophila* larvae are acephalic, since the segments of the head become reduced in size through apoptotic cell death, and are then folded into the interior of the larvae (head involution) where they form the pharynx and dorsal pouch. The musculature also gets reduced to a large extent. Of the muscles of the three gnathal segments only three fibers remain on each side, which form the ventral pharyngeal musculature[6] (Fig. 2B). Anteriorly, these muscles insert at an apodeme formed at the boundary between T1 and the labial segment. Posteriorly, they attach to the tentorium, a crescent shaped apodeme stretching out between the pharynx and dorsal pouch[10] (Fig. 2B). Contraction of the ventral pharyngeal muscles pulls the tentorium forward, thereby reducing the pharynx in length and causing an increase in pressure in the pharynx. The opposite effect is caused by contraction of the dorsal pharyngeal musculature, which is formed by an array of thick, short muscle fibers that upon contraction pull the pharynx roof upward, thereby causing a vacuum that sucks food inside the mouth opening. The dorsal pharyngeal musculature derives from the labrum, and can be homologized with the "dilator of the cibarium" and "dilator of the pharynx" muscles found in adult insects[4] (see Fig. 4A).

Similar to those of the head, the muscle pattern found in the larval tail which comprises segments A8, A9 and A10 is strongly modified from the canonical trunk pattern (Fig. 2C). Tail muscles are associated with the anus (e.g., the GS fibers of A9) and the posterior spiracle (e.g., the dorsal obliques and lateral longitudinals which insert at the base of the spiracle and act as retractors of this structure).

The Somatic Musculature of the Adult Fly

The musculature of the larva is almost completely histolized during metamorphosis. At the same time, a new set of adult muscles is formed from a pool of myoblasts which had remained undifferentiated during embryonic and larval life.[5,11] The adult muscle pattern of the abdominal segments is similar to the larval pattern. Typically, one finds an internal and external layer of dorsal and ventral longitudinal muscles, and sets of lateral transverse muscles (Figs. 1G,H and 2G). By contrast, the segments (including the resident muscles) of the thorax and the newly formed head are highly modified in order to function in locomotion, feeding, mating and sensory reception.

The body wall of the adult fly, similar to that of most other arthropods, is covered by sclerotized cuticle, which has important consequences for the function and patterning of the muscles. In the soft-bodied larva, longitudinal muscles insert at intersegmental apodemes (Fig. 2D). Unhindered by the cuticle, contraction of these muscles causes the segment to shorten. The situation is changed in the adult. Here, large parts of the cuticle are hardened, forming metamerically reiterated sclerites. Sclerites of neighboring segments are separated by rings of soft membraneous cuticle domains. Contraction of longitudinal muscles, inserting at neighboring sclerites, will retract the posterior segment into the anterior one, comparable to the sliding of the cylinders making up a telescope (Fig. 2E,F). Domains of soft cuticle, called pleura, also divide each sclerite into a dorsal part (tergite) and ventral part (sternite). As a result, lateral transverse muscles can compress segments in the dorsal-ventral axis by pulling tergites and sternites towards each other.

The metameric boundaries imposed upon the adult abdomen by the alternating sclerites and membraneous cuticle domain in between deserves special mentioning, because they do not coincide with the intersegmental boundaries of the larval body wall. As illustrated in Figure 2D,E, the larval (primary) intersegmental apodeme, along with a narrow strip of cuticle right in from of the apodeme, become incorporated into a sclerite. Thus, adult sclerites relate to larval segments like parasegments to segments in the embryo! This peculiar phase shifted adult segmentation is found in all homometabolous insects and has been called "secondary segmentation" in the classical literature.[4] Within the sclerite, the former intersegmental apodeme forms a visible ridge called the antecostal ridge.

Pattern of Abdominal Muscles of the Adult Fly

Compared to the abdominal musculature of the larva (and most insects in general), the abdominal musculature of the adult fly is reduced. Dorsally one finds a single layer of short longitudinal muscles, varying in number between 15 and 25 per segment (Fig. 2H). With their anterior end these muscles insert in the middle of a tergite, with their posterior end at the antecostal ridge of the posteriorly adjacent segment. Ventral longitudinal muscles are reduced to two fibers that span the entire length of the segment. Lateral transverse muscles form a dense array of 12-15 thin, parallel fibers that insert at the tergites (dorsally) and sternites (ventrally), respectively.

In newly eclosed adults there exists a small set of temporary muscles that are actually remnants of the larval musculature. In each segment, these temporary muscles include the oblique dorsal muscle (a member of the group of dorsal acute muscles of the larva), and the internal lateral muscle (the former larval segment border muscle).[12]

Muscles of the terminal abdominal segments form a modified pattern adapted to movement of the external genitalia. In females, the abdomen ends with two reduced segments, A8 (carrying the opening of the gonads called vulva) and A9 (anal plate, flanking the anus). In A8, muscles corresponding to dorsal and ventral longitudinals and lateral transversals are still

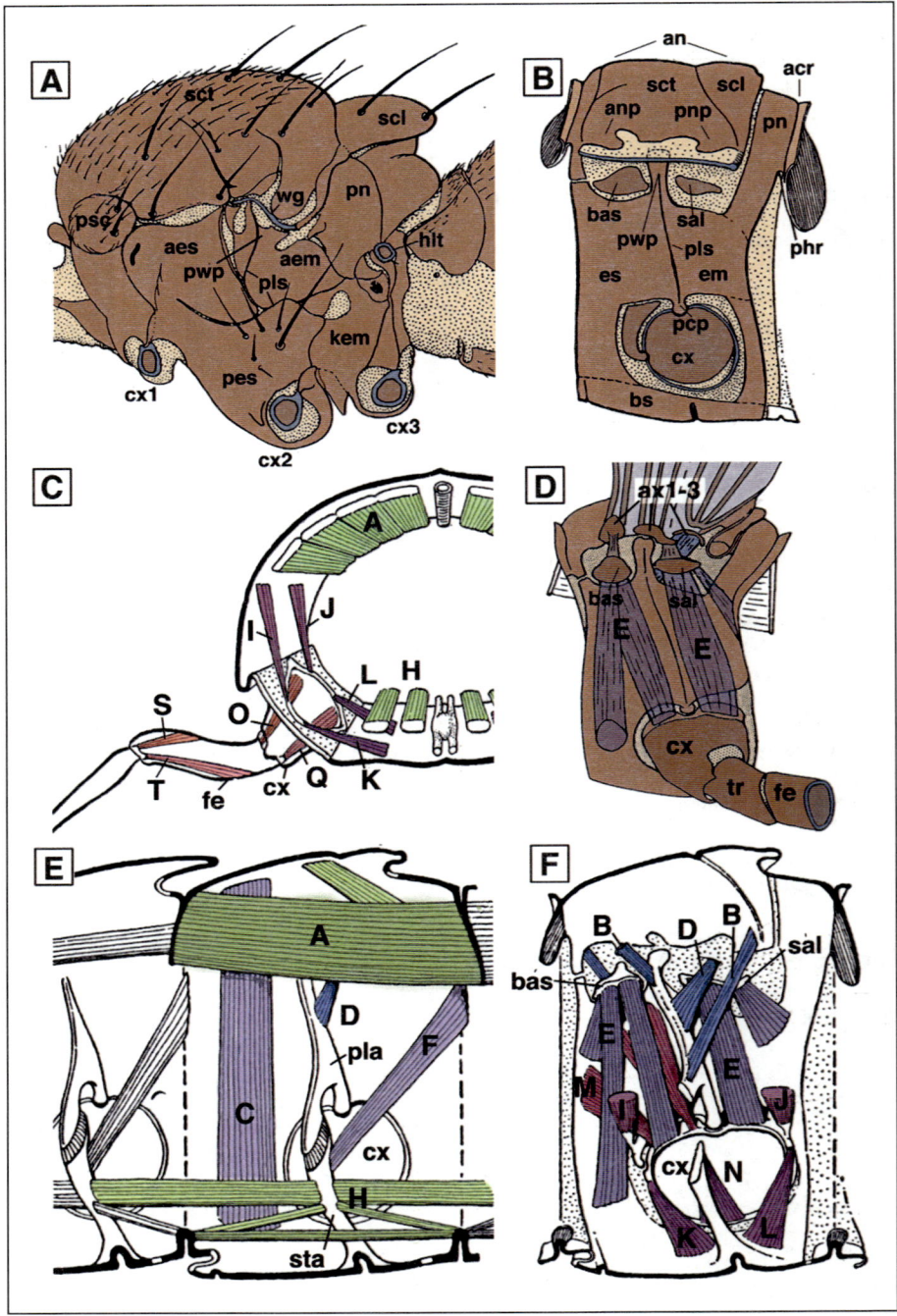

Figure 3. Please see figure legend on the next page.

Figure 3, shown on previous page. Somatic musculature of the *Drosophila* adult's thoracic segments. A,B) External morphology of wing bearing segments in the *Drosophila* adult (A; after Ferris[30]) and generalized insect (B; after ref. 4). Lateral view, anterior to the left. Sclerotized cuticle shaded dark brown, soft cuticle light brown. Attachment sites of appendages (wg wing; hlt haltere; cx coxa of leg) are colored blue. In *Drosophila* and other Dipterans, the wing bearing segment (mesothorax) is increased in size to accommodate the massive flight muscles (see panel D of Fig. 4). The sclerotized dorsal cuticle (alinotum, an, and postnotum, pn) and lateral cuticle (episternum, es, and epimeron, em) of the fly mesothoracic segment occupies much of surface of the thorax. C) Diagrammatic transverse section of thoracic segment of generalized insect, showing arrangement of leg muscles (after ref. 4). Some of the lateral transverse muscles (purple), inserting at strategic positions at the base of the coxa (cx), have aquired the function to promote "I, K", remote "J, L", abduct "I, J" and adduct "K, L" the leg. Intrinsic muscles of the leg segments act as flexors "O, S" or extensors "Q, T" of the distally adjacent segment. D) External view of thoracic segment of generalized insect, illustrating articulation of the wing and leg. Embedded in the base of the wing are three axillary sclerites (ax1-3). The middle one of these articulates with the pleural wing process (pwp). Embedded into the soft cuticle right underneath the wing base are two sclerites called basalar (bas) and subalar (sal). Muscles pulling the wing downward (depressors) insert at the axillary sclerites (axillary muscles, not shown in D) or at the basalar and subalar (basalar and subalar muscles, Snodgrass' group E). E,F) Diagrammatic interior views of thoracic segment of generalized insect illustrating pattern of muscles (anterior to the left). Panel E shows muscles located closer to the sagittal plane; in F, these muscles are omitted, facilitating the view onto the complex pattern of transverse muscles located more laterally, that is, closer to the body wall (after ref. 4). Capital letters refer to topologically and functionally defined muscle classes defined by Snodgrass. A complete key to these muscle classes is given in Figure 4. Other abbreviations: acr, acrotergite; aem, anepimeron (T2); aes, anepisternum (T2); an, alinotum; anp, anterior notal wing process; ax1-3, axillary sclerites; bas, basalar sclerite; bs, basisternum; cx, cut surface of coxa; es, episternum; em, epimeron; fe, femur; hlt, cut surface of haltere; kem, katepimeron (T2); pcp, pleural coxal process; phr, phragma; psc, prescutum (T2); pes, preepisternum (T2); pla, pleural epiphysis; pls, pleural suture; pn, postnotum; pnp, posterior notal wing process; pwp, pleural wing process; sal, subalar sclerite; scl, scutellum; sct, scutum; sta, sternal epiphysis; tr, trochanter; wg, cut surface of wing.

recognizable. A specialized dorsal and ventral muscle (147, 150 in Fig. 2H) projects interiorly to insert at the uterus; the transverse and ventral muscles of A8 (148, 150, 151) move the gonopods, modified sternites flanking the vulva. The musculature of A9 is reduced to two sets of short transverse fibers, the dorsal and ventral rectal muscles.

In male flies, abdominal segments A6 and A7 are strongly reduced. A8 and A9 are modified/fused and carry the genital apparatus, in particular the protrusible penis (aedagus) flanked by the claspers. These structures are specializations of the sternite of A9. A10 is strongly reduced and flanks the anus. The musculature of the male terminal segments has not been mapped in any detail.

Characteristics of the Thoracic Body Wall Muscles

The thoracic segments of insects bear the appendages involved in locomotion—legs and wings—which is strongly reflected in the muscle pattern. Legs are formed as outgrowths of the ventro-lateral thoracic segments (Fig. 3A-D). The sclerotized body wall ventral of the leg forms the basisternum; dorsal of the leg, at a level where soft pleural cuticle prevailed in abdominal segments, the cuticle is sclerotized as well and forms the episternum. The episternal region of each thoracic segment is further modified by specialized ridges and processes which differ considerably in various insect groups. Figure 3 semi-schematically compares the external anatomy of the thorax of a generalized winged insect (3B) with that one in *Drosophila* (3A). Worth special mention is the pleural suture, which extends vertically in the middle of the episternal region between the base of the leg and the wing. The suture demarcates an invagination of the epidermis that continues internally, into the body cavity, where it forms a ridge, called pleural apophysis, at which numerous muscles of the wing and leg insert (Fig. 3E,F). A similar muscle attachment site, the sternal apophysis, is formed in the center of the sternum, ventral of the legs.

The tergum of the wing-bearing thoracic segments underwent an increase in size and complexity as well (Fig. 3A,B). The antecostal sutures, representing the attachment sites for the dorsal longitudinal muscles which in the thorax have become a major component of the indirect flight muscles, are strongly enlarged. They are called phragmata (sing: phragma; Fig. 3B). Likewise, the acrotergites (the sclerotized posterior domain of a dorsal segment) are enlarged. In a wing-bearing segment, the enlarged acrotergite is called postnotum, notum being the commonly used name for the tergum of a thoracic segment. The tergite itself (called alinotum) is also enlarged in size and vaulted upward to receive the volume of the indirect flight muscles.

Articulation of Legs and Wings with the Body Wall

The proximal segment of the fly leg, called coxa, is joined to the epi/basisternum via a cuff of soft cuticle. Stability is provided by a joint, the pleural articulation, where a thin bridge of sclerotized cuticle (pleuro-coxal process) directly joins the coxa to the episternum (Fig. 3A,B). As a result of this articulation, the coxa is relatively free to move in all directions (similar to our shoulder joint). The type of articulation can be different in other insect groups, such that coxal movement is more restricted. The more distal segments of the leg (trochanter/femur, tibia, and tarsus) are articulated in such a way that movement is restricted to one axis (similar to our knee joint).

The wing base is connected to the pleural membrane by soft cuticle. Articulation occurs at three sclerotized points represented by the three axillary sclerites of the wing base, and the anterior/posterior notal processes and pleural wing process of the body (Fig. 3B,D). This type of articulation allows for limited movement of the wing in all axes. Two small sclerites, called basalar and subalar sclerites, are embedded in the pleural membrane ventrally adjacent to where the wing articulates (Fig. 3B,D,F). Muscles depressing the wing either insert at the axillary sclerites of the wing (direct wing depressors), or at the basalar/subalar sclerites (indirect wing depressors; see below).

To describe the types of movements and the way they are effected by muscles the following terms are used[13,14] (Fig. 4C): Protraction, retraction, levation and depression describe a movement of the entire limb. In protraction and retraction, the limb is pulled forward and backward, respectively. In levation and depression, the limb is raised and lowered, respectively. Promotion, remotion, abduction, adduction and rotation refer to movements of the coxa. Promotion and remotion move the coxa forward and backward (leading to protraction/retraction of the whole limb); abduction and adduction raises and lowers the coxa (causing levation and depression of the limb); rotation rolls the coxa along its length axis. Forward rotation is also referred to as pronation; backward rotation supination. Finally, when referring to the simple movement of the distal leg segments, one refers to extension (the angle between the segments is increased) and flexion (the angle is decreased). Movement of the wings is described in similar terms. Raising and lowering the wings is referred to as levation and depression; back- and forward movements are called adduction and abduction. Rotation turns the wing along its length axis.

General Bauplan of the Thoracic Musculature

Snodgrass[4] published a survey of the archetypical pattern of thoracic muscles he was able to identify in all winged insects. This pattern builds upon the simpler pattern, found in nonappendage-carrying segments, that consists of dorsal and ventral longitudinal/oblique and lateral transverse muscles (Fig. 1G,H). In the wing-bearing segments of the thorax, the dorsal musculature (group A according to Snodgrass; (Fig. 3E) is increased in mass. Contraction of these muscles will shorten the notum, which causes wing depression. The ventral longitudinal musculature (group H) is not changed notably from the archetype. The lateral transverse muscles have experienced the most diversification in adaptation to locomotion. The group that comes closest in shape and position to the original lateral transversals is Snodgrass' group E, called epipleural muscles (Fig. 3F). They stretch from the base of the wing (basalar and subalar sclerite) to the base of the leg and variably are used in different insect groups for flight or walking.

Beside the epipleurals, there is a group of transverse muscles that is longer and whose insertion has correspondingly shifted to a more dorsal and ventral level: the tergo-sternal muscles (group C of Snodgrass; Fig. 3E). They act as the second major group of indirect flight muscles, by depressing the thorax and thereby elevating the wing. Finally, the remainder of the thoracic transverse musculature breaks down into nine groups of short fibers grouped strategically around the appendages, and acting to move these appendages. Dorsally one finds the tergo-pleural muscles (group B) and axillary muscles (group D), inserting anterior and posterior to the wing base, and causing the wing to extend and flex, respectively. These muscles also rotate the wing. Seven groups of short transverse fibers move the leg, forming the external leg musculature (Fig. 3C,F). Muscle group I, coming from dorsal (episternum or tergum) and attaching at the anterior side of the coxa (that is, anterior of the pleural articulation) serve as promotors/abductors of the coxa. There is a corresponding group J that inserts at the posterior coxa and acts as the remotors/abductors. Muscle group K and L, originating ventrally of the leg at the anterior and posterior basisternum serve as promoters/ adductors and remotors/adductors, respectively, and also rotate the coxa. Group M (pleurocoxal muscles) extend between the episternum and the dorsal base of the coxa, and abduct the coxa. Group N are the corresponding sterno-coxal adductors. Group P consists in the extracoxal depressors of the trochanter, long vertical fibers that with their proximal end insert at the tergum or more ventral levels of the body wall, and with their distal end at the trochanter.

Intrinsic leg muscles are located inside the coxa, trochanter/femur, and tibia. They are arranged as antagonistic sets, acting to flex or extend the leg at a given joint. Flexion at the coxa-trochanter joint results in the elevation of the leg, which is why the muscles, inserting dorsally in the coxa and trochanter, are called leg elevators (Fig. 3C, group O). Ventral coxal muscles (group Q) act as leg depressors. In most insects, including *Drosophila*, trochanter and femur are fused without articulation into one unit. A single muscle extends along the length of the trochanter into the femur. Contraction of this muscle, called reductor of the femur (group R; Fig. 4B), shortens the trochanter and pulls the base of the femur towards the coxa.

Musculature of the *Drosophila* Legs and Wings

Extrinsic and intrinsic leg muscles with the properties of Snodgrass' groups I – V can be found in *Drosophila* (Fig. 4B,D-F). The indirect flight musculature consists on each side of six median dorsal longitudinals (Snodgrass' group mA; 45a-f) (which act as wing depressors). The indirect wing levators are represented by two dorsolateral obliques (group lA; 46a,b), three long tergo-sternal fibers (group C; 47a-c) and two muscles which were homologized to the tergal remotors of the leg (group J; 48a,b), but which also serve as indirect wing levators.

The direct flight musculature, mainly concerned with extending, flexing and rotating the wing, is recruited from Snodgrass' groups B, D, and E. Two tergo-pleural muscles (group B; 49, 50), called prealar muscles in *Drosophila*, extend from the notum anterior to the wing to the basalar, a small sclerite located in the membraneous cuff that connects the wing base to the notum. Contraction of these muscles extends the wing. Two epipleural muscles (group E; 51, 52; called basalars in *Drosophila*) extend the wing and pull the anterior wing margin down, thereby pronating the wing. The opposite function, i.e., flexion and supination of the wing, is carried out by several axillary muscles (group D; 54, 55) which extend from the episternum posterior and ventral to the wing to the axillary sclerites of the wing base.

Musculature of the Thorax Other Than Wing and Leg Muscles

In the prothorax the muscles are reduced to eight short cervical fibers, likely to be homologues of the dorsal and ventral longitudinal muscles (groups A and H; 20-27), which move the head. The metathorax. beside the massive muscles involved in flight and walking, has a ventral longitudinal muscle (representing group H; 81) and a small transverse muscle that acts as the occlusor of the posterior spiracle. In the metathorax, representatives of the dorsal longitudinal muscles are completely absent. Two small lateral muscles move the haltere (77, 78).

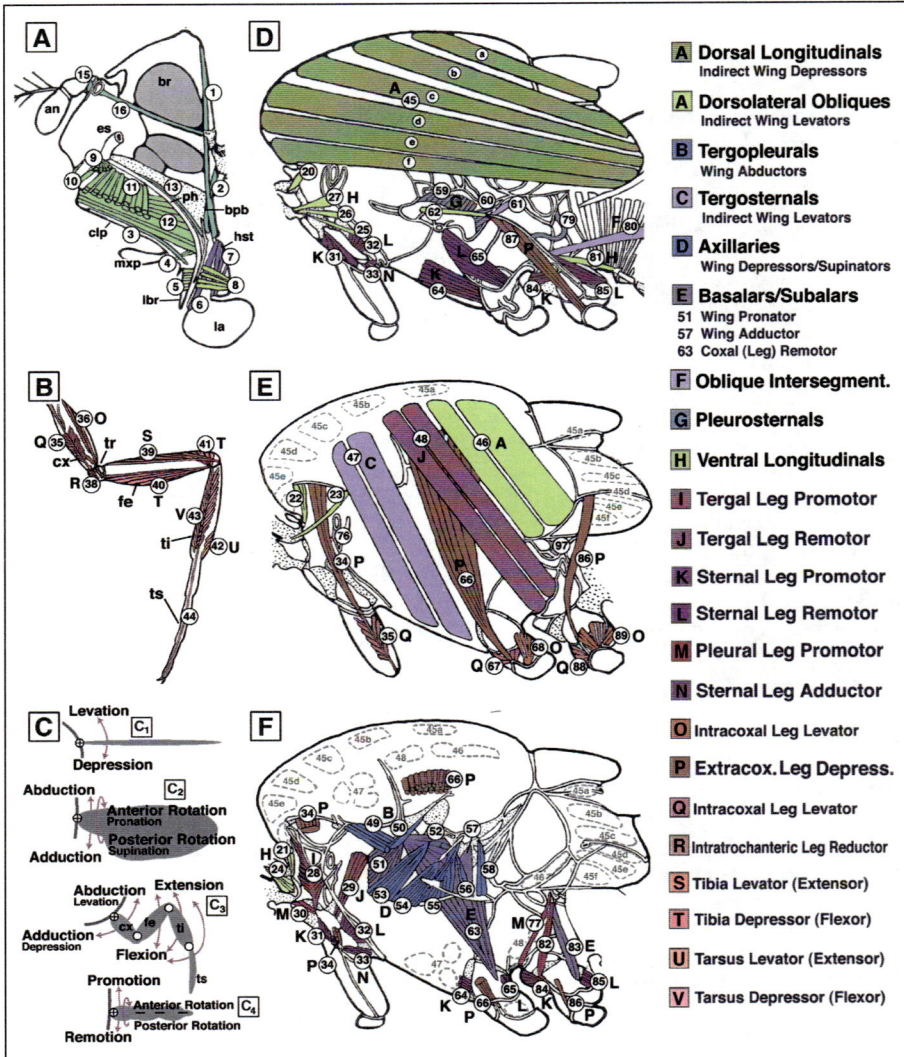

Figure 4. Somatic musculature of *Drosophila* head and thorax (after ref. 12). Numbers of individual muscles correspond to the ones introduced by Miller.[12] Capital letters refer to Snodgrass' system of muscles. A key to this system is shown on the right of this figure. A) Side view of head, illustrating the pattern of muscles associated with the pharynx and proboscis. 1 retractor of rostrum; 2 accessory retractor of rostrum; 3 flexor of proboscis; 4 adductor of maxillary palp apodeme; 5 dilator of epipharynx; 6 retractor of paraphyses of labella (extend labella); 7 retractor of furca (spread labella); 8 transverse muscles of haustellum (extend labella); 9,10 retractors of fulcrum (rotate pharynx upward into head); 11, 12 dilators of pharynx; 13 gracilis (operates salivary pump); 14 (not shown) and 15 muscles of second antennal segment; 16 muscle of frontal pulsatile organ. B) Frontal view of leg showing leg segments with pattern of intrinsic leg muscles. For names of individual muscles see key on the right of figure. C) Movements of wings and legs. C_1 and C_3 show schematic frontal views of wing and leg, respectively; C_2 and C_4 are dorsal views (anterior up). (This figure legend is continued on the next page.)

Figure 4, continued. D-F) Lateral views of thorax illustrating pattern of muscles of neck, wing and legs. Anterior is to the left. The three panels show muscles inserting at different medio-lateral levels. D) contains the muscles located closest to the sagittal plane, that is furthest away from the body wall. These include most prominently the massive dorsal longitudinals (indirect wing depressors). Panel F) shows the short direct wing muscles located close to the body wall; E contains the muscle sets in between those shown in D and F. Muscles are shown in different colors intended to facilitate their recognition as members of Snodgrass' functional-topological classes (identified by capital letters). Hatched outlines in E and F indicate insertions of muscles not shown on these panels. Individual muscles: Prothorax: 20, 21 (not shown), 22, 23, 24: dorsal and lateral longitudinals of T1 (move head); 25, 26, 27 ventral longitudinals of T1 (move head) 28 tergopleural promotor of coxa; 29 pleural remoter and abductor of coxa; 30 pleural promotor 31 sternal anterior rotator (pronator); 32 sternal posterior rotator (supinator and remotor); 33 sternal adductor; 34 extracoxal depressor of trochanter; 35, intracoxal depressor of trochanter; 36 and 37 (the latter not shown) intracoxal levators of trochanter; 38 reductor of femur; 39 levator of tibia; 40, 41 depressors of tibia; 42 levator of tarsus; 43 depressor of tarsus. Mesothorax: 45a-f dorsal longitudinal muscles (indirect wing depressors); 46a, b dorso-lateral oblique muscles (indirect wing levators); 47a-c tergosternal muscles (indirect wing levators); 48a, b tergal remotors of coxa (indirect wing levators); 49, 50 muscles of prealar apophysis (tergopleurals; wing abductors); 51, 52 basalar muscles (wing pronators); 53, 56 muscles of first axillary; 54, 55 muscles of third axillary (wing supinators); 57 subalar muscle (wing adductor); 58 muscle of axillary cord (wing supinator); 59, 60 pleurosternal muscles; 61 lateral intersegmental muscle; 62 ventral longitudinal muscle; 63 pleural remotor of coxa; 64 sternal promotor of coxa; 65 sternal remotor of coxa; 66 extracoxal depressor of trochanter; 67 intracoxal depressor of trochanter; 68 intracoxal levator of trochanter; 76 occlusor of anterior spiracle. Metathorax: 77 muscle of haltere; 79 pleurosternal muscle; 80 intersegmental muscle; 81 ventral longitudinal muscle; 82 pleural promotor of coxa; 83 pleural remotor of coxa; 84 sternal promotor of coxa; 85 sternal remotor of coxa; 86 extracoxal depressor of trochanter; 88 intracoxal depressor of trochanter; 89 intracoxal levator of trochanter; 97 occlusor of posterior spiracle. Other abbreviations: an, antenna; bpb, basiproboscis; br, brain; clp, clypeus; cx, coxa; es, esophagus; fe, femur; hst, haustellum; la, labellum; lbr, labrum; mxp, maxillary palp; ph, pharynx; ti, tibia; tr, trochanter; ts, tarsus.

Musculature of the Head

The segments that form the insect head are strongly modified from the archetypical pattern, which makes it impossible to place individual muscles into the scheme established by Snodgrass[4] for the segmental muscles of the trunk. This is particularly the case for insects with sucking mouthparts, including *Drosophila*. Here, the gnathal segments and preoral segments are strongly reduced and have fused into the proboscis (Fig. 4A). Most of the proboscis is formed by the labial segment. One distinguishes a proximal basiproboscis contiguous with the posterior head capsule, a middle part called mediproboscis (haustellum), and a distal labellum. A small part of the proximo-lateral wall of the proboscis are maxillary, and the anterior proboscis wall is formed by the clypeo-labrum.

Muscles which can be assigned to the labial segment comprise a set of short longitudinal and transverse muscles inside the haustellum which extend and spread the labellum (Fig. 4A, 6-8), and two long slender muscles that, inserting at the posterior basis of the proboscis and the head capsule, retract the proboscis (1, 2). A muscle from the anterior face of the haustellum (3) acts as a flexor (forward motion) of the proboscis. A fourth slender muscle extends from the opening of the salivary duct towards the head capsule (muscle of the salivary pump; 13).

The maxillary segment is reduced to two small external processes, called maxillary palp and lacinia, an internal apodeme situated at the base of the lacinia, and an ill defined strip of epidermis forming part of the posterior head capsule. A set of short muscle fibers extending from the head capsule down towards the apodeme can move the maxillary palp and assist in movement of the haustellum (4).

Anterior in the head one finds two short muscles inserting at the base of the antenna (15). Muscles derived from the labrum move the proboscis and dilate the mouth cavity. These muscles comprise proximally the dilators of the cibarium and pharynx (9-12); functionally and ontogenetically comparable to the dorsal pharyngeal musculature of the larval head), and distally the flexors and compressors of the labrum (5).

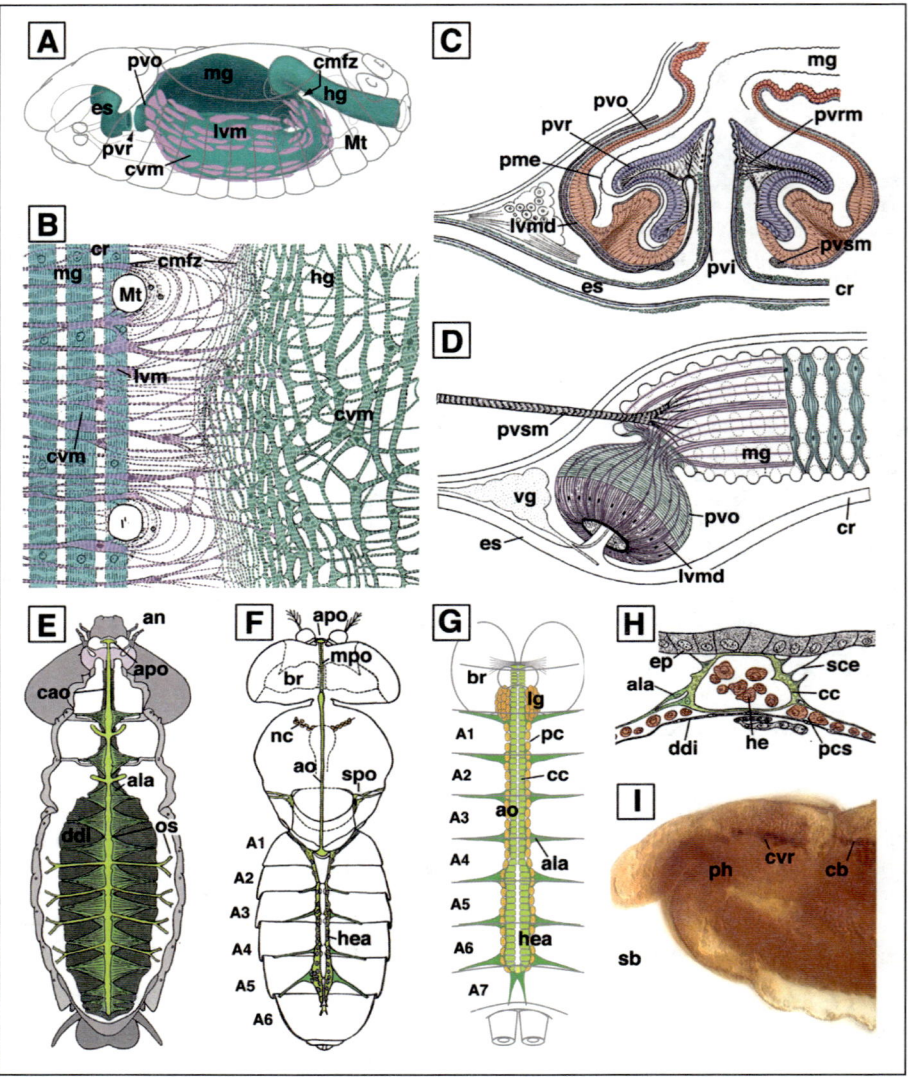

Figure 5. Visceral muscles of the digestive and vascular organs. A) Schematic image of stage 15 *Drosophila* embryo (dorso-lateral view, anterior to the left) showing distribution of precursors of circular visceral fibers (turquois, cvm) and longitudinal visceral fibers (purple, lvm) around the foregut (esophagus, es), midgut (mg) and hindgut (hg; from 6). Precursors of circular muscles are found around all of these segments, except for two narrow gaps coinciding with the location of the presumptive recurrent layer of the proventriculus (pvr) and a narrow domain where the Malpighian tubules (Mt) open into the hindgut (cmfz). Longitudinal fibers are restricted to the midgut (mg) and the outer layer of the proventriculus (pvo). B) Pattern of visceral muscles around midgut (mg) and hindgut (hg). Circular fibers fibers of midgut (cvm) form an inner layer of regular bands; longitudinal fibers (lvm) of variable thickness cover the circular muscle layer. No circular fibers are formed in the distal-most part of the hindgut (cmfz) that receives the Malpighian tubules (Mt). Circular muscles of the hindgut are more irregular in size and orientation than their midgut counterparts, and form thin longitudinal branches that interconnect neighboring fibers. (This figure legend is continued on the next page.)

Figure 5, continued. C,D) Visceral muscles of the proventriculus. Panel C shows sagittal section of the proventriculus, Panel D is a schematic surface view (anterior to the left). The outer (pvo) and inner (pvi) layers of the proventriculus are surrounded by circular fibers (colored turquois); these muscles are lacking from the recurrent layer (pvr). Longitudinal muscles (purple) continue from the midgut (mg) into the outer proventricular layer (pvo). To these outer longitudinal muscles a deep set of longitudinal fibers (lvmd) is added in the pvo. Specialized radial muscles (pvrm) and sphincter muscles (pvsm) are instrumental to carry out the valve-function of the proventriculus. E-G) Schematic dorsal views of the dorsal vessel (cardiac musculature) in *Periplaneta* (E; after 4), adult *Drosophila* (F; after 12) and larval *Drosophila* (G; from ref. 6). Panel H represents a schematic cross section of the heart of the wasp *Phaenoserphus* (after ref. 31). The dorsal vessel comprises a posterior heart (hea) and anterior aorta (ao), both formed by a tube of myo-epithelial cardiocytes (cc), suspended from the dorsal epidermis by segmentally arranged alary muscles (ala). In many insects, mesoderm cells form a connective tissue layer (dorsal diaphragm, ddi) underneath the dorsal vessel; the hemolymph space in between the dorsal diaphragm and the dorsal epidermis represents the pericardial sinus (pcs). Segmentally arranged openings in the dorsal vessel (ostia, os) function as valves that let the hemolymph enter the dorsal vessel. Anteriorly, the aorta continues in many insects (cephalic aorta, cao) to terminate in the anterior pulsatile organ (apo) located in between the antennae (an). In adult *Drosophila*, the cephalic aorta is rudimentary, forming the muscles of the anterior pulsatile organ (mpo). Additional pulsatile organs are found at the base of the appendages (scutellar pulsatile organ, spo, in *Drosophila*). Accessory pulsatile organs have not been described in the *Drosophila* larva. However, a population of *tinman*-positive mesoderm cells (shown in panel I in a stage 15 embryo) derived from the labrum assemble into an ill-defined strand that connects to the aorta, and that may represent a rudimentary cephalic vessel (cvr in I). Other abbreviations: br, brain; cb, cardioblasts; cr, crop; ep, epidermis; he, hemocytes; lg, lymph gland; nc, thoracic nephrocytes; pc, pericardial nephrocytes; ph, pharynx; pme, peritrophic membrane; sce, sustentacular cell fibers; vg, ventricular ganglion (of stomatogastric nervous system).

A number of temporary muscles attach to a sac-like structure, called ptilinum, of the antero-dorsal head of a pupa/young (not shown in Fig. 4A). During eclosion of the adult, the ptilinum is everted by pressure from the hemolymph, thereby pushing forward through the pupal case. Temporary muscles attached to the base of the ptilinum and the anterior head capsule assist in retracting the ptilinum.

Visceral Musculature

The digestive tract of insects consists of a monolayered epithelium surrounded by striated, syncytial muscle fibers. Details about the pattern of visceral muscle have not been worked out for the *Drosophila* larva or adult. The description which illustrates the adult pattern of visceral muscles most thoroughly is the one prepared by Graham-Smith[15] for the Dipteran *Calliphora erythrocephala*. This work, in conjunction with the reviews provided by Miller[12] and Crossley,[16] as well as more recent data on the embryonic pattern of visceral muscles[6,17,18] serves as the basis for the following brief survey.

The digestive tract consists of foregut and hindgut, derived from the terminal ectoderm, and the midgut, derived from the endoderm. The foregut is further subdivided into pharynx, esophagus, and proventriculus; the hindgut comprises (from distal to proximal) the Malpighian tubules, small intestine, large intestine, and rectum (Fig. 5A). As a general rule, the foregut from the level of the esophagus onward, and the hindgut are surrounded by a single layer of circular muscles. The midgut is surrounded by two muscle layers, an inner layer of circular muscles, and an outer layer of longitudinal muscles.

Visceral Musculature of the Foregut

The anterior segments of the digestive tract are the mouth cavity and pharynx (cibarium), which in flies form a specialized chamber adapted to suck up liquid food. Somatic muscle fibers, described in previous sections dealing with the larval and adult head, attach at the walls of the cibarium/pharynx and the head capsule, generating a vacuum in side these cavities upon contraction (Fig. 4A). Visceral muscle fibers are not found. They are equally absent in the salivary gland, a derivative of the labial segment which opens into the pharynx.

The visceral musculature of the esophagus and crop (a blind outpocketing of the esophagus in the adult) is formed by circular fibers. In the adult blowfly, these fibers appear to branch out considerably, forming diagonal and longitudinal side arms that cross the main circular fibers at a more external level. This orthogonal network of visceral fibers parallels the one formed around the hindgut, shown in Figure 5B. Specialized strands of circular bundles of densely packed fibers form sphincter muscles at the entrance to the esophagus, proventriculus, and crop, and prevent the regurgitation of food.

The proventriculus forms a complex organ at the junction between esophagus and midgut. It can be best described as an intussusception of the proximal esophagus into the distal midgut. In a length section (Fig. 5D) one appreciates the three concentrically arranged epithelial cylinders. The inner cylinder (plug) constitutes a continuation of the esophagus, exhibiting a flat epithelial layer surrounded by dense circular visceral muscle fibers. The middle, or recurrent, layer is formed by a cuboidal epithelium that lacks a muscle lining. The outer layer (cardia) is a continuation of the midgut epithelium, surrounded by inner circular muscles and outer longitudinal muscles (Fig. 5C,D).

One of the roles of the proventriculus is the secretion of the peritrophic membrane. This is an extracellular membrane secreted at the apical surface of the cells of the cardia (Fig. 5D). The peritrophic membrane is molded in the narrow cleft between cardial and recurrent epithelium, from where it is pushed posteriorly throughout the mid- and hindgut. Specialized muscles found in the proventriculus are likely to aid in the moulding and movement of the peritrophic membrane. Thus, both in the plug and cardia region, an array of deep longitudinal fibers underlies the circular fibers. Furthermore, radial fibers project from the distal plug outward towards the recurrent epithelium (Fig. 5C). The existence of these specialized visceral fibers has not yet been substantiated in adult *Drosophila*, and is most likely absent in *Drosophila* larvae.

Visceral Musculature of the Mid- and Hindgut

The embryo's midgut is surrounded by inner circular and outer longitudinal fibers. Both are approximately 2 μm in diameter. Circular fibers are much more closely spaced than longitudinal fibers. In the late embryo, the latter from a highly regular array of 20 parallel strands running alongside the midgut at regular intervals (Fig. 5A). The musculature of the midgut is responsible for the peristaltic waves that pass over the gut at regular intervals.

The visceral musculature of the hindgut, like that one of the esophagus, is circular, with irregular diagonal and longitudinal branches given off by the circular strands (Fig. 5B). Specialized arrays of inner longitudinal muscles and radial muscles are found in the rectum. Both longitudinal and circular fibers can also be found on the proximal segment of the Malpighian tubules, the so-called ureter. This complex musculature, which is entirely uncharted in the *Drosophila* larval and adult stage, can be assumed to function during defecation, as well as water reabsorption that occurs in the rectum.

The ovarioles and testes are surrounded by a layer of circular visceral muscles resembling those that are associated with the digestive tract[19-21] (Fig. 6). Muscle fibers are flat and multinucleate, and are ensheathed on all sides by basement membrane. The role of ovarian and testicular muscles is unclear; it is likely that the slow peristaltic waves passing along the length of the ovarioles are required for normal oogenesis and spermatogenesis.

Circulatory System

Evolutionary Remarks and General Structure of the Insect Circulatory Organs

The vascular system has its origin in clefts and lacunae formed between mesodermally derived cells.[22] This can be appreciated best in some coelomate invertebrate groups, such as the nermertines, annelids, or hemichordates. In these animals the mesoderm on either side of the embryo forms a series of hollow vesicles called coelomata (Fig. 7A). The epithelial lining of the coelomata, called mesothelium, is attached on one side (called the somatopleura) to the inner

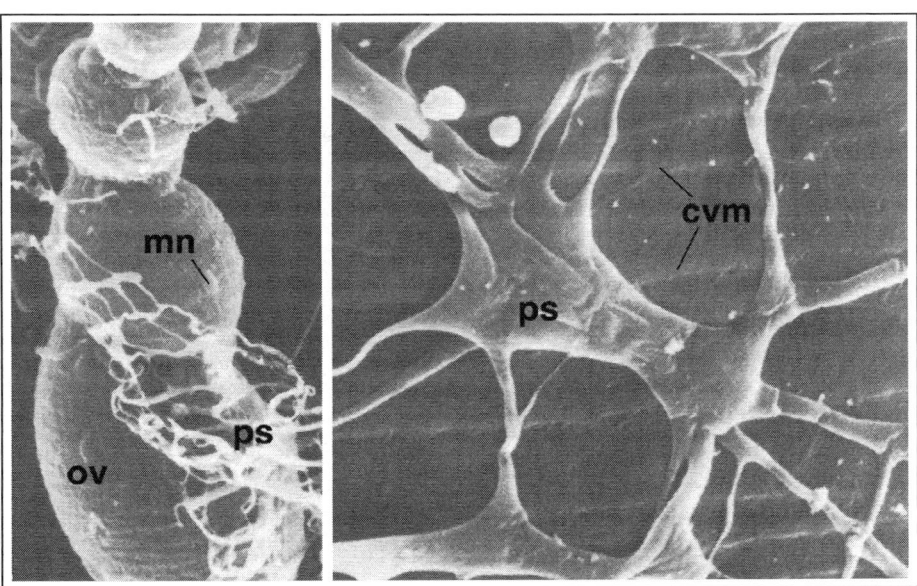

Figure 6. Circular visceral musculature of the *Drosophila* ovary (from ref. 21, with permission). Low power SEM (left) and high power SEM (right) of the ovarian wall (ov), showing meshwork of mesodermal cells that form the peritoneal sheath (ps). Covered by basement membrane, circular muscle fibers (cvm) appear as rather inconspicuous belts underneath the peritoneal sheath. Muscle nuclei (mn) protrude from the otherwise smooth ovarian surface.

surface of the body wall, on the other side (splanchnopleura) to the visceral organs. In the dorsal and ventral midline, as well as in between neighboring vesicles, the splanchnopleural mesothelia come together and form septa. Clefts remaining open within these septa represent blood vessels. The specialized splanchnopleural cells forming these vessels have myofibrils and are contractile; coordinated contractions of the vessels pump the blood throughout the body.

In insects, closed mesodermal coelomata lined by somatopleural and splanchnopleural mesothelia form only transiently. At later developmental stages coelomata dissociate and give rise to an open body cavity called myxocoel or hemocoel. Contractile vascular tubes develop from myoepithelial cells that segregate from the splanchnopleura (Fig. 7B). The most conspicuous tube forms a dorsal vessel that runs throughout the length of the animal. Additional contractile organs develop in the head and thorax.

The dorsal vessel of all insects comprises a wide posterior part (heart) located in the abdomen, and a narrow part (aorta) continuing from the heart through the thorax into the head[4,13,23] (Fig. 5E). Heart and aorta are lined by thin myoepithelial cells, called cardiocytes. Peristaltic waves of contraction move the hemolymph in the heart forward into the aorta. The hemolymph can enter the lumen through paired metamerical valves in the heart wall, called ostia. In most insects, the heart and aorta is a simple, unbranched tube; in some groups, segmental branches (which bear the ostia at their distal openings) occur. Anteriorly, the aorta terminates behind the brain. However, specialized contractile cells (called muscles of the frontal pulsatile organ in *Drosophila*[12]) continue forward, passing through the brain alongside the esophagus, and terminate at the front of the head in contact with a pulsatile ampulla between the antennae. Beside the antennal ampullae, accessory pulsatile organs also variably exist at the base of the wings and legs. The role of these pulsatile organs is to pump hemolymph from the appendages towards the dorsal vessel.

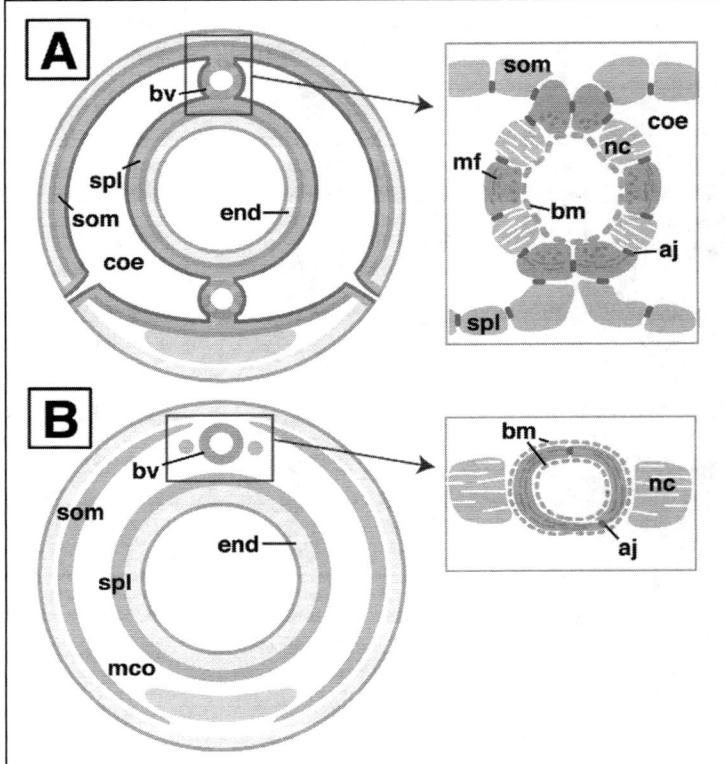

Figure 7. Insect vascular system-phylogenetic perspective. A) Schematic cross section of coelomate protostome (e.g., annelid). The mesoderm forms bilaterally symmetric epithelial sacs (coe = coelomata). Medial lining of coelom, attached to the endoderm (end) represents the splanchnic (= visceral) layer (spl); lateral wall is the somatic layer (som). Clefts between left and right coelomata form blood vessels (bv). Inset to the left shows blood vessel at higher magnification. The blood vessel wall is formed by contractile myo-epithelial cells (mf= myofibrils) and excretory nephrocytes (nc). Note that the basal surface of the myoepithelium, covered by a basement membrane (bm) faces the lumen of the blood vessel; the apical surface, marked by adherens junctions (aj), faces the lumen of the coelom. B) Schematic section of insect. The mesoderm does not form an epithelial layer surrounding a coelom, although the division into inner splanchnic layer (spl) and outer somatic layer (som) is still apparent. The dorsal blood vessel (bv) also shares numerous similarities with its counterpart in coelomate protostomes. Notably, it is formed by a contractile myo-epithelium. A basement membrane lines the lumen of the vessel, but can also be found at its outer surface. Nephrocytes (nc) do not form part of the vessel wall, but are aligned along its outer surface.

Beside the dorsal vessel and accessory pulsatile organs, a number of connective and muscular cells are associated with the insect circulatory system. Numerous small, ill-defined suspensory cells extend short fibers that connect the heart and aorta to the dorsal epidermis (Fig. 5H). A more prominent connection between the dorsal vessel and body wall is provided by the dorsal diaphragm. The diaphragm is formed by paired segmental muscles, the alary muscles, which extend between the cardiocytes and the lateral epidermis. Gaps between alary muscle fibers are filled out by additional, ill-defined connective tissues. In some insect groups, the dorsal diaphragm forms a wall that completely encloses, along with the dorsal epidermis, the dorsal hemolymph compartment (Fig. 5E,H). In these cases we speak of the pericardial sinus.

In most insects, including *Drosophila*, the dorsal diaphragm is much reduced, containing but few fibrillar cells and slender alary muscles (Fig. 5F).

Structure of the *Drosophila* Circulatory System

The adult circulatory structures have been described by Miller (21); information has been added in more recent work by, among others, Jensen[24] (for *Calliphora*), Rizki,[23] Bate,[5] and Molina and Cripps.[25] The adult heart consists of four chambers that extend throughout abdominal segments 1-5 (Fig. 5F). Five pairs of alary muscles insert at the constrictions that mark the boundaries between the heart chambers and suspend the heart from the dorsal epidermis. In each chamber, a pair of ostia, located right anterior to the insertion point of the alary muscles, form valves that allow hemolymph to enter the heart. Posteriorly, the blind ending heart tube is attached to the rectum and dorsal epidermis of abdominal segment 6. Apart from the alary muscles, thin fibrils extend between the cardiocytes and dorsal epidermis. It is not known to what extent these fibrils are but processes of cardiocytes, or formed by a separate type of cell that might correspond to the "suspensory cells" described for other insects (see above). Miller[12] describes occasional small nuclei that he attributes to cells forming the fibrils.

The heart is formed by two rows of cardiocytes that enclose the central lumen. Longitudinal and circular myofibrils fill the cytoplasm of these cells. In addition, a separate layer of longitudinal muscle cells is attached to the ventral wall of the heart. These subcardial longitudinal muscles seem to be absent from the larval heart.

Anterior to the first heart chamber the dorsal vessel narrows and forms the aorta. Structurally, the aorta is built similarly to the heart, but it lacks ostia and alary muscles. Also the ventral longitudinal muscle layer typical for the heart has not been reported for the aorta. The aorta extends throughout the thorax and neck on the dorsal side of the digestive tract, and terminates with an opening into the hemocoel behind the brain. However, as described for insects in general above, a pair of muscles continues anteriorly from the tip of the aorta, through the brain hemispheres, to the frons of the head where it ends in connection with a pulsatile ampulla located between the antennae. This accessory circulatory structure, which has been investigated in more detail in other insects, pumps hemolymph from the antennae into the head capsule. A pair of scutellar pulsatile organs serve a similar function for the wings.[12] These organs are formed by a simple muscle sheet spanning between the scutellum and the bases of the wings (Fig. 5F). Rhythmic contraction draws blood from the bases of the wings into the central hemocoel of the thorax. Additional pulsatile organs, described in other insects to exist in the wings and legs, have not been discovered so far in *Drosophila*.

The larval circulatory system closely resembles the adult one (Fig. 5G). The heart consists of three chambers located in abdominal segments 5-7. Each chamber allows the intake of hemolymph through a pair of ostia. The cellular composition and development of the larval heart has been studied in great detail. Each metameric heart chamber is formed by six pairs of cardiocytes. Two pairs (molecularly marked by the expression of the transcription factor Seven-up) form the ostia, whereas the intervening four pairs form the heart wall proper. Alary muscles attach to the ostia forming cardiocytes and extend towards the epidermis, where they insert at the intersegmental apodemes. The aorta has the same cellular composition as the heart, except that the *seven-up*-positive cardiocytes do not form ostia between them. Unlike the adult, the larva has alary muscles attaching the aorta to the body wall in the same manner as the alary muscles of the heart. The aorta extends throughout segments T1 through A4. Anteriorly it ends behind the brain, in connection with the dorsal pouch which constitutes the involuted head capsule of the larval body.

A number of populations of cells have been described in the embryo which may correspond to precursors of structures added to the adult circulatory system. The transition between the two stages which occurs during metamorphosis has not been described yet in detail. There seems to be general agreement that the posterior part of the larval heart is histolyzed and replaced by adult cardiocytes. Where these cells are located in the embryo is not known. In addition to part of the heart that needs to be replaced, there are a number of other cell types in the adult circulatory system that are not present in the larva. These cells include the ventral longitudinal muscles of the heart, the suspensory cells, the accessory pulsatile organs of the scutellum and antennae, and the preaortic muscles in the head. In a recent investigation of the *Drosophila* head mesoderm[26] (Fig. 5I), a population of *tinman*-positive cells derived from the labrum has been described which in location could well represent the primordium of the preaortic muscles (and may also include precursors of the antennal pulsatile organ). These cells migrate along the roof of the esophagus and attach to the anterior end of the aorta in the late embryo.

In the embryonic trunk, two populations of "pericardial cells" were discovered on the basis of molecular markers. They include the *tinman*-positive and the *even-skipped*-positive pericardial cells.[27] These cells do not differentiate as bona fide pericardial cells, which are ultrastructurally defined as nephrocytes. In fact, neither *tinman*-positive nor *eve*-positive pericardial cells are attached to the late larval heart tube when it is dissected out of the body. However, we surmise that at least some of these cells could represent precursors for adult specific heart structures. For example, the *tin*-positive cells, which adopts a location underneath the cardiocytes in the late embryo, could be the precursors of the adult subcardial longitudinal muscles.

Concluding Remarks

The *Drosophila* musculature, similar to the nervous system, has become a fertile ground for developmental studies that approach fundamental questions of cell fate determination, cellular morphogenesis, and pattern formation. The system is uniquely suited for these undertakings because muscle fibers (at least in the somatic musculature) are relatively few and form an invariant pattern that, as shown in this chapter, has been documented in considerable detail. Areas that need more attention are the visceral musculature and circulatory system, in particular in regard to their postembryonic development and metamorphosis. It will also be helpful to incorporate more than is currently done a comparative outlook into projects addressing patterning and morphogenesis of the musculature. Thus, unraveling the genetic pathway controlling the formation of a certain muscle population, that exists in a highly derived state in Drosophila, may be greatly helped by looking out for the corresponding muscles in other arthropods. It would be especially helpful if the classical attempt of Snodgrass to define an archetypical arthropod muscle pattern could be placed on a more solid footing, by trying to identify markers for subsets of muscles and investigate their expression throughout the arthropod clade.

References

1. Rieger RM, Tyler S, Smith III JPS et al. Platyhelminthes: Turbellaria. In: Harrison FW, Bogitsh BJ eds. Microscopic Anatomy of Invertebrates. New York: Wiley-Liss, 1991:3.
2. Morris J, Nallur R, Ladurner P et al. The embryonic development of the flatworm Macrostomum sp. Dev Gen Evol 2004; 214:220-239.
3. Manton SM. The arthropoda. Habits, functional morphology and evolution. Oxford: Clarendon Press, 1977.
4. Snodgrass RE. Principles of Insect Morphology. New York: McGraw-Hill, 1935:311-315.
5. Bate M. Mesoderm. In: Bate M, Martinez-Arias A, eds. The Development of Drosophila. New York: Cold Spring Harbor Laboratory Press, 1993.
6. Campos-Ortega JA, Hartenstein V. The embryonic development of Drosophila melanogaster. 2nd Ed. Springer, 1997.
7. Haas MS, Brown SJ, Beeman RW. Pondering the procephalon: The segmental origin of the labrum. Dev Genes Evol 2001; 211:89-95.
8. Jürgens G, Hartenstein V. The terminal regions of the body pattern. In: Bate M, Martinez-Arias A, eds. The Development of Drosophila. New York: Cold Spring Harbor Laboratory Press, 1993.
9. Rempel JG. The evolution of the insect head: The endless dispute. Quaest Entomol 1975; 11:7-25.
10. Younossi-Hartenstein AY, Tepass U, Hartenstein V. The embryonic origin of the primordia of the adult Drosophila head. Roux's Arch Dev Biol 1993; 203:60-73.
11. Robertson CW. The metamorphosis of Drosophila melanogaster, including an accurately timed account of the principal morphological changes. J Morphol 1936; 59:351-399.

12. Miller A. The internal anatomy and histology of the imago of Drosophila melanogaster. In: Demerec M, ed. The Biology of Drosophila. New York: Wiley, 1950:420-534.
13. Chapman RF. The insects: Structure and function. Cambridge: Harvard University Press, 1982.
14. Hughes GM. The coordination of insect movements. I. The walking movements of insects. J Exp Biol 1952; 29:267-284
15. Graham-Smith GS. The alimentary canal of Calliphora erythrocephala L., with special reference to its musculature, and to the proventriculus, rectal valve, and rectal papillae. Parasitology 1934; 26:176-248.
16. Crossley AC. The morphology and development of the Drosophila muscular system. In: Ashburner M, Wright TRF, eds. The Genetics and Biology of Drosophila, Vol. 2b. New York: Academic Press, 1979:499-560.
17. Klapper R, Stute C, Schomaker O et al. The formation of syncytia within the visceral musculature of the Drosophila midgut is dependent on duf, sns and mbc. Mech Dev 2002; 110:85-96.
18. Martin BS, Ruiz-Gomez M, Landgraf M et al. A distinct set of founders and fusion-competent myoblasts make visceral muscles in the Drosophila embryo. Development 2001; 128:3331-8.
19. King RC. Ovarian development in Drosophila melanogaster. New York, London, San Francisco: Academic press, 1970.
20. Lindsley DL, Tokuyasu KT. Spermatogenesis. In: Ashburner M, Wright TRF, eds. The Genetics and Biology of Drosophila, Vol. 2b. New York: Academic Press, 1980:226-295.
21. Mahowald AP, Kambysellis MP. Oogenesis. In: Ashburner M, Wright TRF, eds. The Genetics and Biology of Drosophila, Vol. 2b. New York: Academic Press, 1980:141-225.
22. Ruppert EE, Carle KJ. Morphology of metazoan circulatory systems. Zoomorphology 1983; 103:193-208.
23. Rizki TM. The circulatory system and associated cells and tissues. In: Ashburner M, Wright TRF, eds. The Genetics and Biology of Drosophila, Vol. 2b. New York: Academic Press, 1978:397-452.
24. Jensen PV. Structure and metamorphosis of the larval heart of Calliphora erythrocephala. K Dansk Vidensk Selsk Biol Skrift 1973; 20:2-19.
25. Molina MR, Scripps RM. Ostia, the inflow tracts of the Drosophila heart, develop from a genetically distinct subset of cardial cells. Mech Dev 2001; 109:51-59.
26. DeVelasco B, Shen J, Go S et al. Embryonic development of the Drosophila corpus cardiacum, a neuroendocrine gland with similarity to the vertebrate pituitary, is controlled by sine oculis and glass. Dev Biol 2004; 274:280-294.
27. Ward EJ, Skeath JB. Characterization of a novel subset of cardiac cells and their progenitors in the Drosophila embryo. Development 2000; 127:4959-4969.
28. Luther A. Untersuchungen and rhabdocoelen turbellarien IV. Ueber einige repraesentanten der familie proxenetidae. Acta Zool Fenn 1943; 38:3-100.
29. Anderson MS, Halpern ME, Keshishian H. Identification of the neuropeptide transmitter proctolin in Drosophila larvae: Characterization of muscle fiber-specific neuromuscular endings. J Neurosci 1988; 8:242-55.
30. Ferris GF. External morphology of the adult. In: Demerec M, ed. The Biology of Drosophila. New York: Wiley, 1950:368-419.
31. Eastham LES. The postembryonic development of Phaenoserphus viator, aq parasite of the larva of Pterosticus niger, with notes on the anatomy of the larva. Parasitology 1929; 21:1-21.

CHAPTER 3

Mesoderm Formation in the *Drosophila* Embryo

Noriko Wakabayashi-Ito and Y. Tony Ip*

Abstract

All muscle cells develop from the mesoderm, which is the middle germ layer in the early embryo. The mesoderm itself derives from the ventral cells of the blastoderm stage embryo. Therefore, the regulatory events controlling dorsal-ventral development in the oocyte and the early embryo are the earliest events in muscle formation. The first stage in dorsal-ventral development can be traced back to the oocyte, where Gurken-Torpedo signaling establishes dorsal-ventral asymmetry. The ventral half of the oocyte is then allowed to express Pipe, which serves to activate a series of serine proteases. These activation events ultimately lead to the stimulation of the Toll receptor in the ventral side of the early embryo. In the final stage of this Toll maternal cascade there is the formation of the Dorsal protein gradient in the ventral nuclei of the embryo blastoderm. The Dorsal gradient both activates and represses zygotic gene expression to establish mesodermal cell fate and promote mesoderm invagination. The invaginated mesoderm then differentiates into appropriate muscle tissue types according to further positional information.

Introduction

Mesoderm development in the *Drosophila* embryo leads to the formation of various tissues and organs including: somatic and visceral muscles; somatic gonad; heart; fat bodies; macrophages, and lymph glands.[1] The origin of all these tissues can be traced back to the early embryo when patterning of the blastoderm occurs. Blastoderm is the early embryo that has very few distinct morphological features and contains approximately 5000 cells formed from multiple rounds of nuclear division of the fertilized egg. Because the mesoderm comes from the ventral cells of the blastoderm, dorsal-ventral patterning of the blastoderm also determines mesoderm formation. In this chapter, we will discuss how the seemingly homogenous cell population in the blastoderm gives rise to different cell fates, particularly the mesoderm.

At the blastoderm stage the *Drosophila* embryo can be broadly divided into three territories along the dorsal-ventral axis.[2] The dorsal half of the embryo gives rise to the dorsal ectoderm, the lateral cells give rise to the neuroectoderm, and the ventral cells constitute the presumptive mesoderm. These three territories have distinct patterns of gene expression, which coincide with the various cell fates. The morphology of the newly laid *Drosophila* embryo already reflects a dorsal-ventral asymmetry, as it is flattened on the dorsal side and curved on the ventral side. Therefore, dorsal-ventral patterning and mesoderm formation are initiated during oogenesis and are continued in the early embryo.

*Corresponding Author: Y. Tony Ip—Program in Molecular Medicine, University of Massachusetts Medical School, 373 Plantation Street, Room 109, Worcester, Massachusetts 01605, U.S.A. Email: Tony.Ip@umassmed.edu

Muscle Development in Drosophila, edited by Helen Sink. ©2006 Eurekah.com and Springer Science+Business Media.

The dorsal-ventral patterning of the *Drosophila* embryo is established by three sequential phases. The first phase takes place in the oocyte, and involves the Gurken-Torpedo pathway, a TGF-α-like signaling pathway.[3] The second phase involves the localized activation of upstream components of the Toll pathway in the oocyte, which in turn leads to stimulation of the Toll receptor and the Dorsal transcription factor in the ventral cells of the embryo.[4] The third phase incorporates the activation of zygotic genes such as *twist* and *snail* by Dorsal to promote the invagination of the ventral cells to form the middle germ layer, that is mesoderm.[5,6]

Establishment of Dorsal-Ventral Polarity during Oogenesis

The first phase of dorsal-ventral patterning in *Drosophila* development is initiated during oogenesis. This patterning relies on a series of intercellular communications in the egg chamber between the germ line derived oocyte and the surrounding somatically derived follicle cells.[3,7] Extensive mutagenesis screens for recessive female sterile mutations identified a number of genes that affect dorsal-ventral polarity in both the eggshell and the embryo. Among these mutants *gurken* and *torpedo* were found to cause strong ventralization of the eggshell and the embryo, indicating that their wild-type gene products are required for the specification of dorsal fates.[8] Subsequently it was shown that the initial polarizing signal from the oocyte to follicle cells is determined by *gurken*, which encodes a Transforming Growth Factor-α (TGF-α)-like protein.[9] During mid oogenesis the oocyte's nucleus moves to the future dorsal-anterior side of the oocyte[10] and *gurken* mRNA is consequently localized to the dorsal-anterior corner of the oocyte. As a result there is restricted distribution of the Gurken protein. This in turn leads to localized activation of Gurken's receptor Torpedo (the Epidermal Growth Factor Receptor (EGFR) homologue) in dorsal follicle cells[11] (Fig. 1).

Activated Torpedo proceeds to transmit the Gurken-triggered signal in the follicle cells through the mitogen-activated protein kinase (MAPK) signaling pathway.[3,12] The MAPK signaling pathway components Raf and D-MEK are required downstream of Torpedo. A reduction in either Raf or D-MEK activity produces ventralized eggs,[13,14] demonstrating that dorsal cell differentiation depends on a series of phosphorylation events.

The Torpedo signaling in the follicle cells also requires autocrine amplifiers that include Spitz and Rhomboid.[15] The signal from Torpedo activates the transcription of *rhomboid*, which

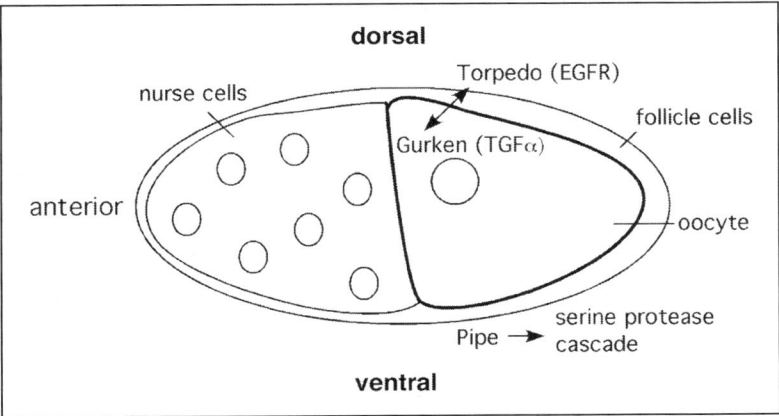

Figure 1. The Gurken-Torpedo pathway defines dorsal-ventral polarity in the oocyte. The location of the oocyte nucleus within the developing oocyte provides positional information, such that the Gurken gene product is active in the dorsal-anterior region. Signaling between the germ line oocyte and the surrounding somatic follicle cells establishes the dorsal fate. The expression of Pipe, an upstream activator of the Toll pathway, is restricted to the ventral region of the oocyte due to the activity of the Gurken-Torpedo pathway in the dorsal region.

encodes the transmembrane protease Rhomboid.[16,17] Removal of Rhomboid activity by using antisense RNA leads to the production of ventralized eggs. On the other hand, ectopic expression of Rhomboid causes the production of dorsalized eggs, and this dorsalizing activity requires Gurken and Torpedo.[18] Rhomboid directly cleaves Spitz, another TGF-α-like ligand, within its transmembrane domain.[17] The diffusible Spitz then collaborates with Torpedo to define dorsal fate in the egg chamber. Activated Torpedo signaling also restricts the expression of *pipe* to ventral follicle cells.[19,20] Pipe (see below) initiates a second signaling cascade, the Toll pathway, in the ventral side of the oocyte and continues in the early embryo.

The Toll Pathway in the Oocyte and Embryo

The Toll pathway takes effect in the early embryo, however the pathway is initiated during oogenesis. Thus, the genes involved are maternal effect genes, which are required within the genome of the mother, but the gene products function in the developing oocyte or embryo. Twelve maternal effect genes have been isolated in this pathway: *pipe, windbeutel, nudel, gastrulation defective, snake, easter, spätzle, Toll, tube, pelle, dorsal,* and *cactus*[21-24] (Fig. 2). Mutations in these genes in the mother perturb the polarity of the embryo, but there is no effect on the shape of the eggshell. Recessive mutations in any one of these genes, except *cactus,* leads to dorsalization of the embryos. Of these genes *pipe, windbeutel,* and *nudel* are expressed in the somatic follicle cells, whereas the rest of genes are expressed in the germ line oocyte.[23]

Pipe is one of the earliest acting components in the Toll pathway that ultimately leads to the production of active ligand for the Toll receptor. While Pipe is predicted to act as a glycosaminoglycan modifying enzyme, its substrate is not known.[25] *Windbeutel* encodes a putative resident protein of the endoplasmic reticulum, and is responsible for the correct subcellular localization of Pipe.[26,27] The hypothetical downstream substrate of Pipe and Windbeutel is postulated to activate an extracellular serine protease cascade in the perivitelline space, the fluid-filled compartment that lies between the vitelline membrane and the embryonic plasma membrane.[25] In this serine protease cascade, the first protease is Nudel, which has both a protease domain and an extracellular matrix association domain.[28] Local activation of Nudel stimulates the next protease, Gastrulation defective.[29-31] Downstream of Gastrulation defective is Snake, which acts upstream of the final protease Easter.[32,33]

Earlier studies suggested the presence of a negative regulator on Easter activity.[34] By analogy to mammalian blood clotting proteases, serine protease inhibitors, serpins, have been hypothesized to be involved in this negative regulation. Indeed, two reports show that one of the *Drosophila* serpins, Serpin27A, acts as an inhibitor that controls spatially the Easter activity in the embryo.[35,36]

The Easter protease is likely the direct activator of Spätzle in the ventral region. Spätzle is synthesized as full-length protein products of sizes ranging from 47 kD to 60 kD.[37] The full length Spätzle proteins are then proteolytically processed to become 23 kD proteins, which correspond to biologically active Spätzle.[37,38] Deletion studies suggested that proteolysis activates Spätzle by releasing an inhibitory effect of the amino-terminal domain on the carboxy-terminal domain, which has the cystine-knot motif found in many vertebrate growth factors.[37,39]

It is likely that the mature form of Spätzle works as a ligand for Toll in ventral cells.[38,40] Activation of Toll then establishes the nuclear gradient of the transcription factor Dorsal (Fig. 2). Toll encodes a transmembrane receptor protein that is distributed uniformly in the embryo, but is normally activated by Spätzle only in ventral cells.[41,42] Intracellular signaling from the Toll receptor to the Dorsal transcription factor requires the proteins MyD88, Tube, Pelle, and Cactus.[21-24,43] MyD88 and Tube are adaptor proteins that contain death domains and are recruited to the Toll receptor after stimulation.[44-47] The complex also recruits Pelle, a death domain containing serine/threonine kinase.[47,48] The increased local concentration of Pelle may result in autophosphorylation and activation.[49,50] The activated Pelle transmits the signal to the Cactus/Dorsal complex. Cactus contains ankyrin repeats and acts as an inhibitor that retains Dorsal in the cytoplasm.[51,52] Signal induced phosphorylation and degradation of Cactus

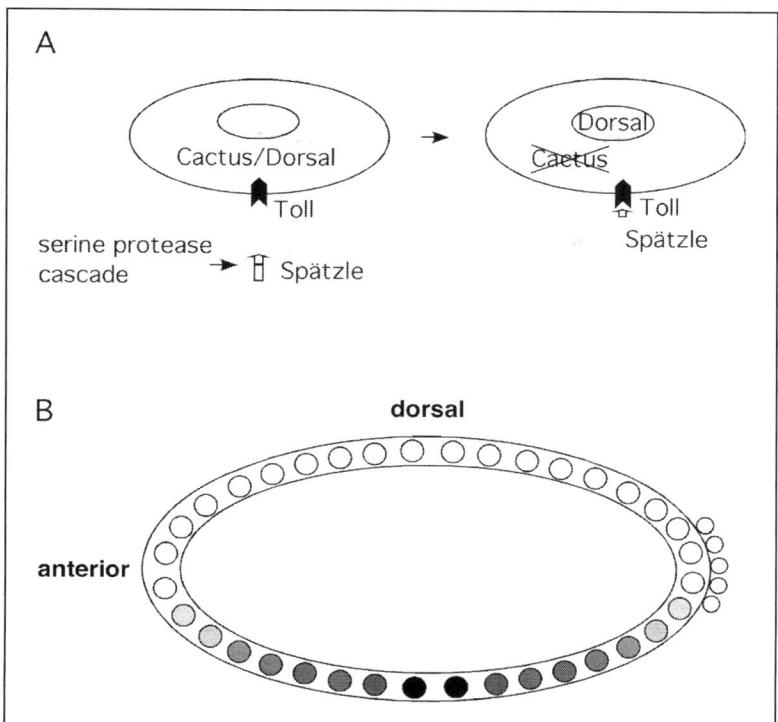

Figure 2. Spatially restricted activation of Toll establishes ventral cell fate in the early embryo. A) The positional information initiated in the oocyte is carried over to the fertilized egg. Initiated by early acting components such as Pipe in the oocyte, the serine protease cascade in the early embryo activates Spätzle by proteolytic processing. The mature Spätzle probably interacts directly with Toll. The signaling pathway causes degradation of Cactus, and allows Dorsal to enter the nuclei. B) Because the early step of activation occurs in ventral regions of the oocyte, the Toll pathway is only active in ventral regions of the embryo. The diffusible components of the pathway activate Toll gradually along the dorsal-ventral axis, leading to the formation of a nuclear gradient of Dorsal protein, with the highest level in ventral nuclei.

allows Dorsal to enter the nuclei, where Dorsal regulates the expression of many zygotic genes to pattern the early embryo.[53,54]

The Toll-Dorsal signaling pathway is well conserved in mammalian cells. Toll homologues in mammals include the Toll-like receptors (TLRs) and interleukin-1 receptor (IL-1R), all of which contain the signature Toll/interleukin-1 receptor (TIR) cytoplasmic domain. Both *Drosophila* and mammalian MyD88 are employed as adaptor proteins for Toll signaling. Tube does not exhibit significant sequence homology to known mammalian proteins,[44] although it may function similarly to other MyD88 homologues by forming adaptor complexes with MyD88 during signaling. Pelle belongs to the IL-1R associated kinase (IRAK) family.[55] The Cactus/Dorsal complex is homologous to the mammalian IκB/NF-κB complex. Phosphorylation of these cytoplasmic complexes leads to degradation of the inhibitors, Cactus and IκB, and the release of the transcription factors, Dorsal and NF-κB, to enter nuclei for transcriptional regulation.[56] The mammalian TLR pathways are required for innate immune response, a biological function seeming very different from that of the *Drosophila* Toll pathway. However, many of the same components of the *Drosophila* Toll pathway are used for antimicrobial response in larvae and adults, providing an evolutionarily conserved function for Toll and TLR signaling.[57]

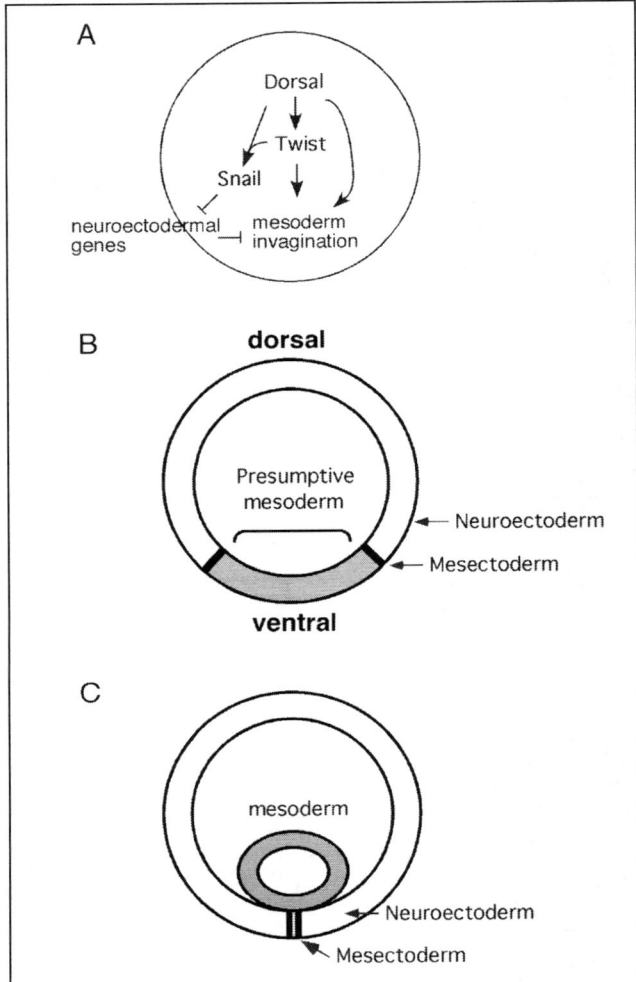

Figure 3. Mesoderm invagination is regulated by the dorsal-ventral determinants. The panels represent cross sections of the early embryo at blastoderm and gastrula stages. A) Dorsal activates the transcription of *twist* and *snail*. Dorsal and Twist activate genes required for the establishment of mesodermal cell fates and for mesoderm invagination. Snail represses genes that should not be expressed in the ventral domain. The Dorsal/Twist/Snail circuitry acts at the top of the regulatory hierarchy to control many aspects of cell fate determination, cell movements, and cell cycle regulation in order to promote mesoderm formation. B) The different cell fates along the dorsal-ventral axis of the blastoderm are as indicated in the panel. C) High levels of nuclear Dorsal promote ventral cell invagination, forming the mesoderm germ layer during gastrulation. The cell movement also brings the mesectoderm to the ventral midline and the lateral cells to the ventral regions to form the neuroectoderm.

Zygotic Gene Activation and Mesoderm Specification

The major role of the maternal Toll pathway is to initiate a third event: transcription of zygotic genes that specify mesodermal cell fate (Fig. 3). The transcriptional regulation is mediated by the last maternal component, Dorsal.[2] Due to the diffusible nature of many upstream components in the perivitelline fluid, the gradual change of the Toll receptor activation along the dorsal-ventral axis leads to differential nuclear translocation of Dorsal. High levels of Dorsal are present in ventral nuclei, and progressively lower levels of the protein are present in lateral nuclei.[58-60] This gradient initiates the formation of the mesoderm, neuroectoderm, and dorsal ectoderm by setting the expression limits of key zygotic genes in the three tissue types.

Among the many zygotic targets of Dorsal, the *twist* and *snail* genes have particularly important roles in mesoderm formation.[61,62] Embryos that lack Twist or Snail do not form mesoderm and die at the end of embryogenesis.[63,64] In *twist* mutant embryos no substantial ventral invagination occurs, while *snail* mutant embryos exhibit a slightly more severe phenotype. Embryos that are double mutant for *twist* and *snail* show the most severe phenotype, such that no cell shape changes occur and no ventral furrow is formed even at later stages.[65] Therefore, the two gene products have similar but somewhat distinct functions in mesoderm formation.

Twist is a basic helix-loop-helix transcription factor and is expressed in the ventral side of the embryo.[66] Twist is required for the activation of many downstream genes coding for proteins essential for mesoderm development, including N-cadherin, PS2α-integrin, muscle specific β3-tubulin, muscle myosin, and the MyoD homologue Nautilus.[67,68] While Dorsal directly activates both Twist and Snail, Twist also functions as a coactivator of Dorsal in the blastoderm to activate the expression of Snail.[69] This Dorsal-Twist cooperation establishes a sharp "on-off" pattern of Snail that coincides with the presumptive mesoderm.[67,70,71] Either expanding or reducing the expression of Snail correlates with changes of ventral cell invagination and the formation of future mesoderm. Moreover, genetic rescue experiments demonstrated that forced expression of Snail in the absence of Twist, but not vice versa, can promote ventral invagination.[72] This suggests that *snail* has a more direct role in regulating the downstream events leading to mesoderm invagination.

The Snail protein has five zinc fingers and works as a sequence-specific, DNA-binding repressor.[62,67,70,71,73] Snail represses the transcription of a number of target genes, such as *single-minded, rhomboid, lethal of scute {l(1)sc}, short gastrulation, Crumb, Delta, E-cadherin,* and members of *Enhancer of split* complex, to restrict their expression outside of the presumptive mesoderm. A simple model is that in the absence of Snail derepression of these target genes interferes with mesodermal cell fate and mesoderm invagination. However, which of the derepressed downstream genes are responsible for blocking invagination is not known.[74] A possible scenario is that in *snail* mutants the cumulative effect of expressing many of these genes in the presumptive mesoderm interferes with invagination.

Mesoderm Invagination

The invagination of ventral cells during gastrulation leads to the formation of the mesoderm layer.[65,75] Before invagination, the ventral cells go through cell autonomous morphological changes. The apical surfaces, which are facing outside the embryo, of the ventral cells first flatten and then constrict. This process transforms the columnar epithelial cells into a wedge shape, which probably helps the cells move into the basal, interior side.[6] In *snail* mutant embryos none of the ventral cells flatten or constrict their apices, suggesting that Snail is essential for most of the changes in apical morphology that normally occur in ventral cells. In *twist* mutant embryos apical flattening is observed in a narrower stripe of cells along the ventral midline and some cells even form transient apical constriction.[65]

There are two other proteins that affect cell shape changes during ventral invagination. In embryos lacking Folded gastrulation or Concertina their early phases of flattening and constriction take place normally, but the late phases of synchronous apical constriction never occur.[76,77] As a consequence, the whole population of ventral cells continues to have apical constriction at a slow rate characteristic of the first phase. The ventral invagination in these mutant embryos is irregular. Although the phenotypes of *folded gastrulation* and *concertina* mutants are similar, *folded gastrulation* is required zygotically whereas *concertina* is a maternal effect gene. *folded gastrulation* encodes a putative secreted molecule and *concertina* a Gα-like protein.[76]

Folded gastrulation is expressed in ventral cells under the control of Twist and Snail.[77] Therefore, one of the functions of Twist and Snail during mesoderm formation is to activate the expression of Folded gastrulation, which may stimulate a pathway that requires Concertina to coordinate the cell shape changes in the ventral cells.

Another important event that modulates mesoderm invagination is the control of mitotic timing in ventral cells.[78-80] Increased expression of String, a Cdc25 homologue, causes cell division to occur earlier in the ventral cells, concomitant with a block in ventral invagination. Normally, the mesodermal cells are poised to go through cell division but are restricted to do so until after invagination. The level of String protein determines the timing of cell division. A novel regulator, Tribbles, has been shown to reduce the level of String protein in ventral cells, such that invagination can proceed normally before cell division occurs.[78-80] The Tribbles-regulated String degradation also receives input from Twist and Snail, although the detailed mechanism is not understood. Once the mesoderm is established, the invaginated cells divide three times rapidly and follow other instructions to develop into various tissues.

Conclusion

We have summarized how dorsal-ventral polarity is established in the early *Drosophila* embryo. Establishment of a ventral cell fate is absolutely essential for the physical event of invagination, which leads to the formation of the mesoderm layer. A few pathways, including Torpedo (EGFR), Toll, and Folded gastrulation, are involved from the time of dorsal-ventral axis formation in the oocyte to the time of ventral invagination in the embryo. The coordination of these signaling events, in conjunction with transcriptional regulation, timing of mitosis, and rearrangement of cytoskeletons, is required for the simple, yet elegant process of mesoderm formation.

References

1. In: Bate M, Martinez Arias A, eds. The Development of Drosophila melanogaster. New York: Cold Spring Harbor Laboratory Press, 1993.
2. Stathopoulos A, Levine M. Dorsal gradient networks in the Drosophila embryo. Dev Biol 2002; 246:57-67.
3. Van Buskirk C, Schupbach T. Versatility in signaling: Multiple responses to EGF receptor activation during Drosophila oogenesis. Trends Cell Biol 1999; 9:1-4.
4. Anderson KV. Pinning down positional information: Dorsal-ventral polarity in the Drosophila embryo. Cell 1998; 95:439-442.
5. Ip YT, Gridley T. Cell movements during gastrulation: Snail dependent and independent pathways. Curr Opin Genet Dev 2002; 12:423-429.
6. Leptin M. Gastrulation in Drosophila: The logic and the cellular mechanisms. EMBO J 1999; 18:3187-92.
7. Roth S. The origin of dorsoventral polarity in Drosophila. Philos Trans R Soc 2003; 358:1317-1329.
8. Schupbach T. Germ line and soma cooperate during oogenesis to establish the dorsoventral pattern of egg shell and embryo in Drosophila melanogaster. Cell 1987; 49:699-707.
9. Neuman-Silberberg FS, Schupbach T. The Drosophila dorsoventral patterning gene gurken produces a dorsally localized RNA and encodes a TGF alpha-like protein. Cell 1993; 75:165-74.
10. Spradling AC. Germline cysts: Communes that work. Cell 1993; 72:649-51.
11. Roth S, Neuman-Silberberg FS, Barcelo G et al. Cornichon and the EGF receptor signaling process are necessary for both anterior-posterior and dorsal-ventral pattern formation in Drosophila. Cell 1995; 81:967-78.
12. Nilson LA, Schupbach T. EGF receptor signaling in Drosophila oogenesis. Curr Top Dev Biol 1999; 44:203-43.
13. Brand AH, Perrimon N. Raf acts downstream of the EGF receptor to determine dorsoventral polarity during Drosophila oogenesis. Genes Dev 1994; 8:629-39.
14. Hsu JC, Perrimon N. A temperaturesensitive MEK mutation demonstrates the conservation of the signaling pathways activated by receptor tyrosine kinases. Genes Dev 1994; 8:2176-87.
15. Wasserman JD, Freeman M. An autoregulatory cascade of EGF receptor signaling patterns the Drosophila egg. Cell 1998; 95:355-64.

16. Bier E, Jan LY, Jan YN. Rhomboid, a gene required for dorsoventral axis establishment and peripheral nervous system development in Drosophila melanogaster. Genes Dev 1990; 4:190-203.
17. Urban S, Lee JR, Freeman M. Drosophila rhomboid-1 defines a family of putative intramembrane serine proteases. Cell 2001; 107:173-82.
18. Ruohola-Baker H, Grell E, Chou TB et al. Spatially localized rhomboid is required for establishment of the dorsal-ventral axis in Drosophila oogenesis. Cell 1993; 73:953-65.
19. Peri F, Technau M, Roth S. Mechanisms of Gurken-dependent pipe regulation and the robustness of dorsoventral patterning in Drosophila. Development 2002; 129:2965-75.
20. James KE, Dorman JB, Berg CA. Mosaic analyses reveal the function of Drosophila Ras in embryonic dorsoventral patterning and dorsal follicle cell morphogenesis. Development 2002; 129:2209-22.
21. Wasserman SA. Toll signaling: The enigma variations. Curr Opin Genet Dev 2000; 10:497-502.
22. Morisato D, Anderson KV. Signaling pathways that establish the dorsal-ventral pattern of the Drosophila embryo. Annu Rev Genet 1995; 29:371-99.
23. LeMosy EK, Hong CC, Hashimoto C. Signal transduction by a protease cascade. Trends Cell Biol 1999; 9:102-7.
24. Amiri A, Stein D. Dorsoventral patterning: A direct route from ovary to embryo. Curr Biol 2002; 12:R532-4.
25. Sen J, Goltz JS, Stevens L et al. Spatially restricted expression of pipe in the Drosophila egg chamber defines embryonic dorsal-ventral polarity. Cell 1998; 95:471-81.
26. Sen J, Goltz JS, Konsolaki M et al. Windbeutel is required for function and correct subcellular localization of the Drosophila patterning protein Pipe. Development 2000; 127:5541-50.
27. Konsolaki M, Schupbach T. Windbeutel, a gene required for dorsoventral patterning in Drosophila, encodes a protein that has homologies to vertebrate proteins of the endoplasmic reticulum. Genes Dev 1998; 12:120-131.
28. Hong CC, Hashimoto C. An unusual mosaic protein with a protease domain, encoded by the nudel gene, is involved in defining embryonic dorsoventral polarity in Drosophila. Cell 1995; 82:785-94.
29. Konrad KD, Goralski TJ, Mahowald AP et al. The gastrulation defective gene of Drosophila melanogaster is a member of the serine protease superfamily. Proc Natl Acad Sci USA 1998; 95:6819-24.
30. Dissing M, Giordano H, DeLotto R. Autoproteolysis and feedback in a protease cascade directing Drosophila dorsal-ventral cell fate. EMBO J 2001; 20:2387-93.
31. LeMosy EK, Tan YQ, Hashimoto C. Activation of a protease cascade involved in patterning the Drosophila embryo. Proc Natl Acad Sci USA 2001; 98:5055-60.
32. Chasan R, Jin Y, Anderson KV. Activation of the easter zymogen is regulated by five other genes to define dorsal-ventral polarity in the Drosophila embryo. Development 1992; 115:607-16.
33. Smith CL, DeLotto R. Ventralizing signal determined by protease activation in Drosophila embryogenesis. Nature 1994; 368:548-51.
34. Misra S, Hecht P, Maeda R et al. Positive and negative regulation of Easter, a member of the serine protease family that controls dorsal-ventral patterning in the Drosophila embryo. Development 1998; 125:1261-7.
35. Hashimoto C, Kim DR, Weiss LA et al. Spatial regulation of developmental signaling by a serpin. Dev Cell 2003; 5:945-50.
36. Ligoxygakis P, Roth S, Reichhart JM. A serpin regulates dorsal-ventral axis formation in the Drosophila embryo. Curr Biol 2003; 13:2097-102.
37. Morisato D, Anderson KV. The spatzle gene encodes a component of the extracellular signaling pathway establishing the dorsal-ventral pattern of the Drosophila embryo. Cell 1994; 76:677-88.
38. Schneider DS, Jin Y, Morisato D et al. A processed form of the Spatzle protein defines dorsal-ventral polarity in the Drosophila embryo. Development 1994; 120:1243-50.
39. DeLotto Y, DeLotto R. Proteolytic processing of the Drosophila Spatzle protein by Easter generates a dimeric NGF-like molecule with ventralising activity. Mech Dev 1998; 72:141-148.
40. Weber AN, Tauszig-Delamasure S, Hoffmann JA et al. Binding of the Drosophila cytokine Spatzle to Toll is direct and establishes signaling. Nat Immunol 2003; 4:794-800.
41. Hashimoto C, Hudson KL, Anderson KV. The Toll gene of Drosophila, required for dorsal-ventral embryonic polarity, appears to encode a transmembrane protein. Cell 1988; 52:269-79.
42. Hashimoto C, Gerttula S, Anderson KV. Plasma membrane localization of the Toll protein in the syncytial Drosophila embryo: Importance of transmembrane signaling for dorsal-ventral pattern formation. Development 1991; 111:1021-8.
43. Wakabayashi-Ito N, Belvin MP, Bluestein DA et al. Fusilli, an essential gene with a maternal role in Drosophila embryonic dorsal-ventral patterning. Dev Biol 2001; 229:44-54.

44. Letsou A, Alexander S, Orth K et al. Genetic and molecular characterization of tube, a Drosophila gene maternally required for embryonic dorsoventral polarity. Proc Natl Acad Sci USA 1991; 88:810-4.
45. Feinstein E, Kimchi A, Wallach D et al. The death domain: A module shared by proteins with diverse cellular functions. Trends Biochem Sci 1995; 20:342-4.
46. Charatsi I, Luschnig S, Bartoszewski S et al. Krapfen/dMyd88 is required for the establishment of dorsoventral pattern in the Drosophila embryo. Mech Dev 2003; 120:219-26.
47. Sun H, Towb P, Chiem DN et al. Regulated assembly of the Toll signaling complex drives Drosophila dorsoventral patterning. EMBO J 2004; 23:100-110.
48. Shelton CA, Wasserman SA. Pelle encodes a protein kinase required to establish dorsoventral polarity in the Drosophila embryo. Cell 1993; 72:515-25.
49. Grosshans J, Schnorrer F, Nusslein-Volhard C. Oligomerisation of Tube and Pelle leads to nuclear localisation of dorsal. Mech Dev 1999; 81:127-38.
50. Shen B, Manley JL. Pelle kinase is activated by autophosphorylation during Toll signaling in Drosophila. Development 2002; 129:1925-33.
51. Geisler R, Bergmann A, Hiromi Y et al. Cactus, a gene involved in dorsoventral pattern formation of Drosophila, is related to the IkB gene family of vertebrates. Cell 1992; 71:613-621.
52. Kidd S. Characterization of the Drosophila cactus locus and analysis of interactions between cactus and dorsal proteins. Cell 1992; 71:623-635.
53. Belvin MP, Jin Y, Anderson KV. Cactus protein degradation mediates Drosophila dorsal-ventral signaling. Genes Dev 1995; 9:783-93.
54. Bergmann A, Stein D, Geisler R et al. A gradient of cytoplasmic Cactus degradation establishes the nuclear localization gradient of the dorsal morphogen in Drosophila. Mech Dev 1996; 60:109-23.
55. Cao Z, Henzel WJ, Gao X. IRAK: A kinase associated with the interleukin-1 receptor. Science 1996; 271:1128-1131.
56. Belvin MP, Anderson KV. A conserved signaling pathway: The Drosophila toll-dorsal pathway. Annu Rev Cell Dev Biol 1996; 12:393-416.
57. Hoffmann JA. The immune response of Drosophila. Nature 2003; 426:33-38.
58. Roth S, Stein D, Nusslein-Volhard C. A gradient of nuclear localization of the dorsal protein determines dorsoventral pattern in the Drosophila embryo. Cell 1989; 59:1189-202.
59. Rushlow CA, Han K, Manley JL et al. The graded distribution of the dorsal morphogen is initiated by selective nuclear transport in Drosophila. Cell 1989; 59:1165-77.
60. Steward R. Relocalization of the dorsal protein from the cytoplasm to the nucleus correlates with its function. Cell 1989; 59:1179-88.
61. Jiang J, Kosman D, Ip YT et al. The dorsal morphogen gradient regulates the mesoderm determinant twist in early Drosophila embryos. Genes Dev 1991; 5:1881-91.
62. Ip YT, Park RE, Kosman D et al. The dorsal gradient morphogen regulates stripes of rhomboid expression in the presumptive neuroectoderm of the Drosophila embryo. Genes Dev 1992; 6:1728-39.
63. Simpson P. Maternal-zygotic gene interactions during formation of the dorsoventral pattern in Drosophila embryos. Genetics 1983; 105:615-632.
64. Grau Y, Carteret G, Simpson P. Mutation and chromosomal rearrangements affecting the expression of snail, a gene involved in embryonic patterning in Drosophila melanogaster. Genetics 1984; 108:347-360.
65. Leptin M, Grunewald B. Cell shape changes during gastrulation in Drosophila. Development 1990; 110:73-84.
66. Thisse B, el Messal M, Perrin-Schmitt F. The twist gene: Isolation of a Drosophila zygotic gene necessary for the establishment of dorsoventral pattern. Nucleic Acids Res 1987; 15:3439-53.
67. Leptin M. Twist and snail as positive and negative regulators during Drosophila mesoderm development. Genes Dev 1991; 5:1568-76.
68. Abmayr SM, Keller CA. Drosophila myogenesis and insights into the role of nautilus. Curr Top Dev Biol 1998; 38:35-80.
69. Shirokawa JM, Courey AJ. A direct contact between the dorsal rel homology domain and Twist may mediate transcriptional synergy. Mol Cell Biol 1997; 17:3345-55.
70. Kosman D, Ip YT, Levine M et al. Establishment of the mesoderm-neuroectoderm boundary in the Drosophila embryo. Science 1991; 254:118-22.
71. Alberga A, Boulay JL, Kempe E et al. The snail gene required for mesoderm formation in Drosophila is expressed dynamically in derivatives of all three germ layers. Development 1991; 111:983-92.
72. Ip YT, Maggert K, Levine M. Uncoupling gastrulation and mesoderm differentiation in the Drosophila embryo. EMBO J 1994; 13:5826-34.

73. Kasai Y, Nambu JR, Lieberman PM et al. Dorsal-ventral patterning in Drosophila: DNA binding of Snail protein to the single-minded gene. Proc Natl Acad Sci USA 1992; 89:3414-3418.
74. Hemavathy K, Meng X, Ip YT. Differential regulation of gastrulation and neuroectodermal gene expression by Snail in the Drosophila embryo. Development 1997; 124:3683-91.
75. Sweeton D, Parks S, Costa M et al. Gastrulation in Drosophila: The formation of the ventral furrow and posterior midgut invaginations. Development 1991; 112:775-89.
76. Parks S, Wieschaus E. The Drosophila gastrulation gene concertina encodes a G alpha-like protein. Cell 1991; 64:447-58.
77. Costa M, Wilson ET, Wieschaus E. A putative cell signal encoded by the folded gastrulation gene coordinates cell shape changes during Drosophila gastrulation. Cell 1994; 76:1075-89.
78. Mata J, Curado S, Ephrussi A et al. Tribbles coordinates mitosis and morphogenesis in Drosophila by regulating string/CDC25 proteolysis. Cell 2000; 101:511-522.
79. Seher TC, Leptin M. Tribbles, a cell-cycle brake that coordinates proliferation and morphogenesis during Drosophila gastrulation. Curr Biol 2000; 10:623-629.
80. Grosshans J, Wieschaus E. A genetic link between morphogenesis and cell division during formation of the ventral furrow in Drosophila. Cell 2000; 101:523-531.

CHAPTER 4

Development of the Cardiac Musculature

Rolf Bodmer*

Abstract

The simplicity of the *Drosophila* heart paired with the genetic versatility of this model system make it ideal for studying general principles of embryonic patterning and the specification of individual cell types and lineages. Over the past decade, a remarkably comprehensive model of the genetic network involved in the sequential steps of cardiac specification has emerged: (1) the heart field is specified and positioned to dorsal mesodermal margin by the combined action of the ectodermally derived inductive signals encoded by *wingless* and *dpp* on the adjacent mesoderm expressing the homeobox transcription factor Tinman. (2) As a consequence, Tinman and the GATA factor Pannier and likely other factors are confined to the cardiac mesoderm, and their combined activity is thought to initiate the formation of the heart. (3) Additional positional information, including patterning by *hedgehog* emanating from the ectoderm, further subdivides the cardiac mesoderm into distinct segmental units of progenitors, which then undergo stereotyped lineages. (4) Furthermore, the distinction between (anterior) aorta and (posterior) heart proper involves the broad regional specification by the hox genes, most notably *abdA*. How these genetically identified factors involved in heart formation activate specific enhancers of target genes to generate exquisitely specific patterns of gene expression confined to the heart field are now being elucidated. The insights from *Drosophila* enabled the discovery of a highly homologous gene network involved in vertebrate heart formation and led to the elucidation of the molecular basis of many forms of congenital heart disease.

Introduction

The *Drosophila* heart is a linear tube lying along the dorsal midline of the body axis. It pumps hemolymph through the larval and adult body cavity in an open circulatory system.[1,2] Metamorphosis during the pupal stages causes the heart to be substantially remodeled.[3-5] The larval heart consists of two major cell types: the inner two rows are contractile muscle cells (myocardial cells), and the outer rows are the pericardial cells that flank them (Fig. 1A). Along the anterior-posterior axis of the heart two morphologically distinct sections are distinguishable. The narrower anterior portion is the aorta, while the wider posterior part is the heart proper. These two sections are separated by a simple valve.[2] The posterior larval heart also contains three segmental pairs of inlet valves, called the ostia (Fig. 1B,C).[2,6-8] A row of crescent-shaped myocardial cells from either side of the embryo form a central cavity, the lumen of the heart. Finally the heart is suspended along to dorsal midline epidermis by means of small somatic muscles.

*Rolf Bodmer—The Burnham Institute, 10901 N. Torrey Pines Road, La Jolla, California 92037, U.S.A. Email: rolf@burnham.org

Muscle Development in Drosophila, edited by Helen Sink. ©2006 Eurekah.com and Springer Science+Business Media.

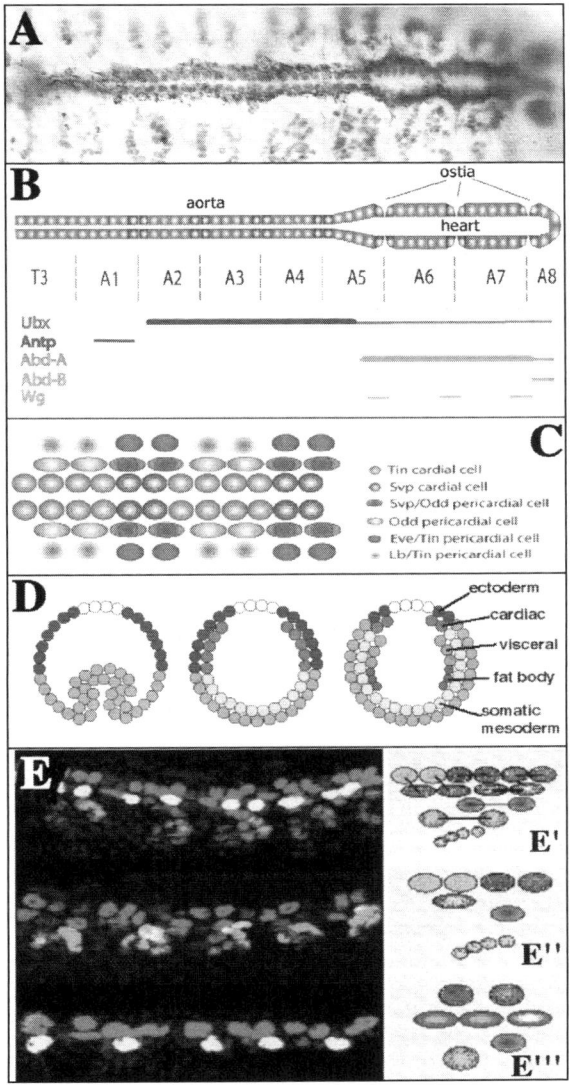

Figure 1. A) Photomicrograph showing the cardiac musculature. The myocardial nuclei are positive for Dmef2 (blue) and the pericardial cells are positive for Pericardin (brown). Note blue myocardial nuclei in between brown percardial cells. B) Myocardium of the heart tube (Tin-expressing cell, red; Svp-expressing cells blue) with segmental register of Hox proteins and Wg. Wg marks the ostia. C) Myocardial and pericardial cell types defined by the combination of transcription factors expression. D) Mesodermal movements during gastrulation involves invagination along the ventral midline (left), dorsal migration (middle, in red is the dorsal medoderm) and subdivision into mesodermal subtypes as indicated (right). Note that the heart forms at the dorsal mesodermal edge (in red). E) Five consecutive hemisegments double-labeled for Tin (red) and Eve (green) in *wild-type* (top), *cycA* mutant (middle) and *cycA;numb* double mutant (bottom). Diagrams of the cardiac lineages in A2-A7 of: E' *wild-type* ; E" *cycA* mutant and E''' *cycA;numb* double mutant. The top row in each panel is the myocardial cells, middle 3 rows are the pericardial cells and green beads indicate the DA1/1 muscle. In E' the bars indicate lineage relationships.[28] A color version of this figure is available online at http://www.Eurekah.com.

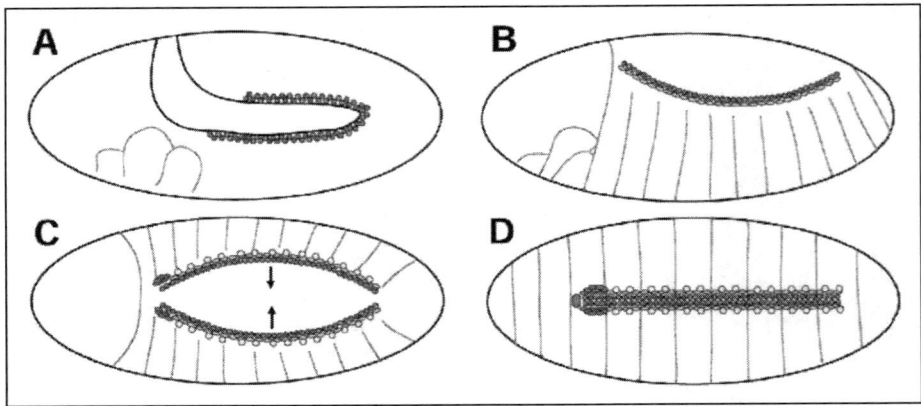

Figure 2. Stages of cardiac mesoderm and heart tube assembly. A) Side view of heart field at late stage 11 germ band extension. B) Side view of heart progenitors at stage 13 germ band retraction. C) Dorsal view of bilateral rows of cells in the process of migration towards the midline (arrows). D) Dorsal view of the heart after assembly.

From Mesoderm to Heart Progenitors

During gastrulation, in the ventral third portion of the *Drosophila* blastoderm embryo, the presumptive mesoderm invaginates in a furrow along the ventral midline and localizes internally (Fig. 1D). Once internalized the mesoderm flattens and spreads dorsally by cell migration, forming a morphologically uniform monolayer of cells in close contact with the ectoderm.[9] Coincidentally, the long germ band folds over and extends anteriorly to bring the posterior end of the embryo in proximity with the head region (Fig. 2). After the mesodermal monolayer reaches the dorsal edge of the ectoderm, molecular markers begin to distinguish the dorsal from the ventral portion of the mesoderm, followed by the first discernable morphological subdivisions (Fig. 1D). The dorsal mesoderm gives rise to the outer dorsal somatic mesoderm and the inner visceral mesoderm. The dorsal mesoderm forms mostly somatic muscles. The inner visceral mesoderm segregates as an epithelium, and serves to guide the migration of the mid-gut primordia that grow in from the embryo's anterior and posterior ends.[10-12] In close temporal sequence the heart progenitors emerge at the dorsal-most edge of the outer layer of the dorsal mesoderm. They form the cardiac mesoderm (Fig. 2).[13,14] The fat body and somatic gonad are also specified at this time.[15,16]

The bilaterally symmetrical heart primordia form at the germ band extended stage (Fig. 2). Then following germ band retraction (Fig. 2) the two rows of cardiac mesoderm move towards the dorsal midline (Fig. 2). Here they assemble into a linear tube (Fig. 2). As the myocardial cells align in register with one another they assume an apical-basal (dorsal-ventral) polarity.[17,18] While the molecular-genetic basis of cardiac morphogenesis has not yet been studied in detail, many of the molecular determinants of cardiogenesis have been identified. Their homologues in vertebrates have functions that are remarkably conserved.[14,19-23]

Cardiac Lineages

The lineages that contribute to the cell types forming the heart tube and surrounding pericardial cells have been studied in considerable detail (Figs. 1D,E).[24-28] Within the emerging cardiac anlagen (primordial cells) at the dorsal mesoderm edge (Fig. 2) groups of progenitors express distinct combinations of transcription factors (Fig. 1D), and undergo stereotyped patterns of cell division to generate cardiac cell diversity (Fig. 1E).

The cardiac progenitors emerge as a two to three cell-wide unit of dorsal mesodermal cells at mid-stage 11, when most if not all express the Tinman (Tin) homeobox transcription factor

(Fig. 1B,C). They extend from the third thoracic segment (T3) to the eighth abdominal segment (A8).[13,29] Additional markers are then expressed in subsets of prospective heart cells (Fig. 1B,C). The *even-skipped* (*eve*) segmentation homeobox gene is expressed in two pericardial cells with large nuclei in each hemisegment.[30] The tandem homeobox genes *ladybird* (*lbe*)[31] is also expressed in two pericardial cells. The Seven-up COUP steroid receptor (Svp)[32] is expressed in two myocardial cells and also two pericardial cells. The two myocardial cells coexpress the *dorsalcross* T-box encoding genes,[33] whereas two of the four pericardial cells coexpress the zinc finger protein Odd-skipped (Odd) and the remaining two are Svp-positive.[34] *Tin* is expressed in two myocardial cells and at least four pericardial cells. Two of the pericardial cells coexpress *eve* and the other two coexpress *lbe*.[27,35] The myocardial cells share characteristics with other muscle forming mesodermal cells such as expression of *Dmef2*, a MADS-box transcription factor (Fig. 1A),[36-39] muscle myosin heavy chain[40] and beta-3-tubulin.[41]

Interestingly, some of the heart precursors divide symmetrically, whereas others exhibit asymmetric cell division patterns. Two sets of asymmetric lineages have been described in the cardiac mesoderm based on cell division patterns[24] clonal analysis.[27,34] and cell cycle mutant phenotypes.[28] The Svp lineage generates two myocardial cells and two pericardial cells per hemisegment, whereas the Eve lineage generates two Eve pericardial cells and two founder cells for dorsal somatic muscles. Thus, cardiac and somatic muscle associated cells do not always derive from separate lineages. A common feature of the cardiac asymmetric lineages is that they distinguish between myogenic and nonmyogenic sibling cell fates. In the case of the Eve lineage, one Eve progenitors per hemisegment generates two Eve pericardial cells and a somatic muscle founder cell (DO2/8 i.e., muscle letter/number hereon). In contrast the other Eve progenitor in the hemisegment gives rise to another muscle founder cell (DA1/1) and an unidentified sibling. The final division of the immediate precursor of the Eve pericardial cells is then symmetrical. The other heart associated cell fates in the main portion of the heart (A2-8) undergo symmetric cell divisions, which is in contrast to the 'outflow track' (T3-A1), where most lineages appear to be asymmetric.[27,28]

Studies of cell cycle mutants suggested some interesting features of these cardiac lineages.[28] Most cells in the developing embryo undergo only three additional, partially synchronous rounds of cell division after the blastoderm stage, with the notable exception of the CNS.[42,43] The last round of global cell division, mitosis 16, is blocked in *cyclinA* (*cycA*) mutants.[44,45] *CyclinB* (*cycB*) mutants have no obvious effect during these divisions, but *cycA,cycB* double mutants synergistically arrest at cycle 15.[44] In embryos mutant for *string* (*stg*) (*cdc25* in yeast), cell division is already arrested at cycle 14.[42,46,47] The CDK inhibitors Dacapo and Fizzy-related negatively regulate Cyclin levels and are required to exit the cell cycle.[48,49] Heart precursors continue to differentiate in cell cycle mutants and in embryos in which cell cycle inhibitors are overexpressed. The cell fate of symmetrically dividing progenitors of either myocardial- or pericardial-only lineages is not altered by premature arrest of cell division. In contrast, arrested progenitors of the asymmetric Eve and Svp lineages always adopt a myogenic, as opposed to pericardial fate (Fig. 1E, middle panel). In cycle 15 arrested *cycA;cycB* double mutants, only two of the normally six myocardial cells are formed. One expresses *tin* and the other *svp*. This suggests that in each hemisegment two higher order myocardial progenitors are initially specified, one giving rise to the *tin*-expressing myocardial cells and the other to the *svp*-positive myocardial and pericardial cells. In conclusion, there are probably 5-6 cardiac progenitors specified in each hemisegments.

The molecular basis for specifying alternative cell fates during asymmetric cell division has been elucidated in some detail, and involves asymmetric segregation of intracellular determinants between the progenitor and its daughter cells. The presence of the membrane-associated *numb* gene product is necessary and sufficient for specifying a daughter cell fate that is different from its Numb devoid sibling. This occurs by inhibiting the activation of the transmembrane Notch receptor and a four-pass transmembrane protein encoded by *sanpodo*.[26,50-59] In the *Drosophila* heart, the asymmetric *eve* and *svp* lineages are under the control of *numb*, *Notch* and *sanpodo*.[25,27,28,34,60]

In contrast to the asymmetric lineages, constitutive activation or repression of the *Notch* pathway has no apparent effect on the symmetric myocardial-only or pericardial-only lineages. Following *numb* loss-of-function or ectopic Notch activation there are twice the number of *svp* and *eve* pericardial cells at the expense of the *svp* myocardial or *eve* muscle founder cells. In contrast the number of *tin*-expressing myocardial cells and the other pericardial cells are unaffected. When *numb* is overexpressed the opposite phenotype is observed in *svp* and *eve* lineages, but again without affecting cell fate of the symmetrical lineages. Hence regulation of Notch signaling dramatically affects cell fate in asymmetric lineages by preventing a myogenic fate, but even forced expression is incapable of altering the myogenic- or pericardial-only fate of symmetrically dividing progenitors. These conclusions are also supported by experiments in which cell cycle mutants are combined with lineage mutants (Fig. 1E, lower panel).[28] Remarkably, a switch from *Notch* insensitivity to *Notch* responsiveness can occur even within the same lineage: the initial *svp* progenitor first divides symmetrically then asymmetrically, whereas the Eve pericardial cells progenitor first divides asymmetrically and then symmetrically. It may very well be that a convenient way to increase complexity of an organism during evolution is through regulating *numb/Notch* responsiveness during phases of cell proliferation.

Specification and Positioning of the Cardiac Mesoderm

The mesoderm, which is a prerequisite of heart formation, depends on formation of the embryonic dorsal-ventral body axis.[61] The first mesoderm-specific genes that are activated in the ventral third of the blastoderm embryo are *twi* and *snail*, encoding a basic helix-loop-helix and a zinc finger protein respectively.[62,63] Twi acts as a positive determinant of mesoderm, whereas Snail acts by repressing neuroectodermal gene expression within the mesodermal anlagen.[64-69] At the blastoderm stage Twi turns on: Tin;[13] Zinc Finger Homeodomain Protein-1, a nuclear protein containing a homeodomain and several zinc fingers;[70,71] Heartless (a.k.a. DFR1), a membrane protein of the Fibroblast Growth Factor Receptor family;[72-75] *Drac1*, a GTPase;[76] Pointed P2, an ETS transcription factor;[77] and Dmef2, the *Drosophila* member of the Myocyte Enhancer Factor-2 family of MADS-box-containing nuclear proteins.[36-39] Twi and its targets are likely to provide a mesoderm-specific context for appropriate differentiation of mesodermal derivatives.

During gastrulation the mesodermal anlagen invaginates along the ventral midline. The mesoderm then flattens into a monolayer that spreads dorsally in close apposition to the ectoderm, a process requiring Heartless function.[73-75] If the mesoderm does not come into contact with the dorsal ectoderm, heart and visceral muscle formation is severely reduced. As outlined below, secreted signals emanating from the dorsal ectoderm are crucial for inducing cardiogenesis (reviewed in ref. 78). In contrast to Heartless' role in cell migration, Tin is crucial for mesodermal regionalization and specification of dorsal fates.

Although *tin* is first expressed uniformly in the presumptive trunk mesoderm, directly dependent on Twi,[13,79] it is restricted to the dorsal portion of the mesoderm at stage 10. *tin* mutants primarily lack derivatives of the dorsal mesoderm, including the heart, the visceral and a set of dorsal somatic body wall muscles.[12,35] Thus, specification of 'dorsal mesoderm' by Tin is crucial for subsequent subdivisions and differentiation. Dorsal mesoderm specification, including *tin* expression, also depends on the product of *decapentaplegic* (*dpp*). *dpp* encodes a secreted factor that is a member of the Transforming Growth Factor-β (TGF-β, superfamily, and is required for formation of the dorsal-ventral axis.[61,80-82] *dpp* is expressed in the dorsal ectoderm and signals across germlayers to pattern the underlying dorsal mesoderm (reviewed in ref. 78).[83-85] From mid-stage 11 on *tin* is further restricted to the heart progenitors in the trunk region (Fig. 2A), and is maintained through adulthood.[5] Enhancer elements of *tin* have been found that direct expression in the early mesoderm, the dorsal mesoderm and cardiac mesoderm.[79,85,86]

Although *dpp* is required for heart formation and cardiogenesis, the restriction of *tin* expression to the heart-forming region at the dorsal mesodermal margin requires additional positional information. This pattering information is provided by the striped expression of the segmentation gene *wingless*. This gene encodes a secreted glycoprotein of the Wnt class, and is essential for cardiac specification.[87,88] Genetic evidence suggests that the dynamic expression patterns of *dpp*, *wingless* and *tin* is not only essential but also instructive for initiating cardiogenesis and correct positioning of the heart progenitors.[89] Recently it was shown that the cardiac restricted expression of *pannier*, which encodes a GATA transcription factor, with *tin* provides heart-forming competence within the mesoderm.[90-93]

In *Drosophila*, *wingless* is expressed and required in two phases of heart development.[31,87,89] The first phase is mediated by the canonical *wingless* pathway.[88] This contrasts with vertebrates, where it appears to be noncanonical Wnt signaling that promotes cardiogenesis.[94] It remains to be determined whether in flies the second phase of cardiogenic *wingless* signaling is noncanonical. *dpp* also has two phases of expression: first in broad dorsal ectodermal band, and later in a narrow stripe along the dorsal ectodermal margin above the cardiac progenitors. It is tempting to speculate that *dpp*, like *wingless*, also acts in two phases to initiate heart development. *pannier* is a target of *dpp* in the ectoderm and is, in turn, required for maintaining the late phase of *dpp* expression.[93,95,96] This led to the model where the convergence of *wingless* and *dpp* signaling initiates and maintains *tin* and *pannier* expression in the cardiac mesoderm, then these transcription factors provide the appropriate cardiogenic context for heart development to proceed.[21,89,93] Therefore the genetic cascade of heart specification is:

Pannier

Twist → Tinman (meso); Wingless, Dpp (ecto) → ↓↑ → cardiogenic fate

Tinman

pannier is a likely direct target of Tin.[97] Yet *pannier* overexpression throughout the mesoderm results in transient, ectopic *tin* expression. This suggests that both of these transcription factors are required to maintain each other's expression. *tin* and *pannier* may in fact act synergistically in specifying cardiac development, since mesodermal overexpression of both but not either alone induces robust and sustained expression of cardiac-specific markers, such as the basic Helix-Loop-Helix transcription factor encoded by *dHand* and the ABC-class transmembrane protein encoded by *dSUR*.[93] This is consistent with in vitro experiments using vertebrate cell lines, in which it was shown that the vertebrate homologs of these genes activate target genes synergistically (e.g., refs. 98-100). Taken together, the following sequence of events is a probable scenario of cardiogenesis: at stage 9/10 *dpp* maintains *tin* expression in the dorsal mesoderm and *pannier* expression in the dorsal ectoderm. At early to mid-stage 11, *tin* and *dpp/wingless* initiate *pannier* in the cardiogenic region, and *pannier* and *dpp/wingless* maintain *tin* expression in the heart field. Later, *tin* and *pannier* maintain each other, and ectodermal *pannier* maintains late *dpp* expression.

A set of T-box genes, encoded by *neuromancer I* (a.k.a. *H15*)[101] and its neighbor *neuromancer II*, may also participate in determining cardiogenesis.[101a] Interestingly, in vitro protein binding and transactivation studies in vertebrates suggest a synergistic relationship between Nkx, Gata and T-box transcription factors.[102,103] In the future, it will be interesting to see if the combination of *tin*, *pannier* and *neuromancer* genes is not only essential but also sufficient in promoting heart formation.

Specification of Cardiac Cell Types

The next question of interest is how individual cell fates are specified within the heart field, a process which may occur in parallel with the overall cardiac initiation. Four groups of heart progenitors emerge within the prospective cardiogenic region in stereotyped, segmentally repeated clusters, expressing either *eve*, *lbe*, *svp* or *odd*.[30,31,58,60,104] The first clusters to emerge within a segmental unit are the *eve* cells, at a time when *tin* is still expressed in the entire dorsal mesoderm. The specification of the cell types expressing mesodermal *eve* and the regulation of this expression has been extensively studied.[25,24,71,104-108] In addition to *dpp, wingless, tin* and *twi*, specification of the *eve* clusters also requires Ras-mediated input from Epidermal Growth Factor and Fibroblast Growth Factor Receptor tyrosine kinases. Eve is not only a marker for the cells it is expressed in but is also required as an identity gene for their correct differentiation.[60,71,104] To study *eve*'s role in the mesoderm, the enhancer responsible for mesodermal *eve* expression was selectively eliminated in a genomic rescue construct.[109] Flies that are null mutant for *eve* but contain this 'eve meso minus' rescue construct as a transgene are viable but lack the mesodermal *eve* clusters and exhibit defective heart function.

Anterior to the *eve* clusters in each segment emerge the *lbe* clusters.[31] In *lbe* loss-of-function mutant the *eve* clusters are enlarged, while mesodermal overexpression of *lbe* suppresses *eve*.[60,104] Conversely, 'eve meso minus' rescue embryos have expanded *lbe* expression, and mesodermal overexpression of *eve* eliminates *lbe* expression in the heart. This suggests that *eve* and *lbe* repress each other and are thus present in nonoverlapping clusters.

lbe is also required in a group of ectodermal cells that contact the outflow track of the heart.[110] This is reminiscent of the neural crest cells that pattern the cardiac outflow track in vertebrates.[111] Posterior to the Eve cells are the *svp*-expressing cell clusters near the segmental boundaries (see also Fig. 1B,C). *svp* is first coexpressed with *tinman*, but then suppresses it in the *svp* cells.[5,6,7] Unlike *lbe* and *eve*, however, *svp* gain- or loss-of-function does not influence cardiac *eve* or *lbe* expression. The fourth group consists of two sets of *odd*-expressing pericardial cells,[34] one of which also expresses *svp*. The function of *odd* in the heart has not yet been determined.

How the genetic information leading to precise positioning of cardiac cell types is integrated at the transcriptional level has been studied in detail using *eve*'s mesodermal enhancer (eme).[104,107,109,112] eme contains multiple consensus or in vitro binding sites for all the transcription factors that are required genetically for *eve* expression in the cardiac mesoderm, such as for Pangolin/dTCF (*wingless* pathway), Mad (*dpp* pathway), Pointed (ETS factor, ras pathway), Twi and Tin.[104,107,112] Overall, these sites seem to act in an additive fashion, such that mutating progressively more *tin* sites eventually eliminates reporter gene expression. Therefore the genetic network that provides cardiogenic competence to the dorsal edge of the mesoderm also acts combinatorially at a single enhancer.

The above activator inputs appear to explain well why mesodermal *eve* is expressed within the cardiac mesoderm, but they do not explain how *eve* transcription is confined to a rather small segmental cluster within the heart. Since Lbe acts genetically as a repressor of *eve*, it may do so by direct interaction with the mesodermal *eve* enhancer. Indeed, Lbe not only binds to the enhancer in vitro, but mutating the corresponding sites leads to an expansion of reporter gene expression to encompass the entire heart field in vivo.[104] In addition, eme without Lbe consensus sites is no longer repressible by Lbe as expected when Lbe directly regulates *eve* in the mesoderm. In summary, the precision of mesodermal *eve* expression is achieved by a combination of activators and repressors converging on a small enhancer.

Formation of the neighboring *eve* and *lbe* clusters in each segment requires a similar combination of activating genetic inputs (*dpp, wingless, tin,* etc.) and they exclude each other from their territory by mutual repression. It is unclear, however, what mechanism positions the *eve* progenitors posterior to *lbe*. An emerging candidate is *hedgehog*, which is expressed in stripes adjacent and posterior to those of *wingless*, and is already known to be required for heart formation via its role in maintaining *wingless* expression.[87,88] To test for additional cardiac patterning functions, *hedgehog* has been eliminated while maintaining *wingless*. Interestingly, in such *hedgehog* off–*wingless* on embryos the *eve* cell clusters do not form and the *lbe* clusters are expanded to all *tin*-expression cardiac progenitors (J. Liu, L. Qian and R. Bodmer, unpublished data). Overexpression of *hedgehog* and activation of the ras pathway expands *eve* and reduces *lbe* in the cardiogenic mesoderm[106] (J. Liu, L. Qian and R. Bodmer, unpublished data). Moreover, manipulation of *hedgehog* alters the cardiac levels of *rhomboid* mRNA, which codes for a protease that regulates Epidermal Growth Factor Receptor-mediated Ras-activation.

These findings suggest that *hedgehog*, via augmenting ras signaling, activates the mesodermal *eve* cells nearby in the underlying mesoderm, and inhibits the *lbe* clusters. These then form further away. The mutual repression between *eve* and *lbe* then maintains the stereotyped anterior-posterior arrangement of these cell fates within each segment. In conclusion, we now understand some of the basic genetic principles by which individual heart progenitors and their progeny are specified, positioned and distinguished from their neighbors.

Many more genes are likely involved in cardiogenesis, either during specification or at a later differentiation step. Two gene functions have recently been identified that when mutated cause a dramatic increase in the myocardial cell population of the heart: the ETS gene *pointed* and the *muscle-specific homeobox* (*msh*) gene.[27,60] *msh* and *pointed* may restrict the number of heart progenitors to the most dorsal region of the mesoderm as well as play a role in specifying dorsal skeletal muscle founders. *msh* seems to act in a cross-repressive network with *lbe* and *eve* to specify spatially restricted cardiac (and skeletal) cell fates; and in the case of *pointed*, it has been suggested that its role in determining myocardial cell number is likely in addition to its function as an effecter of the ras pathway. In order to find additional gene functions that participate in cardiogenesis, heart-specific gain- and loss-of-function screens have recently been carried out. In a gain-of-function screen based on mesoderm-specific overexpression 84 out of 2,293 gene functions were found to affect *lbe* or *eve* expression.[108] In a RNAi-based loss-of-function screen 132 out of 5,849 genes altered embryonic heart morphology.[113]

Hox Gene Function in the Heart

In addition to the segmental specification described earlier, the embryonic heart is divided into three regions (Fig. 2): the outflow track (T3-A1), the aorta (A2-A5) and the wider posterior heart (A6-A8), which also contains six *wingless*-expressing hemolymph inlet valves known as ostia (Figs. 1 and 2).[5] Three studies have recently elucidated the role of the *hox* genes in this broad regionalization of the heart (reviewed in ref. 29).[8,114,115]

The principle *hox* gene player involved in discriminating between the aorta and the posterior heart is *abdominal-A* (*abd-A*). The expression pattern of other *hox* genes in the embryonic heart is arranged in an anterior-posterior pattern (Fig. 1B,C), reminiscent of their nested expression pattern during segmental identity specification in the early embryo. *Antennapedia* (*Antp*) is expressed primarily in the outflow track A1 region, *Utrabithorax* (*Ubx*) at high levels in the aorta (A2-A5), *abdominal-A* (*abd-A*) is confined to the posterior heart (A6-A8), and *Abdominal-B* (*Abd-B*) is expressed in the four posterior-most myocardial cells.

abd-A acts as the homeotic selector for posterior heart identity. In *abd-A* mutants expression of *troponinC-akin1* (*tina1*) and *wingless* in the posterior heart is abolished. In addition the posterior heart looks like the aorta and expresses *Ubx* at high levels. Conversely, in embryos in which *abd-A* is expressed uniformly in the mesoderm or exclusively in the heart the anterior portion of the heart is now wider in diameter expressing *tina1* and forming *wingless* expressing ostia throughout.

Since *Ubx* is prevalent more anteriorly than *abdA* (Fig. 2), it may be the homeotic selector gene for a 'posterior aorta' fate. As expected, loss-of-*Ubx*-function does not alter posterior heart differentiation. The *Ubx* mutant aorta exhibits only minor abnormalities in epithelial polarity,[8] and there is no alteration of *tina1* or *svp* expression. In *Ubx*, *abd-A* double mutants no significant additional changes are observed in the myocardium, although not enough outflow track markers are available to assess a possible anterior transformation. In contrast, the pericardial lymph gland primordia in A1 are expanded along the entire aorta in *Ubx* mutants,[116] consistent with a mesoderm-autonomous homeotic function of *Ubx*. However, when *Ubx* is overexpressed uniformly in the mesoderm *abd-A* expression in the heart is not altered, but surprisingly *tina1* is expanded anteriorly as with *abd-A* overexpression. This suggests that in this overexpression assay *Ubx* can partially substitute for its relative *abd-A*. In contrast the pan-mesodermal overexpression of *abd-A*'s posterior relative *Abd-B* (Fig. 1B,C) completely suppresses cardiogenesis, and in *Abd-B* mutants additional myocardial cells are formed more posteriorly.[113] Taken together, these findings suggest that the *hox* genes function autonomously within the heart progenitors to specify correct regional identities along the anterior-posterior axis.

References

1. Miller A. The internal anatomy and histology of the imago of Drosophila melanogaster. In: Demerec M, ed. Biology of Drosophila. New York: Wiley, 1950:420-534.
2. Rizki TM, Rizki RM. Larval adipose tissue of homoeotic bithorax mutants of Drosophila. Dev Biol 1978; 65:476-482.
3. Jensen PV. Structure and metamorphosis of the larval heart of Calliphora erythrocephala. K Dansk Vidensk Selsk Biol Skrift 1973; 20:2–19.
4. Curtis NJ, Ringo JM, Dowse HB. Morphology of the pupal heart, adult heart, and associated tissues in the fruit fly, Drosophila melanogaster. J Morphol 1999; 240:225-235.
5. Molina MR, Cripps RM. Ostia, the inflow tracts of the Drosophila heart, develop from a genetically distinct subset of cardial cells. Mech Dev 2001; 109:51-59.
6. Gajewski K, Choi CY, Kim Y et al. Genetically distinct cardial cells within the Drosophila heart. Genesis 2000; 28:36-43.
7. Lo PC, Frasch M. A role for the COUP-TF-related gene seven-up in the diversification of cardioblast identities in the dorsal vessel of Drosophila. Mech Dev 2001; 104:49-60.
8. Ponzielli R, Astier M, Chartier A et al. Heart tube patterning in Drosophila requires integration of axial and segmental information provided by the Bithorax Complex genes and hedgehog signaling. Development 2002; 129:4509-4521.
9. Leptin M, Grunewald B. Cell shape changes during gastrulation in Drosophila. Development 1990; 110:73-84.
10. Borkowski OM, Brown NH, Bate M. Anterior-posterior subdivision and the diversification of the mesoderm in Drosophila. Development 1995; 121:4183-4193.
11. Tepass U, Hartenstein V. Epithelium formation in the Drosophila midgut depends on the interaction of endoderm and mesoderm. Development 1994; 120:579-590.
12. Bodmer R. The gene tinman is required for specification of the heart and visceral muscles in Drosophila. Development 1993; 118:719-729.
13. Bodmer R, Jan LY, Jan YN. A new homeobox-containing gene, msh-2, is transiently expressed early during mesoderm formation of Drosophila. Development 1990; 110:661-669.
14. Bodmer R. Heart development in Drosophila and its relationship to vertebrate systems. Trends Cardiovasc. Med 1995; 5:21-28.
15. Azpiazu N, Lawrence PA, Vincent JP et al. Segmentation and specification of the Drosophila mesoderm. Genes Dev 1996; 10:3183-3194.
16. Riechmann V, Irion U, Wilson R et al. Control of cell fates and segmentation in the Drosophila mesoderm. Development 1997; 124:2915-2922.
17. Rugendorff A, Younossi-Hartenstein A, Hartenstein V. Embryonic origin and differentiation of the Drosophilia heart. Roux's Arch Dev Biol 1994; 203:266-280.
18. Chartier A, Zaffran S, Astier M et al. Pericardin, a Drosophila type IV collagen-like protein is involved in the morphogenesis and maintenance of the heart epithelium during dorsal ectoderm closure. Development 2002; 129:3241-3253.
19. Harvey RP. NK-2 homeobox genes and heart development. Dev Biol 1996; 178:203-216.
20. Bodmer R, Venkatesh TV. Heart development in Drosophila and vertebrates: Conservation of molecular mechanisms. Dev Genet 1998; 22:181-186.
21. Lockwood WK, Liu M, Su M-T et al. A genetic model for cardiac pattern formation and cell fate determination. In: Haddad G, Xu T, eds. In Genetic Models In Cardiorespiratory Biology. Lung Biology Series, 2001:179-201.
22. Cripps RM, Olson EN. Control of cardiac development by an evolutionarily conserved transcriptional network. Dev Biol 2002; 246:14-28.
23. Zaffran S, Frasch M. Early signals in cardiac development. Circ Res 2002; 91:457-469.
24. Carmena A, Gisselbrecht S, Harrison J et al. Combinatorial signaling codes for the progressive determination of cell fates in the Drosophila embryonic mesoderm. Genes Dev 1998; 12:3910-3922.
25. Park M, Yaich LE, Bodmer R. Mesodermal cell fate decisions in Drosophila are under the control of the lineage genes numb, Notch, and sanpodo. Mech Dev 1998; 75:117-126.
26. Vervoort M, Dambly-Chaudiere C, Ghysen A. Cell fate determination in Drosophila. Curr Opin Neurobiol 1997; 7:21-28.
27. Alvarez AD, Shi W, Wilson BA et al. Pannier and pointedP2 act sequentially to regulate Drosophila heart development. Development 2003; 130:3015-3026.

28. Han Z, Bodmer R. Myogenic cells fates are antagonized by Notch only in asymmetric lineages of the Drosophila heart, with or without cell division. Development 2003; 130:3039-3051.
29. Lo PC, Frasch M. Establishing A-P polarity in the embryonic heart tube: A conserved function of Hox genes in Drosophila and vertebrates? Trends Cardiovasc Med 2003; 13:182-187.
30. Frasch M, Hoey T, Rushlow C et al. Characterization and localization of the even-skipped protein of Drosophila. EMBO J 1987; 6:749-759.
31. Jagla K, Jagla T, Heitzler P et al. Ladybird, a tandem of homeobox genes that maintain late wingless expression in terminal and dorsal epidermis of the Drosophila embryo. Development 1997; 124:91-100.
32. Mlodzik M, Hiromi Y, Weber U et al. The Drosophila seven-up gene, a member of the steroid receptor gene superfamily, controls photoreceptor cell fates. Cell 1990; 60:211-224.
33. Reim I, Lee HH, Frasch M. The T-box-encoding Dorsocross genes function in amnioserosa development and the patterning of the dorsolateral germ band downstream of Dpp. Development 2003; 130:3187-3204.
34. Ward EJ, Skeath JB. Characterization of a novel subset of cardiac cells and their progenitors in the Drosophila embryo. Development 2000; 127:4959-4969.
35. Azpiazu N, Frasch M. Tinman and bagpipe: Two homeo box genes that determine cell fates in the dorsal mesoderm of Drosophila. Genes Dev 1993; 7:1325-1340.
36. Nguyen HT, Bodmer R, Abmayr SM et al. D-mef2: A Drosophila mesoderm-specific MADS box-containing gene with a biphasic expression profile during embryogenesis. Proc Natl Acad Sci USA 1994; 91:7520-7524.
37. Bour BA, O'Brien MA, Lockwood WL et al. Drosophila MEF2, a transcription factor that is essential for myogenesis. Genes Dev 1995; 9:730-741.
38. Lilly B, Galewsky S, Firulli AB et al. D-MEF2: A MADS box transcription factor expressed in differentiating mesoderm and muscle cell lineages during Drosophila embryogenesis. Proc Natl Acad Sci USA 1994; 91:5662-5666.
39. Lilly B, Zhao B, Ranganayakulu G et al. Requirement of MADS domain transcription factor D-MEF2 for muscle formation in Drosophila. Science 1995; 267:688-693.
40. Zhang S, Bernstein SI. Spatially and temporally regulated expression of myosin heavy chain alternative exons during Drosophila embryogenesis. Mech Dev 2001; 101:35-45.
41. Damm C, Wolk A, Buttgereit D et al. Independent regulatory elements in the upstream region of the Drosophila beta 3 tubulin gene (beta Tub60D) guide expression in the dorsal vessel and the somatic muscles. Dev Biol 1998; 199:138-149.
42. Foe VE. Mitotic domains reveal early commitment of cells in Drosophila embryos. Development 1989; 107:1-22.
43. Campos-Ortega JA, Hartenstein V. The Embryonic Development of Drosophila Melanogaster. Springer Verlag 1997; 405.
44. Knoblich JA, Lehner CF. Synergistic action of Drosophila cyclins A and B during the G2-M transition. EMBO J 1993; 12:65-74.
45. Dong X, Zavitz KH, Thomas BJ et al. Control of G1 in the developing Drosophila eye: Rca1 regulates Cyclin A. Genes Dev 1997; 11:94-105.
46. Edgar BA, O'Farrell PH. Genetic control of cell division patterns in the Drosophila embryo. Cell 1989; 57:177-187.
47. Edgar BA, O'Farrell PH. The three postblastoderm cell cycles of Drosophila embryogenesis are regulated in G2 by string. Cell 1990; 62:469-480.
48. Lane ME, Sauer K, Wallace K et al. Dacapo, a cyclin-dependent kinase inhibitor, stops cell proliferation during Drosophila development. Cell 1996; 87:1225-1235.
49. Sigrist SJ, Lehner CF. Drosophila fizzy-related down-regulates mitotic cyclins and is required for cell proliferation arrest and entry into endocycles. Cell 1997; 90:671-681.
50. Uemura T, Shepherd S, Ackerman L et al. Numb, a gene required in determination of cell fate during sensory organ formation in Drosophila embryos. Cell 1989; 58:349-360.
51. Rhyu MS, Jan LY, Jan YN. Asymmetric distribution of numb protein during division of the sensory organ precursor cell confers distinct fates to daughter cells. Cell 1994; 76:477-491.
52. Spana EP, Kopczynski C, Goodman CS et al. Asymmetric localization of numb autonomously determines sibling neuron identity in the Drosophila CNS. Development 1995; 121:3489-3494.
53. Brewster R, Bodmer R. Origin and specification of type II sensory neurons in Drosophila. Development 1995; 121:2923-2936.
54. Guo M, Jan LY, Jan YN. Control of daughter cell fates during asymmetric division: Interaction of Numb and Notch. Neuron 1996; 17:27-41.
55. Spana EP, Doe CQ. Numb antagonizes Notch signaling to specify sibling neuron cell fates. Neuron 1996; 17:21-26.

56. Lu B, Rothenberg M, Jan LY et al. Partner of Numb colocalizes with Numb during mitosis and directs Numb asymmetric localization in Drosophila neural and muscle progenitors. Cell 1998; 95:225-235.

57. Dye CA, Lee JK, Atkinson RC et al. The Drosophila sanpodo gene controls sibling cell fate and encodes a tropomodulin homolog, an actin/tropomyosin-associated protein. Development 1998; 125:1845-1856.

58. Skeath JB, Doe CQ. Sanpodo and Notch act in opposition to Numb to distinguish sibling neuron fates in the Drosophila CNS. Development 1998; 125:1857-1865.

59. O'Connor-Giles KM, Skeath JB. Numb inhibits membrane localization of Sanpodo, a four-pass transmembrane protein, to promote asymmetric divisions in Drosophila. Dev Cell 2003; 5:231-243.

60. Jagla T, Bidet Y, Da Ponte JP et al. Cross-repressive interactions of identity genes are essential for proper specification of cardiac and muscular fates in Drosophila. Development 2002; 129:1037-1047.

61. St Johnston D, Nusslein-Volhard C. The origin of pattern and polarity in the Drosophila embryo. Cell 1992; 68:201-219.

62. Thisse B, Stoetzel C, Gorostiza-Thisse C et al. Sequence of the twist gene and nuclear localization of its protein in endomesodermal cells of early Drosophila embryos. EMBO J 1988; 7:2175-2183.

63. Boulay JL, Dennefeld C, Alberga A. The Drosophila developmental gene snail encodes a protein with nucleic acid binding fingers. Nature 1987; 330:395-398.

64. Simpson P. Maternal-zygotic gene interactions involving the dorsal-ventral axis in Drosophila embryos. Genetics 1983; 105:615-632.

65. Nusslein-Volhard C, Wieschaus E, Kluding H. Mutations affecting the pattern of the larval cuticle in Drosophila melanogaster. 1. Zygotic loci on the second chromosome. Roux's Arch Dev Biol 1984; 183:267-282.

66. Alberga A, Boulay JL, Dennefeld C. The snail gene required for mesoderm formation in Drosophila is expressed dynamically in derivatives of all three germlayers. Development 1987; 111:983-992.

67. Thisse B, Stoetzel C, Messal ME et al. Genes of the Drosophila dorsal groupcontrol the specific expression of the zygotic gene twist in presumptive mesodermal cells. Genes Dev 1987; 1:709-715.

68. Leptin M. Twist and snail as positive and negative regulators during Drosophila mesoderm development. Genes Dev 1991; 5:1568-1576.

69. Kosman D, Ip YT, Levine M et al. Establishment of the mesoderm-neuroectoderm boundary in the Drosophila embryo. Science 1991; 254:118-122.

70. Lai ZC, Fortini ME, Rubin GM. The embryonic expression patterns of zfh-1 and zfh-2, two Drosophila genes encoding novel zinc-finger homeodomain proteins. Mech Dev 1991; 34:123-134.

71. Su MT, Fujioka M, Goto T et al. The Drosophila homeobox genes zfh-1 and even-skipped are required for cardiac-specific differentiation of a numb-dependent lineage decision. Development 1999; 126:3241-3251.

72. Shishido E, Higashijima S, Emori Y et al. Two FGF-receptor homologues of Drosophila: One is expressed in mesodermal primordium in early embryos. Development 1993; 117:751-761.

73. Shishido E, Ono N, Kojima T et al. Requirements of DFR1/Heartless, a mesoderm-specific Drosophila FGF-receptor, for the formation of heart, visceral and somatic muscles, and ensheathing of longitudinal axon tracts in CNS. Development 1997; 124:2119-2128.

74. Gisselbrecht S, Skeath JB, Doe CQ et al. Heartless encodes a fibroblast growth factor receptor (DFR1/DFGF-R2) involved in the directional migration of early mesodermal cells in the Drosophila embryo. Genes Dev 1996; 10:3003-3017.

75. Beiman M, Shilo BZ, Volk T. Heartless, a Drosophila FGF receptor homolog, is essential for cell migration and establishment of several mesodermal lineages. Genes Dev 1996; 10:2993-3002.

76. Luo L, Liao YJ, Jan LY et al. Distinct morphogenetic functions of similar small GTPases: Drosophila Drac1 is involved in axonal outgrowth and myoblast fusion. Genes Dev 1994; 8:1787-1802.

77. Klambt C. The Drosophila gene pointed encodes two ETS-like proteins which are involved in the development of the midline glial cells. Development 1993; 117:163-76.

78. Bodmer R, Frasch M. Genetic determination of Drosophila heart development. In: Rosenthal N, Harvey R, eds. Heart Development. London, New York: Academic Press, San Diego, 1999:65-90.

79. Yin Z, Xu XL, Frasch M. Regulation of the twist target gene tinman by modular cis-regulatory elements during early mesoderm development. Development 1997; 124:4971-4982.

80. Spencer FA, Hoffmann FM, Gelbart WM. Decapentaplegic: A gene complex affecting morphogenesis in Drosophila melanogaster. Cell 1982; 28:451-461.

81. Ferguson EL, Anderson KV. Localized enhancement and repression of the activity of the TGF-beta family member, decapentaplegic, is necessary for dorsal-ventral pattern formation in the Drosophila embryo. Development 1992; 114:583-597.

82. Francois V, Solloway M, O'Neill JW et al. Dorsal-ventral patterning of the Drosophila embryo depends on a putative negative growth factor encoded by the short gastrulation gene. Genes Dev 1994; 8:2602-2616.

83. Staehling-Hampton K, Hoffmann FM, Baylies MK et al. Dpp induces mesodermal gene expression in Drosophila. Nature 1994; 372:783-786.

84. Frasch M. Induction of visceral and cardiac mesoderm by ectodermal Dpp in the early Drosophila embryo. Nature 1995; 374:464-467.

85. Xu X, Yin Z, Hudson JB et al. Smad proteins act in combination with synergistic and antagonistic regulators to target Dpp responses to the Drosophila mesoderm. Genes Dev 1998; 12:2354-2370.

86. Venkatesh TV, Park M, Ocorr K et al. Cardiac enhancer activity of the homeobox gene tinman depends on CREB consensus binding sites in Drosophila. Genesis 2000; 26:55-66.

87. Wu X, Golden K, Bodmer R. Heart development in Drosophila requires the segment polarity gene wingless. Dev Biol 1995; 169:619-628.

88. Park M, Wu X, Golden K et al. The wingless signaling pathway is directly involved in Drosophila heart development. Dev Biol 1996; 177:104-116.

89. Lockwood WK, Bodmer R. The patterns of wingless, decapentaplegic, and tinman position the Drosophila heart. Mech Dev 2002; 114:13-26.

90. Ramain P, Heitzler P, Haenlin M et al. Pannier, a negative regulator of achaete and scute in Drosophila, encodes a zinc finger protein with homology to the vertebrate transcription factor GATA-1. Development 1993; 119:1277-1291.

91. Winick J, Abel T, Leonard MW et al. A GATA family transcription factor is expressed along the embryonic dorsoventral axis in Drosophila melanogaster. Development 1993; 119:1055-1065.

92. Gajewski K, Fossett N, Molkentin JD et al. The zinc finger proteins Pannier and GATA4 function as cardiogenic factors in Drosophila. Development 1999; 126:5679-5688.

93. Klinedinst SL, Bodmer R. Gata factor Pannier is required to establish competence for heart progenitor formation. Development 2003; 130:3027-3038.

94. Pandur P, Lasche M, Eisenberg LM et al. Wnt-11 activation of a noncanonical Wnt signalling pathway is required for cardiogenesis. Nature 2002; 418:636-641.

95. Ashe HL, Mannervik M, Levine M. Dpp signaling thresholds in the dorsal ectoderm of the Drosophila embryo. Development 2000; 127:3305-3312.

96. Herranz H, Morata G. The functions of pannier during Drosophila embryogenesis. Development 2001; 128:4837-4846.

97. Gajewski K, Zhang Q, Choi CY et al. Pannier is a transcriptional target and partner of Tinman during Drosophila cardiogenesis. Dev Biol 2001; 233:425-436.

98. Sepulveda JL, Belaguli N, Nigam V et al. GATA-4 and Nkx-2.5 coactivate Nkx-2 DNA binding targets: Role for regulating early cardiac gene expression. Mol Cell Biol 1998; 18:3405-3415.

99. Sepulveda JL, Vlahopoulos S, Iyer D et al. Combinatorial expression of GATA4, Nkx2-5, and serum response factor directs early cardiac gene activity. J Biol Chem 2002; 277:25775-25782.

100. Lee Y, Shioi T, Kasahara H et al. The cardiac tissue-restricted homeobox protein Csx/Nkx2.5 physically associates with the zinc finger protein GATA4 and cooperatively activates atrial natriuretic factor gene expression. Mol Cell Biol 1998; 18:3120-3129.

101. Griffin KJ, Stoller J, Gibson M et al. A conserved role for H15-related T-box transcription factors in zebrafish and Drosophila heart formation. Dev Biol 2000; 218:235-247.

101a.Qian L, Liu J, Bodmer R. Neuromancer Tbx20-related genes (H15/midline) promote cell fate specification and morphogenesis of the Drosophila heart. Dev Biol 2005; 279(2):509-24.

102. Garg V, Kathiriya IS, Barnes R et al. GATA4 mutations cause human congenital heart defects and reveal an interaction with TBX5. Nature 2003; 424:443-447.

103. Stennard FA, Costa MW, Elliott DA et al. Cardiac T-box factor Tbx20 directly interacts with Nkx2-5, GATA4, and GATA5 in regulation of gene expression in the developing heart. Dev Biol 2003; 262:206-224.

104. Han Z, Fujioka M, Su M et al. Transcriptional integration of competence modulated by mutual repression generates cell-type specificity within the cardiogenic mesoderm. Dev Biol 2002; 252:225-240.

105. Carmena A, Buff E, Halfon MS et al. Reciprocal regulatory interactions between the Notch and Ras signaling pathways in the Drosophila embryonic mesoderm. Dev Biol 2002; 244:226-242.

106. Buff E, Carmena A, Gisselbrecht S et al. Signalling by the Drosophila epidermal growth factor receptor is required for the specification and diversification of embryonic muscle progenitors. Development 1998; 125:2075-2086.

107. Halfon MS, Carmena A, Gisselbrecht S et al. Ras pathway specificity is determined by the integration of multiple signal-activated and tissue-restricted transcription factors. Cell 2000; 103:63-74.

108. Bidet Y, Jagla T, Da Ponte JP et al. Modifiers of muscle and heart cell fate specification identified by gain-of-function screen in Drosophila. Mech Dev 2003; 120:991-1007.

109. Fujioka M, Emi-Sarker Y, Yusibova GL et al. Analysis of an even-skipped rescue transgene reveals both composite and discrete neuronal and early blastoderm enhancers, and multi-stripe positioning by gap gene repressor gradients. Development 1999; 126:2527-2538.

110. Zikova M, Da Ponte JP, Dastugue B et al. Patterning of the cardiac outflow region in Drosophila. Proc Natl Acad Sci USA 2003; 100:12189-12194.

111. Kelly RG, Buckingham ME. The anterior heart-forming field: Voyage to the arterial pole of the heart. Trends Genet 2002; 18:210-216.

112. Knirr S, Frasch M. Molecular integration of inductive and mesoderm-intrinsic inputs governs even-skipped enhancer activity in a subset of pericardial and dorsal muscle progenitors. Dev Biol 2001; 238:13-26.

113. Kim YO, Park SJ, Balaban RS et al. A functional genomic screen for cardiogenic genes using RNA interference in developing Drosophila embryos. Proc Natl Acad Sci USA 2004; 101:159-64.

114. Lo PC, Skeath JB, Gajewski K et al. Homeotic genes autonomously specify the anteroposterior subdivision of the Drosophila dorsal vessel into aorta and heart. Dev Biol 2002; 251:307-319.

115. Lovato TL, Nguyen TP, Molina MR et al. The Hox gene abdominal-A specifies heart cell fate in the Drosophila dorsal vessel. Development 2002; 129:5019-5027.

116. Mastick GS, McKay R, Oligino T et al. Identification of target genes regulated by homeotic proteins in Drosophila melanogaster through genetic selection of Ultrabithorax protein-binding sites in yeast. Genetics 1995; 139:349-363.

CHAPTER 5

Development of the Somatic Gonad and Fat Bodies

Mark Van Doren*

Abstract

The development of the *Drosophila* fat body and gonads represent excellent models for studying cell type specification, patterning and morphogenesis during organ formation. Moreover, these organs are critical for the proper homeostasis of one generation of the species, while ensuring the production of the next generation. The specification of the fat body and somatic gonad from segmentally repeated domains within the mesoderm highlights how the embryonic patterning systems are used to generate the different cell types that make up distinct organs. The fat body and somatic gonad are formed, at least in part, from analogous domains within the mesoderm. Distinct segmental identities specified by the homeotic genes are utilized to distinguish between the fat body and somatic gonad, and pattern cell types within these tissues. These tissues also rely on information from sex determination to exhibit different functions, and even form distinct organs, in the two sexes. Finally, complex cell-cell interactions and morphogenetic movements are required to form fat body and gonads with the proper tissue architecture.

Specification of the Fat Body and Somatic Gonad

The development of the fat body and somatic gonad (also known as gonadal mesoderm) have been described classically at the descriptive level elsewhere (e.g., ref. 1 and references therein), and so I will restrict this discussion to work where these tissues have been followed using molecular markers. One conclusion from the molecular work is that the majority of fat body and somatic gonad arise from analogous primordia within the mesoderm, which is why these two distinct tissues are considered together in the context of this chapter.

The origins of the fat body have been traced using both terminal differentiation markers such as *alcohol dehydrogenase*[2] and *Drosophila collagen gene 1*,[3] as well as with markers that are involved in specifying fat body cell identity such as *seven-up*[3] and *serpent (srp)*.[4-7] This has established that the fat body primordium can first be identified by stage 10 of embryogenesis (stages as in ref. 1) as bilateral clusters of mesodermal cells in each of parasegments (PS) 4-13. These clusters consist of primary fat body, found in PS 4-9 and PS 13, as well as secondary fat body found in PS 4-13[7] (Fig. 1) The secondary fat body has a similar developmental potential as the primary fat body but is specified independently.

The first molecular markers for the somatic gonad were the retrotransposon *412*[8] and the enhancer trap line *68-77*,[9,10] which are specifically expressed in the somatic gonad at the time of gonad formation (stage 14 of embryonic development). However, since the expression of

*Mark Van Doren—Department of Biology, Johns Hopkins University, Baltimore, Maryland 21218, U.S.A. Email: vandoren@jhu.edu

Muscle Development in Drosophila, edited by Helen Sink. ©2006 Eurekah.com and Springer Science+Business Media.

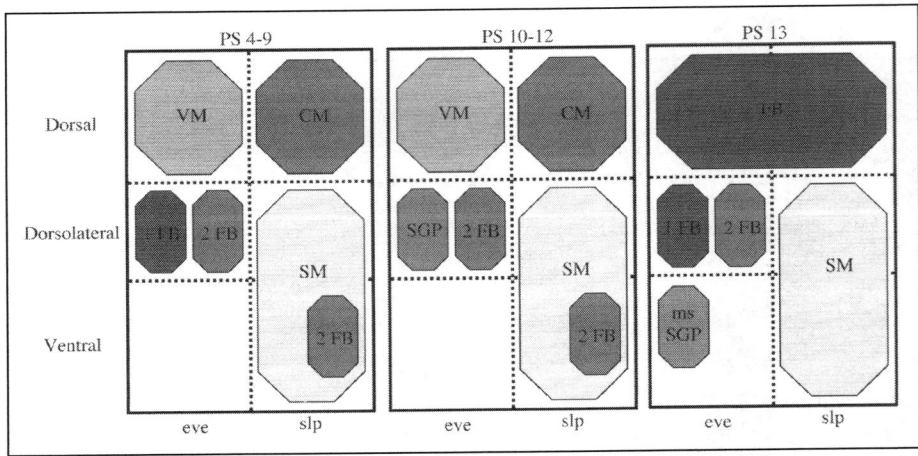

Figure 1. Cell specification and mesodermal address. Different mesodermal cell types are specified in a segmentally repeated fashion using a combination of anterior-posterior and dorsal-ventral patterning information within each segment (after Riechman et al, 1998[7]). Homeotic genes allow different parasegments to form distinct subsets of mesodermal cell types (see also Fig. 2). VM, visceral mesoderm, CM, cardiac mesoderm, 1 FB, primary fat body, 2 FB, secondary fat body, SM, somatic muscles, SGP, somatic gonadal precursors, msSGP, male-specific somatic gonadal precursors.

these markers is complex at earlier time points, the origin of the somatic gonadal precursors was only first unambiguously determined by examining expression of *eyes absent*[11] (*eya*, this gene was originally named *clift*[11] but is officially called *eyes absent* by Flybase). At stage 11, EYA identifies bilateral clusters of somatic gonadal precursors developing in each of PS 10, 11 and 12[11] (Fig. 1).

As with other mesodermal and ectodermal tissues in *Drosophila* (e.g., refs. 12,13), the fat body and somatic gonad are specified in a segmentally repeated fashion and utilize a combination of anterior-posterior and dorsal-ventral patterning information to specify cell identity (Fig. 1). We will refer to this information collectively as a "mesodermal address". The anterior-posterior information is based on the segmentally repeated unit, the parasegment. This is further divided into an "even-skipped" domain and a "sloppy paired" domain within each parasegment.[6] In addition, homeotic genes provide distinct and different identities among the 14 parasegments, allowing them to give rise to different cell types. Dorsal-ventral patterning information divides the mesoderm into discrete regions along this axis, including the dorsal, dorso-lateral, and ventral domains. The mesodermal address of the primary fat body, and a majority of the somatic gonad, is within the even-skipped domain of the dorso-lateral mesoderm.[6,7,14] However, additional fat body and gonadal precursors come from other regions as discussed below.

Broad mesodermal expression of the transcription factors *tinman (tin)* and *zinc finger homeodomain protein-1* are required for the establishment of the dorso-lateral domain of the mesoderm.[15] This is somewhat confusing since later *tin* expression is restricted to the dorsal domain in response to *decapentaplegic (dpp)* signaling from the ectoderm,[12] and is critical for specifying the dorsal mesoderm address. Yet it is clearly the early *tin* expression that is critical for the dorso-lateral domain, since fat body and somatic gonadal precursors are still present in *dpp* mutants that lack the late dorsal *tin* expression. Further, dorso-lateral fates can even be expanded in *dpp* mutants,[15] indicating that *dpp* may be important for blocking dorso-lateral fates in the dorsal most region. A second signal, possibly acting through the Epidermal Growth Factor Receptor, may play a similar role in blocking dorso-lateral fates in the most ventral regions. Thus one model is that the dorso-lateral domain is specified by default, as mesoderm that does not receive either the dorsal or ventral signals.

The development of the primary fat body and somatic gonad appear to be mutually exclusive. Within PS 4-9 and PS 13, the *even-skipped* domain of the dorso-lateral mesoderm gives rise to fat body, while in PS 10-12 this same region gives rise to somatic gonad.[7,14] *srp* normally acts to promote fat body development, and is both necessary and sufficient for fat body specification.[16] *srp* also acts to block formation of the somatic gonad; the somatic gonad is expanded into additional parasegments in *srp* mutants,[7,14] and is repressed by ectopic expression of *srp*.[16]

The homeotic gene *abdominal-A (abdA)* is essential for specification of the somatic gonad in PS10-12,[8] and acts primarily by blocking *srp*. In *abdA* mutants, *srp* is now expressed in the somatic gonad domain of PS 10-12, promoting fat body development in this region at the expense of the somatic gonad.[7,14] In contrast when *abdA* is ectopically expressed extra somatic gonadal precursors are formed[10,17] and *srp* expression is repressed.[14] Somatic gonad appears to be the "default" state for this mesodermal address as somatic gonad forms throughout PS4-13 in *abdA, srp* double mutants.[14] Thus *abdA* mainly acts to block *srp* at this stage, and is not directly required for initial somatic gonad specification. *abdA* may, however, play a positive role later on in somatic gonad patterning and morphogenesis.

Although the above description holds true for the primary fat body cluster and the main body of somatic gonad, there are still additional cells contributing to both tissues that arise from other regions of the mesoderm (Fig. 1). Groups of secondary fat body cells form immediately adjacent to the primary fat body cells in PS 4-9 and PS 13, and these same cells form adjacent to the somatic gonad in PS 10-12.[7] Thus, the secondary fat body is not subject to the fat body vs. somatic gonad distinction as is the primary fat body. In addition, a second set of secondary fat body precursors form with an apparently different mesodermal address (i.e., the sloppy paired domain of the ventral mesoderm) in PS 4-12. Finally, the entire dorsal domain of PS 13 forms fat body, in place of the visceral mesoderm and cardiac mesoderm found in this address in PS 4-12.[7]

Additional cells also contribute to the gonad, at least in one of the sexes. These cells are termed the "male specific somatic gonadal precursors" and are first observed in PS 13. There they appear to form in the even-skipped domain of the ventral mesoderm.[18] These cells are initially formed in both males and females, but ultimately only contribute to the male gonad as described below.

Homeotic genes such as *abdA* are critical for determining which segments of the dorso-lateral mesoderm will give rise to fat body, and which to somatic gonad. However, homeotic genes also play a role in dividing the somatic gonadal precursors into distinct identities (Fig. 2). Molecular markers indicate that the somatic gonad is sub-divided into anterior and posterior cell fates. *escargot* is specifically expressed in anterior somatic gonad,[19] while markers such as *Dwnt2*,[20] the bluetail transposon,[10] and *eya* (at later stages[10]) are expressed in the posterior gonad. It has been proposed that *abdA* alone specifies anterior somatic gonad, while a combination of *abdA* and *Abdominal-B (AbdB)* specify posterior somatic gonad.[10] Indeed, ectopic expression of *abdA* is sufficient to induce the formation of additional somatic gonad that has an anterior identity.[10] Furthermore, loss of *AbdB* function causes a loss of posterior gonad mesoderm identity, even though a similar amount of total somatic gonad is still present.[10,48] Initial expression of ABDB is found in PS 13 and more posterior parasegments.[21,22] However, ABDB expression is eventually found in more anterior parasegments,[21,22] including the posterior regions of the somatic gonad.[48] This supports a role for ABDB in patterning the posterior somatic gonad. These data are consistent with the model that *abdA* alone specifies anterior somatic gonad, while a combination of *abdA* and *AbdB* specifies posterior somatic gonad. Alternatively, *abdA* may be solely required for allowing somatic gonad specification in PS 10-12 (see above), while the presence or absence of *AbdB* determines anterior vs. posterior somatic gonad.

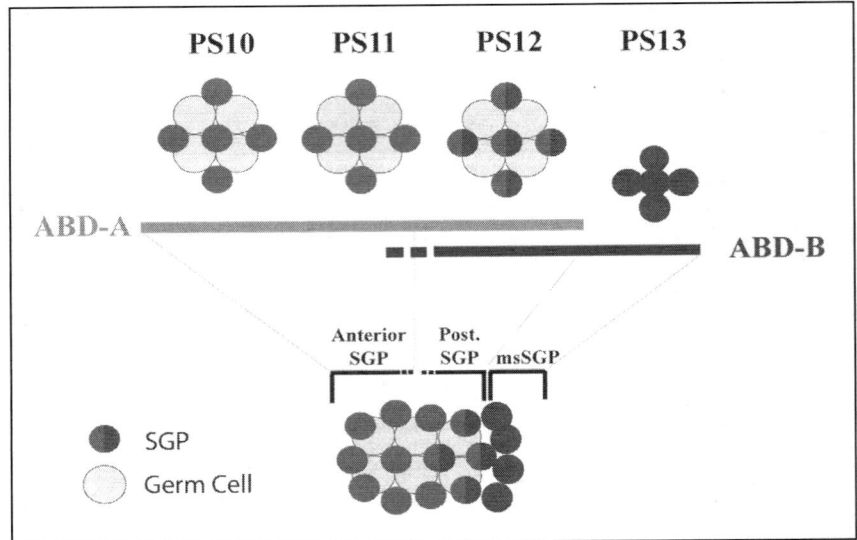

Figure 2. A model for patterning of the somatic gonad by homeotic genes. The action of *abdA* alone patterns anterior (e.g., PS 10) SGP identity, while a combination of *abdA* and *AbdB* specify posterior (e.g., PS 12) SGP identity. Male specific somatic gonadal precursors (msSGPs) form in PS 13 and are regulated by *AbdB* alone.

AbdB is also essential for the distinct cell types that arise in PS 13. *AbdB* is necessary for the specification of the male specific somatic gonadal precursors in PS 13 (Fig. 2), and is sufficient to induce male specific somatic gonadal precursors in more anterior parasegments when ectopically expressed.[48] By analogy, *AbdB* may also be responsible for the formation of the dorsal domain of fat body, in place of visceral and cardiac mesoderm, in this parasegment (Fig. 1). One apparent paradox is that *AbdB* limits male specific somatic gonadal precursors fate to PS 13, but also acts in PS 12 to pattern the posterior somatic gonadal precursors; why don't male specific somatic gonadal precursors form in PS 12? This may reflect differences in the timing of when these cell types express ABDB, or in different thresholds of ABDB activity required.

Morphogenesis and Function of the Fat Body

Morphogenesis
Once the fat body primordia are specified the cells must undergo the complicated morphogenetic movements that create the larval fat body. It is a common theme in mesoderm development that cell types, like the fat body, are specified in a segmentally repeated fashion and must then meet up with like cell types from neighboring segments to form a contiguous tissue. This appears to be an elaborate example of classical cell sorting[23] where virtually every different cell type in the mesoderm is simultaneously seeking out like cell types from different segments. Although little is known about this process at the molecular level, it likely involves cell movement, where cells are able to "sample" different neighboring cell types, along with differential cell adhesion, where like cell types are able to recognize one another via distinct cell surface adhesion properties.[24] The cell movement involved might be random, but it could also be directed, with different mesodermal cell types attracted to each other or to particular locations.

The primary and secondary clusters of fat body from PS 4-13 join together and, with additional cell division, create a continuous band of lateral fat body.[25] In addition, posterior fat body, likely that originating in the dorsal domain of PS 13, extends anteriorly to form the dorsal process (or dorsal extension). Finally, in the anterior and posterior regions, the lateral fat body on either side of the embryo extend ventrally toward the midline to form the anterior and posterior plates.[25] During metamorphosis and early adult stages, the embryonic/larval fat body is discarded and the adult fat body forms.[26] Little is known about the cellular and molecular mechanisms controlling morphogenesis of the larval fat body or the origins and development of the adult fat body.

Function

The "fat body" is perhaps an unfortunate name for such an essential and dynamic tissue. Certainly, this tissue is involved in fat storage and energy metabolism, however this tissue is also essential for a number of other critical biological functions. Related to its role in energy metabolism, the fat body is involved in nutrient sensing, and signals to regulate the growth of tissues in response to nutrient availability.[27,28] Perhaps also linked to energy metabolism, the fat body plays an additional role in regulating the lifespan of the animal.[29]

The fat body also acts as the *Drosophila* "liver", and is critical for a number of related functions, including steroidogenesis. Indeed, even the transcriptional regulatory network that regulates liver-specific gene expression in mammals is conserved in the *Drosophila* fat body (ref. 4 and references therein). The fat body is a major player in the innate immune response, producing anti-microbial peptides in response to bacterial infection.[30] This tissue also produces a protective response to a wide variety of other stresses, including heat shock, mechanical shock, dehydration and irradiation.[31]

Finally, the fat body exhibits additional parallels to the somatic gonad: sexually dimorphic development to nurture the germ cells. The female fat body, but not the male fat body, produces yolk proteins that are transported to the ovary to eventually be taken up by the developing eggs. These proteins, probably with associated lipids and other factors, create the essential stockpile of nutrients that will feed the embryo during the early stages of development. The fat body also controls sex-specific behavioral outputs from the central nervous system.[32] Clearly then, fat storage is only one of a number of important functions for this essential tissue.

Morphogenesis of the Gonad

The formation of the embryonic *Drosophila* gonad is an excellent model for studying morphogenesis and organ formation (Fig. 3). Several distinct cell types, including both germ cells and somatic cells, must come together in a stereotypical fashion to form an organized, patterned, and sexually dimorphic gonad. This process involves a number of types of morphogenetic movement, including cell migration, cell sorting, tissue morphogenesis, and the formation of specialized cell-cell contacts. Correct formation of the gonad is critical for continued germ cell development, and the production of the next generation of the species.

Embryonic Gonad Formation

One aspect of gonad formation involves the migration of the germ cells from their site of origin, the posterior pole of the embryo, to reach the somatic gonad. Although guidance mechanisms that control germ cell migration have been identified, much remains to be learned about how this process is regulated, and how the germ cells are able to specifically recognize and associate with the somatic gonad. Since this volume focuses on mesoderm development, and germ cell migration has been excellently reviewed elsewhere,[33] we will focus on the morphogenesis of the somatic gonad, and the later interactions between the somatic gonad and the germ cells.

Once the somatic gonad is correctly specified, the three clusters of somatic gonadal precursors from PS 10, 11 and 12 must come together to form a contiguous tissue, a process

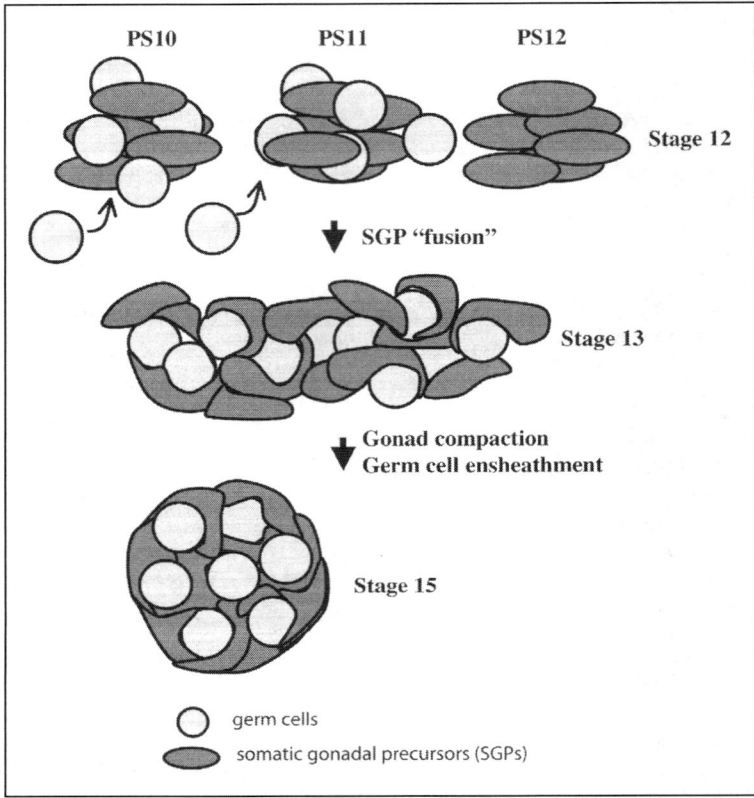

Figure 3. Gonad morphogenesis. A schematic representation of gonad morphogenesis (as it occurs in females), including the steps of somatic gonadal precursors (SGP) cluster fusion, gonad compaction and germ cell ensheathment.

that we will refer to as somatic gonadal precursor cluster "fusion" (not to be confused with myoblast fusion, where the individual cells actually fuse to form multinucleate myotubes). During stage 12, as the germ band retracts, groups of somatic gonadal precursors from separate parasegments move toward one another to form one band of cells. This process is likely to involve cell sorting, including both cell movement and differential cell adhesion, as described for the fat body above. As this occurs, the germ cells complete their migration to the somatic gonad, and become intermingled with the somatic gonadal precursors. Thus, by the beginning of stage 13, the somatic gonadal precursors and germ cells form a loosely associated group that is still spread over three segments of the embryo.[8,10]

During the next stage (stage 14), the somatic gonadal precursors and germ cells undergo a dramatic rearrangement in tissue morphology to form a compact, spherical gonad in PS 10.[8,10] This process is likely to involve additional cell sorting, where the germ cells and somatic gonadal precursors maximize their contacts with one another and minimize their contacts with the surrounding mesoderm. However, this process also involves coordinated changes in individual cell shape and cell-cell interactions[34] that are likely to contribute to the changes in overall tissue morphology. Somatic gonadal precursors are likely to provide the "driving force" for gonad compaction since a well compacted gonad can form in the absence of germ cells.[8,35]

In addition, regionalized cell identity is preserved during the course of gonad compaction and the cells expressing posterior somatic gonadal precursor markers prior to compaction are found at the posterior of the formed gonad. Thus, either these spatial relationships are preserved during gonad morphogenesis, or identity is reassigned in the coalesced gonad.

During the course of gonad formation, a defined tissue architecture is established, especially in terms of the relationships between the somatic gonadal precursors and the germ cells. Germ cells are not observed on the outside of the gonad,[36] and there is also very little germ cell-germ cell contact within the gonad.[34] Instead each germ cell is ensheathed by somatic gonadal precursors, which dramatically alter their individual cell shapes to surround the spherical germ cells[34] Germ cell ensheathment begins at stage 13, before gonad compaction, and is complete in the coalesced gonad at the end of stage 14. However, this remains an active process since the germ cells begin dividing in the gonads at stage 15, but daughter germ cells remain fully ensheathed by the somatic gonadal precursors. Gonad compaction can occur in the absence of germ cell ensheathment, and *vice versa*, and so these two processes appear to be independent aspects of gonad morphogenesis.[34]

Molecular Control of Gonad Morphogenesis

How are the dramatic morphogenetic movements of the somatic gonadal precursors and germ cells regulated at the molecular level to allow for proper gonad formation? One molecule that plays a key role in gonad morphogenesis is the cell surface adhesion protein E-cadherin[34] (E-cadherin is also known as *shotgun* in *Drosophila*). E-cadherin expression is up-regulated in somatic gonadal precursors at the time of somatic gonadal precursor "fusion" (stage 12/13). This is dependent on the somatic gonadal precursor identity factor *eya*, providing a link between somatic gonadal precursor specification and the molecules that regulate gonad morphogenesis. Loss of E-cadherin function affects both gonad compaction and germ cell ensheathment. Thus, this molecule is essential for the formation of the proper gonad architecture, including the specialized somatic gonadal precursor-germ cell contacts. Although E-cadherin is required for germ cell ensheathment by the somatic gonadal precursors, it is difficult to imagine how this homophillic adhesion molecule promotes the heterotypic interaction between these two cell types. Other heterophillic adhesion molecules might provide the specificity necessary to preferentially promote somatic gonadal precursor-germ cell interactions.

E-cadherin may also play an earlier role in the cell sorting events of somatic gonadal precursor cluster fusion, as suggested by the increase in E-cadherin expression in somatic gonadal precursors at this stage. However, revealing such a role for E-cadherin might require stronger loss-of-function conditions, which cannot be generated due to technical limitations. E-cadherin mutants do show a defect in the ability of the male specific somatic gonadal precursors to join the main body of the male gonad, supporting a role for E-cadherin in cell sorting or "fusion" events.

Another factor required for both gonad compaction and germ cell ensheathment is *fear of intimacy*[37] *(foi)*. Since *foi* affects gonad formation without affecting cell identity, it is likely to be specifically involved in the morphogenetic movements required for gonad formation. Interestingly, E-cadherin protein levels are dramatically decreased in *foi* mutants, indicating that *foi* may control gonad morphogenesis by affecting E-cadherin expression or function.[34] FOI is a cell-surface, transmembrane protein with homology to a family of metal ion transporters, suggesting that gonad morphogenesis, and E-cadherin, are regulated by the concentration of a key metal ion.[37]

Unlike *foi* and E-cadherin, *traffic jam (tj)* is not required for gonad compaction, but is required for the formation of correct gonad architecture.[38] In *tj* mutants, a rounded gonad forms that contains both somatic gonadal precursors and germ cells, but the spatial relationship between these cell types is altered; the somatic gonadal precursors fail to intermingle with the germ cells, and instead remain exclusively on the outside of the gonad. Although the ensheathment of the germ cells by the somatic gonadal precursors has not been specifically examined in these mutants, it is likely to also be affected. This is further supported by the finding that *tj* is required for proper germ cell-soma contact in the gonad at all stages of development in both males and females. Although *tj* encodes a protein of the Maf family of transcription factors, it does not seem to affect the specification of cell types in the gonad, and may therefore, like *foi* and E-cadherin, be more specifically involved in regulating gonad morphogenesis.

In the adult gonad *tj* behaves as a negative regulator of the expression of several adhesion molecules, including E-cadherin.[38] The *tj* mutant phenotype earlier during gonad formation might be caused by an overexpression of such adhesion molecules. It is not surprising that too much E-cadherin could be as detrimental to gonad morphogenesis as too little, since a balance of adhesion molecules is essential for tissue architecture. This is further illustrated by experiments where E-cadherin is overexpressed in the germ cells, which also prevents proper germ cell-soma interaction.[34]

Thus, while only a few molecules affecting gonad formation have so far been identified, it is not surprising that the regulation of cell surface adhesion is already emerging as an essential theme in this process. Genetic screens for additional genes regulating gonad formation (Jenkins, Weyers and Van Doren, unpublished) will hopefully provide the tools necessary for a more comprehensive understanding of the molecular events controlling gonad morphogenesis.

Gonad Sexual Dimorphism

Another fascinating aspect of gonad morphogenesis is that the embryonic gonad has the potential to form two dramatically different organs, an ovary or a testis, depending on the sex of the animal. In light of this a critical question is: How does the sex determination pathway regulate the sex-specific development and morphogenesis of the gonad? It has long been known that there is a subtle difference in the number of germ cells that populate the male vs. the female embryonic gonad.[36,39] This difference becomes even more dramatic soon after gonad formation[36] as the testis grows much faster than the developing ovary. However, little was known about when sexual dimorphism is manifested in the somatic gonad, and how the sex determination pathway regulates sexual dimorphism in the gonad.

Recently it was found that the somatic gonad is already sexually dimorphic at the time of gonad formation.[18] An additional group of somatic cells, the male-specific somatic gonadal precursors, forms part of the posterior of the male gonad at the time of gonad formation (stage 14), but is not found in female gonads. One molecular marker that allowed the discovery of these cells is the Sox100B protein.[40] Interestingly, Sox100B is homologous to mammalian Sox 9, a gene that is essential for the male pathway of gonad development in humans, and likely other vertebrate species. Thus the male-specific expression of Sox100B in *Drosophila*[18] provides evidence for a common pathway controlling sexual dimorphism in different species, even though these species have different mechanisms for initial sex determination.

The male-specific somatic gonadal precursors are initially present in both males and females (stage 12 and 13), but eventually disappear in females at the time of gonad formation (stage 14). These cells undergo sex-specific programmed cell death in the female, which is regulated by the cell death gene *head involution defective (hid)*. When programmed cell death is blocked, the male-specific somatic gonadal precursors can join the gonad in both sexes, indicating that programmed cell death is the primary mechanism controlling the sexually dimorphic development of these cells at this stage. Thus, one mechanism by which the sex determination pathway can regulate sexual dimorphism is by regulating the sex-specific activity of programmed cell death genes such as *hid*.[18]

This is but one example of how sexual dimorphism is manifested in the developing somatic gonad. Other sexually dimorphic cell types in the somatic gonad are likely to be regulated through distinct mechanisms. Furthermore, the choice between the male and female germ cell identity, which is essential for proper development of sex-specific gametes, is also dependent on sexual dimorphic development of the soma. Clearly, much remains to be learned about how the sex determination pathway is able to promote the distinct patterns of somatic and germ cell development in the testis and ovary.

Later Gonad Development

Little is known about the development of the gonads from the time of embryonic gonad formation until appearance of the adult gonad after metamorphosis. It is clear that the testis develops sooner than the ovary, and is already functioning to support the early stages of spermatogenesis in the early larvae (reviewed in ref. 41). Each testis is comprised of a single soma-derived stem cell niche that controls both the germline stem cells, which divide to produce the cells that enter spermatogenesis, and the somatic stem cells, which produce the somatic cyst cells that associate with differentiating germ cells and regulate spermatogenesis. It is unknown which somatic cells of the embryonic gonad contribute to the stem cell niche, the somatic stem cells, or any of the other somatic cell types that make up the adult testis.

Ovary morphogenesis is thought to be actively initiated only during the late (third instar) larval period as the individual ovarioles that will form the adult ovary begin to be organized.[42,43] The germ cells are encompassed by the developing ovarioles, and a somatic stem cell niche is formed in each ovariole to regulate the female germline stem cells.[44] Again, little is known about the relationships between the somatic cells of the embryonic gonad, and the numerous somatic cell types that make up the adult ovary.

During metamorphosis (pupal stages), the developing gonads must properly attach to the rest of the reproductive tract. The reproductive tract is derived from the genital imaginal discs, which themselves develop in a sexually dimorphic manner to create the distinct reproductive structures specific for the male or female (reviewed in ref. 45). There is evidence that the gonad and genital discs do not merely join together, but that extensive tissue interactions and cell migrations may occur between these two structures. For example, the pigment cell layer that forms the outer covering of the testis is thought to derive from the gonad, but these cells extend to cover the proximal portion of the reproductive tract (seminal vesicle) derived from the genital disc. Furthermore, there is evidence that the muscle sheath surrounding the testis is actually derived from the genital disc, and these cells migrate from the disc, underneath the testis pigment cell layer, to eventually encompass the entire testis.[20] This muscle cell migration is likely to be responsible for the elongation and coiling of the testis at this stage of development, a process that is known to be regulated by the genital disc.[46] As discussed extensively elsewhere in this volume, muscle is usually derived from the mesoderm, yet the genital disc is an ectodermally derived tissue. How, then, can the genital disc give rise to the testis muscle cell layer? It was recently shown that the male genital disc recruits neighboring mesodermal cells to form part of the disc.[47] It is intriguing to speculate that the muscle cell layer of the testis is derived from mesodermal cells of the male genital disc, and that they themselves had to be sex-specifically recruited into the disc from the surrounding mesoderm. This example highlights the intricate aspects of mesoderm development and organogenesis that can be studied using gonad formation as a model.

Conclusion

The fat body and gonad will continue to provide excellent models for organogenesis during development. A great deal has been learned about how cell identity is established in these tissues. The challenge now is to understand how this identity is translated into the proper form and function of these organs. The identification of *E-cadherin*, *foi*, and *traffic jam* as critical players in gonad formation provides a basis for beginning to understand the molecular mechanisms controlling this process. A further understanding of how gonad and fat body morphogenesis is regulated is likely to illuminate general principles of organogenesis that are broadly applicable to other systems.

References

1. Campos-Ortega J, Hartenstein V. The embryonic development of Drosophila melanogaster. New York: Springer-Verlag 1997.
2. Lockett TJ, Ashburner M. Temporal and spatial utilization of the alcohol dehydrogenase gene promoters during the development of Drosophila melanogaster. Dev Biol 1989; 134:430-437.
3. Hoshizaki DK, Blackburn T, Price C et al. Embryonic fat-cell lineage in Drosophia melanogaster. Development 1994; 120:2489-2499.
4. Abel T, Michelson AM, Maniatis T. A Drosophila GATA family member that binds to adh regulatory sequences is expressed in the developing fat body. Development 1993; 119:623-633.
5. Rehorn KP, Thelen H, Michelson AM et al. A molecular aspect of hematopoiesis and endoderm development common to vertebrates and Drosophila. Development 1996; 122:4023-4031.
6. Riechmann V, Irion U, Wilson R et al. Control of cell fates and segmentation in the Drosophila mesoderm. Development 1997; 124:2915-2922.
7. Riechmann V, Rehorn KP, Reuter R et al. The genetic control of the distinction between fat body and gonadal mesoderm in Drosophila. Development 1998; 125:713-723.
8. Brookman J, Toosy A, Shashidhara L et al. The 412 retrotransposon and the development of gonadal mesoderm in Drosophila. Development 1992; 116:1185-1192.
9. Warrior R. Primordial germ cell migration and the assembly of the Drosophila embryonic gonad. Dev Biol 1994; 166:180-194.
10. Boyle M, DiNardo S. Specification, migration, and assembly of the somatic cells of the Drosophila gonad. Development 1995; 121(6):1815-1825.
11. Boyle M, Bonini N, DiNardo S. Expression and function of clift in the development of somatic gonadal precursors within the Drosophila mesoderm. Development 1997; 124(5):971-982.
12. Frasch M. Induction of visceral and cardiac mesoderm by ectodermal Dpp in the early Drosophila embryo. Nature 1995; 374:464-467.
13. Azpiazu N, Lawrence PA, Vincent J-P et al. Segmentation and specification of the Drosophila mesoderm. Genes Dev 1996; 10:3183-3194.
14. Moore LA, Broihier HT, Van Doren M et al. Gonadal mesoderm and fat body initially follow a common developmental path in Drosophila. Development 1998; 125(5):837-844.
15. Broihier HT, Moore LA, Van Doren M et al. zfh-1 is required for germ cell migration and gonadal mesoderm development in Drosophila. Development 1998; 125(4):655-666.
16. Hayes SA, Miller JM, Hoshizaki DK. Serpent, a GATA-like transcription factor gene, induces fat-cell development in Drosophila melanogaster. Development 2001; 128(7):1193-1200.
17. Greig S, Akam M. The role of homeotic genes in the specification of the Drosophila gonad. Curr Biol 1995; 5:1057-1062.
18. DeFalco TJ, Verney G, Jenkins AB et al. Sex-specific apoptosis regulates sexual dimorphism in the Drosophila embryonic gonad. Dev Cell 2003; 5(2):205-216.
19. Gonczy P, Viswanathan S, DiNardo S. Probing spermatogenesis in Drosophila with P-element enhancer detectors. Development 1992; 114(1):89-98.
20. Kozopas KM, Samos CH, Nusse R. DWnt-2, a Drosophila Wnt gene required for the development of the male reproductive tract, specifies a sexually dimorphic cell fate. Genes Dev 1998; 12(8):1155-1165.
21. Celniker SE, Keelan DJ, Lewis EB. The molecular genetics of the bithorax complex of Drosophila: Characterization of the products of the Abdominal-B domain. Genes Dev 1989; 3:1424-1436.
22. Delorenzi M, Bienz M. Expression of Abdominal-B homeoproteins in Drosophila embryos. Development 1990; 108:323-329.
23. Townes PL, Holtfreter J. Directed movements and selective adhesion of embryonic amphibian cells. J Exp Zool 1955; 128:53-120.
24. Steinberg MS. Reconstruction of tissue by dissociated cells. Science 1963; 180:401-408.
25. Hartenstein V, Jan YN. Studying Drosophila embryogenesis with P-lacZ enhancer trap lines. Roux's Archives of Dev Biol 1992; 201:194-220.
26. Rizki TM. Fat body. In: Wright T, Ashburner M, eds. Genetics and Biology of Drosophila. New York: Academic Press, 1978.
27. Martin JF, Hersperger E, Simcox A et al. Minidiscs encodes a putative amino acid transporter subunit required nonautonomously for imaginal cell proliferation. Mech Dev 2000; 92:155-167.
28. Colombani J, Raisin S, Pantalacci S et al. A nutrient senosr mechanism controls Drosophila growth. Cell 2003; 114:656-658.
29. Rogina B, Reenan RA, Nilsen SP et al. Extended life-span conferred by contransporter gene mutations in Drosophila. Science 2000; 290:2137-2140.
30. Hoffmann JA. The immune response of Drosophila. Nature 2003; 426:33-38.

31. Ekengren S, Tryselius Y, Dushay MS et al. A humoral stress response in Drosophila. Curr Biol 2001; 11:714-718.
32. Dauwalder B, Tsujimoto S, Moss J et al. The Drosophila takeout gene is regulated by the somatic sex-determination pathway and affects male courtship behavior. Genes Dev 2002; 16(22):2879-2892.
33. Starz-Gaiano M, Lehmann R. Moving towards the next generation. Mech Dev 2001; 105(1-2):5-18.
34. Jenkins AB, McCaffery JM, Van Doren M. Drosophila E-cadherin is essential for proper germ cell-soma interaction during gonad morphogenesis. Development 2003; 130:4417-4426.
35. Geigy R. Action de l'ultra-violet sur le pole germinal dans l'oeuf de Drosophila melanogaster. Rev Suisse Zool 1931; 38:187-288.
36. Sonnenblick BP. The early embryology of Drosophila melanogaster. In: Demerec M, ed. Biology of Drosophila. New York: Wiley, 1950:62-167.
37. Van Doren M, Mathews WR, Samuels M et al. Fear of intimacy encodes a novel transmembrane protein required for gonad morphogenesis in Drosophila. Development 2003; 130:2355-2364.
38. Li MA, Alls JD, Avancini RM et al. The large Maf factor traffic jam controls gonad morphogenesis in Drosophila. Nat Cell Biol 2003; 5:994-1000.
39. Poirié M, Niederer E, Steinmann-Zwicky M. A sex-specific number of germ cells in embryonic gonads of Drosophila. Development 1995; 121:1867-1873.
40. Loh SHY, Russell S. A Drosophila group E sox gene is dynamically expressed in the embryonic alimentary canal. Mech Dev 2000; 93:185-188.
41. Fuller M. Spermatogenesis. In: Bate M, Martinez Arias A, eds. The Development of Drosophila melanogaster. New York: Cold Spring Harbor Press, 1993:1:71-147.
42. King RC. Ovarian Development in Drosophila melanogaster. New York: Academic Press, 1970.
43. Godt D, Laski FA. Mechanisms of cell rearrangement and cell recruitment in Drosophila ovary morphogenesis and the requirement of bric a brac. Development 1995; 121(1):173-187.
44. Spradling AC. Developmental genetics of genesis. In: Bate M, Martinez Arias A, eds. The Development of Drosophila melanogaster. New York: Cold Spring Harbor Press, 1993:1:1-70.
45. Sanchez L, Guerrero I. The development of the Drosophila genital disc. Bioessays 2001; 23(8):698-707.
46. Stern C. The growth of testes in Drosophila. II. The nature of interspecific differences. J Exp Zool 1941; 87:159-180.
47. Ahmad SM, Baker BS. Sex-specific deployment of FGF signaling in Drosophila recruits mesodermal cells into the male genital imaginal disc. Cell 2002; 109(5):651-661.
48. DeFalco T, Le Bras S, Van Doren M. Abdominal-B is essential for proper sexually dimorphic development of the Drosophila gonad. Mech Dev 121:1323-1333.

CHAPTER 6

Development of the Larval Visceral Musculature

Hsiu-Hsiang Lee, Stephane Zaffran and Manfred Frasch*

Abstract

The visceral mesoderm of *Drosophila* forms the thin layers of muscle fibers surrounding the digestive tract. Both during their development and after differentiation, these muscle tissues have crucial roles in the morphogenesis and functioning of the gut tube. The visceral muscles of the foregut, midgut and hindgut are generated from several distinct primordia that arise at precisely defined locations in the early mesoderm. Recent studies have greatly advanced our knowledge of the genetic and molecular patterning mechanisms that delineate these primordia during early mesoderm development. Many of the cellular processes and gene interactions that govern important spatial and functional subdivisions within these primordial tissues, as well as the differentiation of visceral myoblasts into gut muscles, have also been clarified. Herein, we present a summary of our current state of knowledge of *Drosophila* visceral muscle development and focus on the genetic and molecular networks that control this developmental process.

Introduction

The larval visceral musculature consists of a thin layer of muscles that surround the digestive tract, and perform slow peristaltic contractions. Unlike vertebrate smooth muscles, *Drosophila* visceral muscles do exhibit striations; however overall their ultrastructure shows many hallmarks of slow and super-contracting muscles that are able to exert constant tension. These features include: irregular and perforated Z-discs; imprecisely aligned filaments; sparse sarcoplasmic reticulum and T-tubules, and an arrangement of ~12 thin filaments around each thick filament (rather than six as in skeletal and flight muscles).[1]

The visceral musculature of the midgut, which represents the main portion of the digestive tract, consists of a mesh with an inner layer of circular and an outer layer of longitudinal muscles (Fig. 1A,B). The circular muscles are binucleate syncytia, which span one half of the circumference of the gut tube and form integrin-mediated muscle attachments with the fibers from the other side along the dorsal and ventral midlines of the tube.[2,3] The longitudinal midgut muscles consist of long, multinucleate fibers.[3] However, unlike the midgut, the foregut and hindgut are surrounded only by circular muscles (Fig. 1A).

The origin of these various types of visceral muscles in the early mesoderm has been clarified by virtue of various molecular markers, which allowed the establishment of a fate map at a stage when the mesoderm has spread out as a single-celled layer underneath the ectoderm.[4-7] At this stage (embryonic stage 10), there are eleven roughly rectangular cell clusters in the dorsal

*Corresponding Author: Manfred Frasch—Brookdale Department of Molecular, Cell and Developmental Biology, Box 1020, One Gustave L. Levy Place, New York, New York 10029, U.S.A. Email: manfred.frasch@mssm.edu

Muscle Development in Drosophila, edited by Helen Sink. ©2006 Eurekah.com and Springer Science+Business Media.

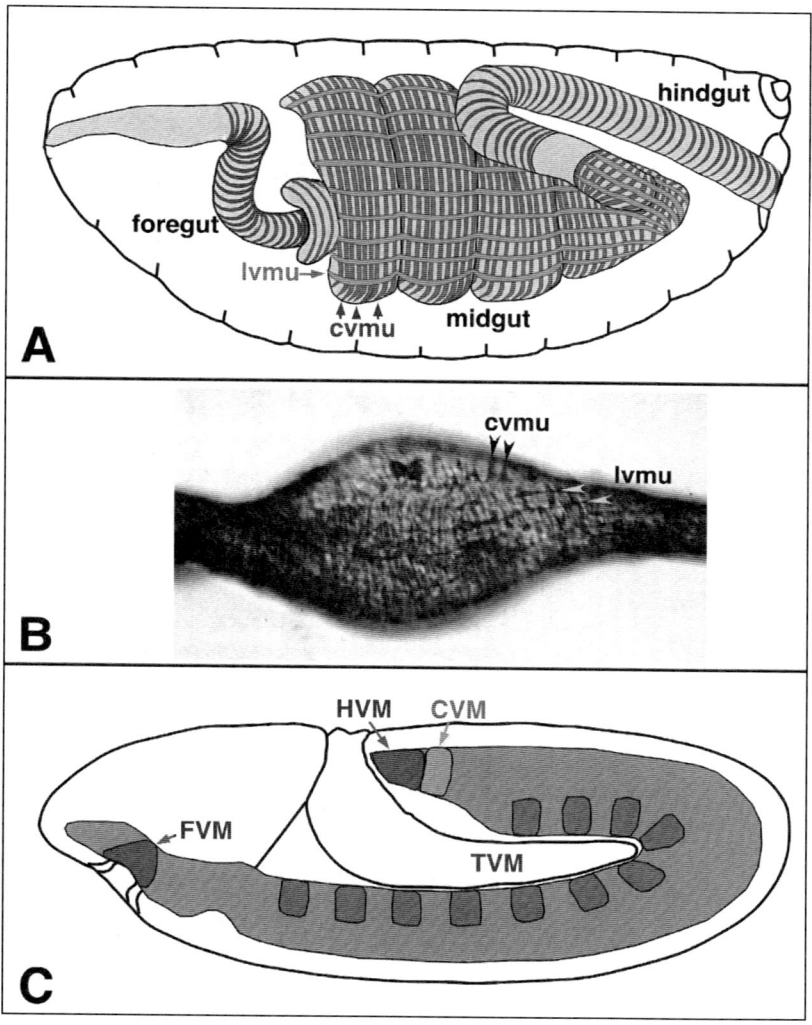

Figure 1. Morphology and developmental origin of visceral muscles in the *Drosophila* embryo. A) Schematic drawing of visceral muscles in a stage 16 embryo. Foregut and hindgut are surrounded by a layer of circular muscle fibers (purple), whereas the midgut is surrounded by an inner layer of circular (cvmu; red) and an outer layer of longitudinal (lvmu; green) muscle fibers. B) Portion of embryonic midgut stained for Myosin Light Chain to visualize circular (cvmu) and longitudinal (lvmu) gut muscle fibers. C) Schematic lateral view of stage 10 embryo depicting the primordia of the caudal visceral mesoderm (CVM) as well as the visceral mesoderm of the foregut (FVM), hindgut (HVM), and trunk (TVM). A color version of this figure is available online at http://www.Eurekah.com.

mesoderm between parasegments 2 and 12, which contain all of the primordial cells of the circular visceral muscles as well as a portion of the cells that will contribute to longitudinal gut muscles (Figs. 1C and 2A,B). Upon their ingression, these cell clusters give rise to the trunk visceral mesoderm. The primordia of the foregut and hindgut visceral musculatures are located close to the anterior and posterior margins of the mesoderm, where they are associated with the stomodeal and proctodeal invaginations that give rise to the foregut (esophagus and proventriculus)

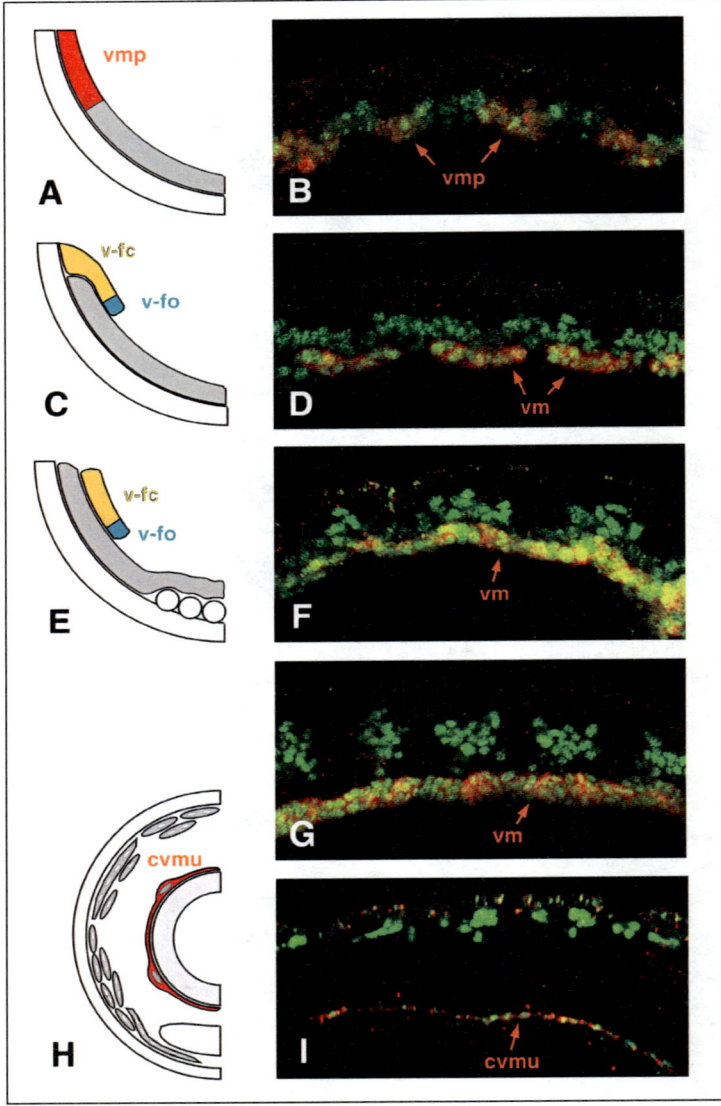

Figure 2. Formation of circular midgut musculature from trunk visceral mesoderm primordia. Left column shows schematic drawings of cross sections (left halves of germband/embryo) and right column shows longitudinal optical sections of embryos stained for *bap3*-lacZ (red) + Mef2 (green). *bap3*-lacZ marks the trunk visceral mesoderm precursors and their descendents, whereas Mef2 marks both visceral and somatic mesoderm derivatives. (Hence, the tissues labeled green only correspond to somatic mesoderm/muscles.) A,B) Stage 10. The mesoderm consists of a monolayer and the trunk visceral mesoderm primordia (vmp) are seen as dorsal metameric cell clusters. C,D) Stage 11. The visceral mesodermal cells (vm) are ingressing towards the interior, subdivided into visceral muscle founders (v-fo) and fusion-competent myoblasts (v-fc), and the cell clusters expand along the anterior-posterior axis. E,F) Stage12. The vm is fully internalized and neighboring cell clusters have merged into a continuous band of cells. G) Stage 13. The vm separates from the segmented somatic mesoderm. H,I) Stage 15. The circular visceral muscle (cvmu) syncytia form a thin layer around the midgut.

and hindgut, respectively (Fig. 1C). An area lying anteriorly adjacent to the hindgut visceral mesoderm, termed caudal visceral mesoderm, contains the progenitors of the longitudinal midgut muscles (Fig. 1C). Initially, the caudal visceral mesoderm anlage forms at the posterior-most end of the mesoderm,[8] suggesting that the hindgut and caudal visceral mesoderm anlagen invert during the course of the rearrangements taking place during early germ band extension.

Morphogenesis and Functions of Trunk Visceral Mesoderm and Circular Midgut Muscles

At embryonic stage 11 the cell clusters of the presumptive trunk visceral mesoderm start ingressing as continuous sheets of cells. During this process the expanding somatic and cardiac mesoderm appears to migrate between the ectoderm and the ventral and lateral margins of the trunk visceral mesoderm primordia, eventually displacing them towards the interior where they form a second dorsal mesodermal cell layer (Fig. 2A-F). At the same time, the trunk visceral mesoderm cells rearrange along the anterior-posterior axis, which results in the elongation of the cell clusters and, ultimately, contact formation among individual clusters from different parasegments (Figs. 2D,F and 3A,B). In addition, the cells along the ventral margins of the cell clusters become morphologically distinct and form arched columnar rows. These particular cells correspond to visceral muscle progenitors[9,10] (Figs. 2C,E and 5B), which after one round of cell division generate visceral muscle founder cells that play key roles in organizing circular midgut muscle development. At late stage 11, the anterior-posterior coalescence of the metameric cell clusters is completed and a continuous band of trunk visceral mesoderm is formed (Fig. 2F). During stage 12 (Fig. 2G), the endoderm of the anterior and posterior midgut rudiments migrates along the trunk visceral mesoderm to meet in the center of the embryo. In addition, the circular visceral muscle founder cells undergo myoblast fusion with more dorsally located cells of the trunk visceral mesoderm, termed fusion-competent cells, to form binucleate syncytia.[9,10] These syncytia are initially arranged in a palisade-like row of columnar cells along the length of the midgut endoderm (see Fig. 3 C,D) and during the closure of the endodermal midgut tube they extend dorso-ventrally to span the entire left and right half of the gut tube, respectively (Fig. 2H,I).

Aside from the peristaltic function of the differentiated midgut musculature, the developing trunk visceral mesoderm also has essential functions for proper morphogenesis of the midgut. In particular, the mesenchymal to epithelial transition and proper migration of the endoderm requires the presence of trunk visceral mesoderm as a substrate.[4,11,12] Subsequently, the visceral mesoderm is required to control the formation of constrictions, individual loops, and gastric caeca (four anterior appendix-like tubes) at specific positions along the anterior-posterior axis of the developing midgut.[13] A role of the visceral musculature in terminal differentiation of endodermal cells and their function in nutrient absorption has also been reported.[14]

Morphogenesis of Longitudinal Midgut, Foregut and Hindgut Muscles

The cells of the caudal visceral mesoderm undergo long-range migration towards the anterior before they form longitudinal midgut muscles. During stage 11, the first cells from this primordium leave their original position, coalesce into two posterior clusters laterally, and then migrate onto the posterior cells of the trunk mesoderm on either side of the embryo (Fig. 3A,B). This migration process continues, predominantly along the dorsal and ventral margins of the trunk visceral mesoderm, until the caudal visceral mesodermal cells have spread out evenly along the future midgut (stage 13).[6] At this stage, these cells function as the founder cells of the longitudinal midgut muscles and fuse with the remaining fusion-competent cells from the trunk visceral mesoderm (Fig. 3D) into multinucleate syncytia.[9,10] Finally, the longitudinal fibers distribute evenly around the circumference of the midgut and align in parallel along the antero-posterior axis (Fig. 1A). Like the circular muscles, the developing longitudinal midgut muscles are also required for normal midgut constrictions.[15] In addition, the migrating

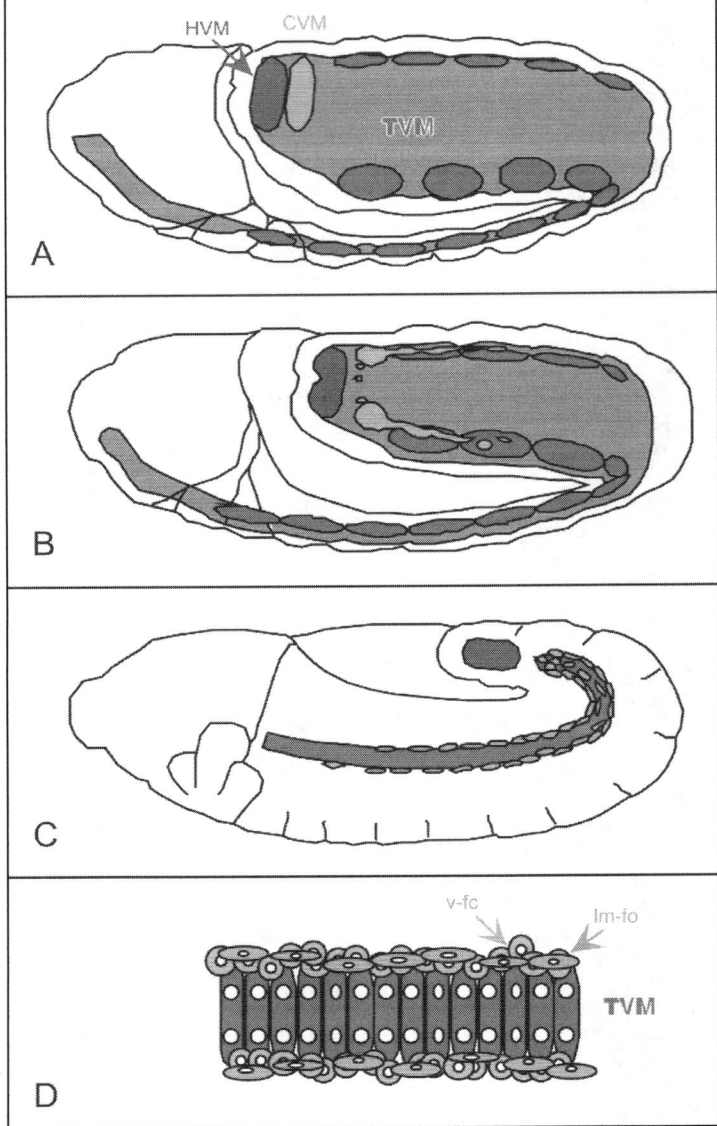

Figure 3. Formation of longitudinal midgut muscles from caudal visceral mesoderm primordia. A) Schematic drawing of a stage 10 embryo (dorso-lateral view). The caudal visceral mesoderm (CVM) is positioned just anteriorly to the hindgut visceral mesoderm (HVM) at this stage. The bilateral cell clusters of the trunk visceral mesoderm (TVM) are also depicted. B) Stage 11. The cells of the CVM have rearranged into two bilateral clusters from which they start migrating onto the TVM on either side of the embryo. C) Mid-stage 12. The CVM cells (at this stage considered longitudinal muscle founders) migrate along the dorsal and ventral surfaces of the TVM. D) Late-stage 12, partial high magnification view. The longitudinal muscle founders (lm-fo) have spread out along the TVM and begin fusing with the remaining fusion-competent cells (v-fc) from the TVM. The circular visceral muscle founders have already fused into binucleated syncytia.

caudal visceral mesodermal cells appear to induce the formation of specific cell types (marked by the expression of the homeobox gene *even-skipped*) in the dorsal somatic or cardiac mesoderm along their path or migration.[8]

The foregut and hindgut visceral mesoderm envelopes the developing ectodermal tube of the foregut and hindgut, respectively, and forms the circular muscles of these portions of the digestive tract.[16]

Genetic and Molecular Mechanisms Controlling Visceral Mesoderm Development

Trunk Visceral Mesoderm and Circular Midgut Muscles

Early Mesoderm Patterning and the Formation of Visceral Mesoderm Primordia in the Trunk

The early mesoderm of *Drosophila* embryos can be divided into three parts, cephalic, trunk and caudal mesoderm, along the anterior-posterior axis. The trunk mesoderm, like the ectoderm, is composed of metameric units, called parasegments (PS).[17-19] The anterior and posterior boundaries of each mesodermal parasegment coincide with those of each ectodermal parasegment. The cells within each mesodermal parasegment can be subdivided into anterior (A) and posterior (P) domains, which are located underneath the anterior and posterior ectodermal compartments, respectively[18] (note that Riechmann et al termed the A and P domains slp and eve domains, respectively[19]) (Fig. 4A). The primordia of different mesodermal tissues, such as visceral, cardiac and somatic muscles, as well as fat body, are derived from mesodermal cells located at precisely defined mesodermal positions with regard to the antero-posterior and dorso-ventral axis.[17-19] Notably, the dorsal/anterior quadrant of each parasegment (i.e., the dorsal portion of each P domain) contains the precursors of the trunk visceral mesoderm. In contrast, the dorsal posterior quadrant (i.e., the dorsal portion of each A domain) forms cardiac and dorsal somatic muscle progenitors, and the ventral quadrants form various ventral somatic muscle progenitors as well as fat body and gonadal mesodermal precursors (Fig. 4A).

The *bagpipe* (*bap*) gene, which encodes a NK family homeodomain-containing transcription factor, is required to specify trunk visceral mesoderm during early mesoderm development.[4] *bap* is expressed in the circular muscle precursors of foregut, midgut and hindgut. In the trunk mesoderm, *bap* expression initiates within 11 dorsally located segmental domains on either side of the embryo and fades when these clusters of cells merge to form continuous bilateral bands of the trunk visceral mesoderm.[4] These *bap*-expressing cells define the above-described trunk visceral mesoderm primordia and include both founder cells and fusion-competent cells. The phenotype observed in late stage *bap* mutant embryos, which consists of a reduction or loss of midgut visceral muscles and midgut constrictions, indicates that the function of *bap* is required for the development of the visceral muscles of the midgut. Indeed, in *bap* deficiency embryos, several trunk visceral mesoderm markers, including the Immunoglobulin-like domain protein Fasciclin III (Fas III), are not expressed, although the primordial cells still ingress. Moreover, at later stages, *bap*-deficient trunk visceral mesoderm precursors fuse with somatic muscle founders, thus becoming incorporated into somatic muscle fibers instead of forming midgut muscle fibers.[7] Analogous, albeit weaker phenotypes are seen in embryos homozygous for the only available point mutant allele of *bap*, the hypomorphic *bap*,[208] which display strongly reduced Fas III expression and only partial transformations from visceral to somatic muscle identities.[4] Together, the expression and phenotypic data make *bap* the earliest known regulator that specifically controls trunk visceral mesoderm development.

Since *bap* mRNA and proteins vanish before the visceral mesoderm differentiates and acquires its final morphology, *bap* must activate downstream genes that regulate subsequent differentiation events. One candidate target gene is *biniou* (*bin*), which belongs to the FoxF subfamily of forkhead domain genes.[7,20] Bin is expressed in all types of visceral musculature

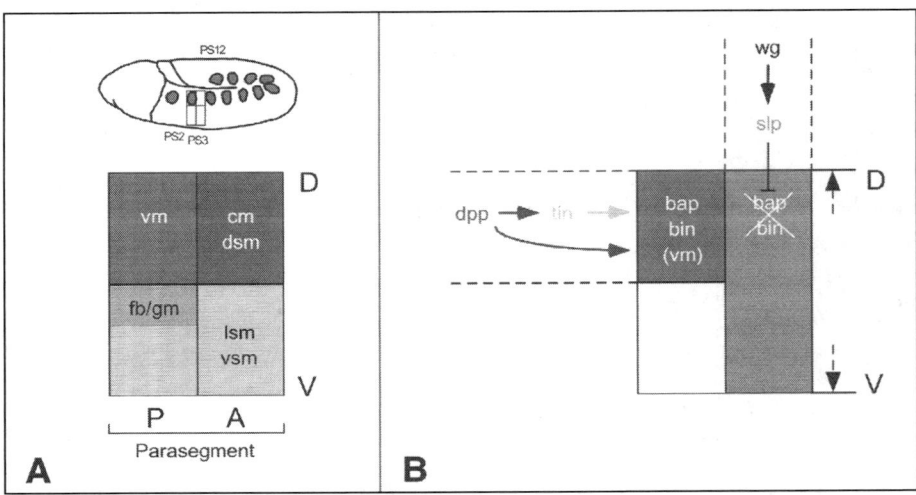

Figure 4. Mesoderm patterning and the specification of trunk visceral mesoderm primordia. A) The 11 segmental clusters of trunk visceral mesoderm primordia (red), corresponding to PS2-PS12, are located in the dorsal mesoderm on either side of the embryo. Each parasegment (PS) can be subdivided into an anterior (A) and a posterior (P) domain. The primordia of different mesodermal tissues originate from specifically defined positions according to their antero-posterior and dorsal-ventral location within each parasegment. B) In the trunk mesoderm, the specification of visceral mesoderm primordia, which express *bap* and *bin*, is controlled by combinatorial inputs from different ectodermal cues and mesodermal regulators. The Dpp signals from dorsal ectoderm (continuous along the A/P axis) activate uninterrupted *tin* expression in dorsal mesoderm, and then work together with Tin to activate *bap* and *bin* expression in the dorsal mesoderm. Simultaneously, segmented Wg signals from ectoderm induce striped *slp* expression in the mesoderm, and this mesodermal Slp abrogates the activitation of *bap* by Dpp + Tin, resulting in segmental expression patterns of *bap* and *bin* that are complementary to that of *slp*. Abbreviations: D, dorsal; V, ventral; vm, visceral mesoderm; cm, cardiac mesoderm; dsm, lsm and vsm, dorsal, lateral and ventral somatic mesoderm; fb, fat body (only the primary fb primordia are shown[65]); gm, gonadal mesoderm. A color version of this figure is available online at http://www.Eurekah.com.

precursors, including those of the circular muscles of the foregut, midgut and hindgut, as well as the longitudinal muscles of the midgut. Like *bap*, *bin* is expressed bilaterally in the 11 segmented domains of trunk visceral mesoderm primordia. However, unlike *bap*, *bin* expression persists through all stages of visceral muscle development.[7] The phenotype of disrupted trunk visceral mesoderm formation and absence of midgut muscles observed in *bin* mutants is identical to that of *bap* mutants, showing that *bin* also plays an important role during early visceral mesoderm development.[7] Additional genetic studies have shown that *bap* is required to activate *bin* expression in the trunk visceral mesoderm primordia and, conversely, that *bin* is necessary to maintain *bap* expression during stage 11 through a positive feedback loop.[7] The observed partial rescue of Fas III expression in the visceral mesoderm of *bap* mutants upon forced *bin* expression specifically in the visceral mesoderm suggests that *bin* is one of the major downstream target genes of *bap*. Bin acts together with Bap to specify the visceral mesoderm primordia during early visceral mesoderm development as well as downstream of *bap* during subsequent stages.[7] After the disappearance of Bap, Bin continues to maintain the cell identities of the visceral mesoderm and to sustain the distinction between visceral and somatic mesoderm, presumably by activating essential patterning and differentiation genes.

The early expression and function of *bap* in the trunk visceral mesoderm primordia have made it a priority to define the upstream regulators that determine the early pattern of metameric *bap* expression in the dorsal mesoderm. These studies showed that this pattern is achieved through combinatorial mechanisms, which require differential combinations of inductive signals from the ectoderm (Fig. 4B). Decapentaplegic (Dpp), a member of the Transforming Growth Factor β (TGF-β) superfamily and a homologue of vertebrate Bone Morphogenetic Proteins (BMPs), is secreted from the dorsal ectoderm all along the anterior-posterior axis of the embryonic trunk[21] and signals towards the underlying cells of the dorsal mesoderm.[22,23] One of the earliest mesodermal functions of Dpp is to maintain the expression of *tinman* (*tin*), another NK homeobox gene, in the dorsal mesoderm.[23] Thickveins (Tkv), the type-I Dpp receptor, Medea, a *Drosophila* member of the Smad protein family of BMP signal effectors, as well as Tin itself, are required to transduce Dpp signals and to activate *tin* as a direct target in the dorsal mesoderm.[24,25] Subsequently, Tin protein cooperates with Dpp signals to activate *bap* expression, and Bap may in turn cooperate with Dpp to activate *bin* expression[4,7,23](Lee and Frasch, unpublished data). Wingless (Wg), a *Drosophila* homolog of vertebrate Wnt molecules, is expressed in transverse stripes in cells of the A compartments of the ectoderm and acts as a negative signal during this process.[18] The Wg signals act indirectly upon *bap* during this event, via inducing the striped expression of the forkhead domain gene *sloppy-paired* (*slp*) in the A domains (hence also termed slp domains) of the mesoderm.[26] The induction of *slp* in the mesoderm (and ectoderm) by Wg involves the binding of the Wg downstream effector, Pangolin (Pan)/dTCF, to several sites within an enhancer of *slp*.[26] Mesodermal Slp acts as a negative regulator of *bap*,[19] which apparently serves to abrogate *bap* activation by Dpp and Tin in a segmental fashion. Recent data have demonstrated that these positive and negative inputs are directly integrated at the level of the *bap* enhancer, which contains essential binding sites for Tin, Smad proteins, and Slp (Lee and Frasch, unpublished data). Together, these observations suggest that the segmental on/off switch of *bap* along the antero-posterior axis is achieved through a molecular switch in which the binding of Tin + Smads to the *bap* enhancer determines the "on" state, whereas additional binding of Slp determines the "off" state of enhancer activity.

As discussed elsewhere, ectodermal Wg and mesodermal Slp are required not only to prevent the formation of visceral mesoderm primordia, but also to promote the development of cardiac and dorsal somatic mesoderm in these same domains.[18,19,26-29]

Early Differentiation Events of the Trunk Visceral Mesoderm

The earliest steps of visceral muscle differentiation are observed through the detection of structural proteins at late stage 11, during and shortly after the formation of the continuous bands of trunk visceral mesoderm. Two of the differentiation markers that have been characterized in more detail are β3 Tubulin (β3Tub60D) and Fasciclin III (Fas III), both of which require *bap* and *bin* for their normal expression. β3Tub60D is a cytoskeletal component that contributes to the normal cytoarchitecture of visceral mesoderm cells.[14] Loss-of-function of β*3Tub60D* causes defects in midgut morphogenesis, and a failure of gut function. Despite normal movements of the food through the gut tube, mutant larvae are deficient in absorbing nutrients across the gut wall, presumably due to defects in the underlying endodermal layer of the gut.[14] β*3Tub60D* is activated in the trunk visceral mesoderm by two (likely functionally redundant) enhancers, vm1 and vm2, that are located in the major intron.[14,30-32] vm1, which has been dissected molecularly, contains an optimal NK homeodomain binding sequence that binds both Tin and Bap, as well as two Bin binding sites.[32] The Tin/Bap and Bin binding sites are required, in a partially redundant manner, for the activity of vm1 in the visceral mesoderm. In addition, certain "weakened" derivatives of vm1, such as one that turns out to lack the Tin/Bap site and one of the two Bin sites, revealed a contribution of the homeotic protein Ultrabithorax (Ubx) to vm1 enhancer activation in parasegment 7.[30,32] This activity of Ubx is mediated by two Ubx binding sites in vm1 that are distinct from the Tin/Bap site. Because β*3Tub60D* and full-length vm1 are expressed uniformly in all parasegments of the visceral mesoderm, the significance of this Hox gene input in vivo is presently not clear.

The presence of internalized and at least partially differentiated trunk visceral mesoderm cells is also required for normal guidance of posterior salivary gland migration.[33] Hence, the visceral mesoderm is not only crucial for the movement of the gut but also for other tissues that develop in close contact with it. Salivary gland migration does not require any *bin*-dependent events of visceral mesoderm differentiation, although it requires *tin* activity as well as PS2 integrin, which is present in all mesodermal cells at this stage.[33] These data provide the first hints that *bin* might control only a subset of differentiation events downstream of *tin* in the visceral mesoderm. While some differentiation events downstream of *tin* are known to require a combination of *bap* and *bin*, others may require either *bap* or *bin* alone for their execution.

Specification of Visceral Muscle Progenitors and Fusion

After the establishment of the *bap* and *bin* expression patterns, the precursor pool of the trunk visceral mesoderm cells becomes subdivided into two different types, founder cells and fusion-competent cells. This subdivision is critical for later differentiation events during visceral muscle development and the generation of binucleated circular visceral muscle fibers. The visceral muscle founder cells form two rows of cells located at the ventral margin of visceral mesoderm, whereas the remaining cells located more dorsally become fusion-competent cells[9,10] (Fig. 5C). The founder cells derive from visceral muscle progenitor cells via a round of cell division that also duplicates the number of presumptive fusion-competent cells (Fig. 5B). The earliest signs of a distinction between the two cell types are seen shortly after the establishment of the *bin* and *bap* expression patterns, when cells at the ventral margin of each visceral mesoderm precursor cluster become positive for activated (diphospho) MAPK.[34] Shortly thereafter, just prior and during the ingression of the visceral mesoderm precursors, the ventral row of cells initiates expression of the *Tbx1*-related T-box gene *org-1* as well as the myoblast-fusion regulator and somatic founder myoblast marker *dumbfounded* (*duf*, aka *kin of irregular chiasm C/kirre*).[9,10] At the same time, these cells, which are now *bona fide* visceral muscle progenitors, acquire their distinct columnar morphology.[9,10,35,36] The founder cells derived from this row of progenitors additionally express the homeobox gene *H2.0*, the bHLH gene *hand*, and up-regulate *bin* expression.[7,37,38] By contrast, the remaining, dorsally located cells remain less ordered and express markers for fusion-competent myoblasts, including the myoblast-fusion regulator *sticks-and-stones* (*sns*).[9,10]

The subdivision of the visceral mesoderm precursors into founder cells and fusion-competent cells and the specification of the ventral row of visceral mesoderm precursors as visceral muscle progenitors is triggered by a novel signaling pathway, which involves Jelly belly (Jeb) as a signal and Alk as its receptor[35,36,39] (Fig. 5A-D). *jelly belly*, which encodes a secreted molecule containing a type A LDL receptor motif, is expressed in somatic myoblast precursors in the ventro-lateral mesoderm, ventrally-adjacent to the visceral mesoderm primordia[40] (Fig. 5A,B). The receptor tyrosine kinase Alk, which is the *Drosophila* homolog of the human proto-oncogene *anaplastic lymphoma kinase* (*ALK*),[41] is expressed in all cells of the trunk visceral mesoderm primordial,[42] simultaneously with *bap* and *bin* (Lee and Frasch, unpublished data). Embryos that are mutant for either *jeb* or *Alk* show the same mesodermal phenotype, namely absence of midgut musculature.[35,40] The visceral muscle phenotype in *jeb* mutants can be rescued with a constitutively activated version of *Alk*, further suggesting that these two molecules act in the same genetic pathway and that *Alk* functions downstream of *jeb* in visceral muscle development.[35] Indeed, cell culture based binding assays demonstrated that Jeb binds to Alk with high affinity and specificity, and binding triggers dose-dependent MAPK activation.[35,36] In the aggregate, the genetic and biochemical data support the notion that Jeb and Alk form a ligand-receptor pair, which functions to specify the cell fate of visceral founders from "naïve" visceral mesoderm precursor cells.

Within the ventral row of visceral mesoderm precursors, which contain increased amounts of surface-bound Jeb, the Jeb signals are transmitted through the Ras/Raf/MAPK cascade downstream of the receptor. In turn, this signaling cascade induces the expression of *org-1*, *duf*, and probably other target genes, which impart the identity as visceral muscle progenitors on these cells. Conversely, the expression of fusion-competent marker such as *sns* is down-regulated by active Jeb/Alk in these cells. The observation that activated Ras can rescue the absence of founder cell markers and of visceral muscles in *jeb* mutants[35] indicates that the Ras/Raf/MAPK pathway

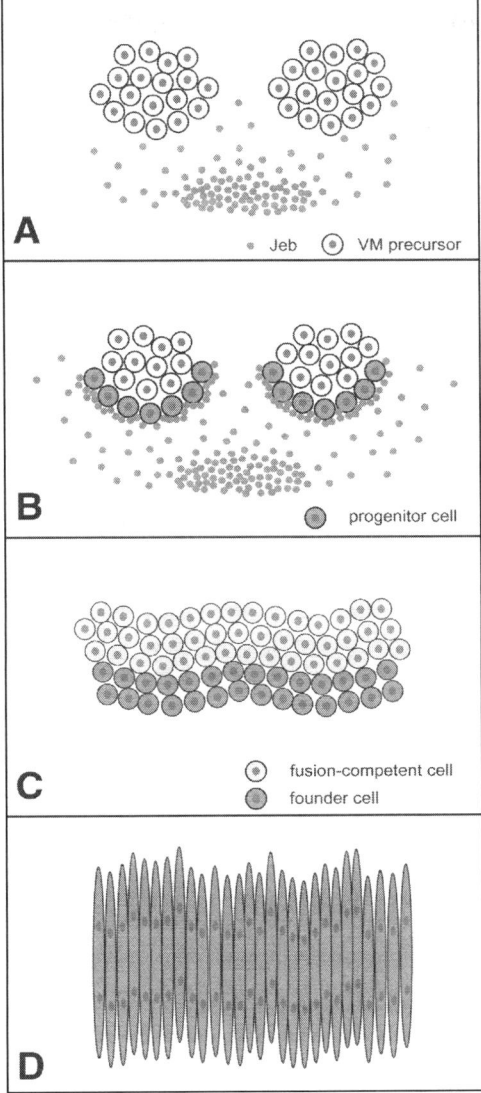

Figure 5. Visceral founder cell induction and the formation of binucleated circular visceral muscles from trunk visceral mesoderm primordia. A) During early visceral mesoderm development, cells in the ventral lateral mesoderm synthesize and secrete Jeb (red) to the visceral mesoderm precursors, which express the receptor Alk. B) Jeb molecules binding to their receptor Alk on the visceral mesoderm precursors can be found highly concentrated at the ventral boundary of the visceral mesoderm primordia. Subsequently, Jeb/Alk signals specify a single row of ventral visceral mesoderm primordia to become visceral muscle progenitors. C) As the cells divide, the visceral mesoderm primordia increase in cell number and expand longitudinally to form a continuous band of trunk visceral mesoderm. Each visceral progenitor will give rise to two founder cells located at the ventral side of the visceral mesoderm. The cells located at more dorsal portion of visceral mesoderm become fusion-competent cells. D) One visceral founder cell fuses with one visceral fusion-competent cell to generate a circular visceral muscle fiber with two nuclei. A color version of this figure is available online at http://www.Eurekah.com.

is sufficient for specification events downstream of Alk. However, it remains possible that additional effectors that have been described for human ALK, such as PI3 kinase and STATs, are involved in other processes such as Jeb-dependent cell rearrangements (see ref. 40).

Although Jeb protein is found highly concentrated along the ventral boundaries of the visceral mesoderm primordia, lower levels can be detected across the whole visceral mesoderm, suggesting there are additional mechanisms restricting the Jeb/Alk signaling to only the ventral-most cell row. It has been shown that Ras and Notch signaling activities play opposing roles in the specification of somatic progenitors and nonprogenitors in somatic muscle development.[43,44] A similar mechanism might be active in visceral muscle development, as genetic studies with loss- and gain-of-function of Notch signaling activities indicate that Notch signaling is able to antagonize the activity of Jeb/Alk/Ras signals.[39] Although in Notch mutants the initial induction of *org-1* and *duf* in a single, ventral row of cells is normal, Notch signaling is required to prevent a subsequent expansion of the expression of these founder cell markers into more dorsally located cells that are to become fusion-competent cells (Lee and Frasch, unpublished data).

Cell lineage studies in *jeb* mutant embryos, in which the cells derived from the visceral mesoderm primordia were marked with *bap-lacZ*, showed that the perduring β-gal protein becomes incorporated into the somatic muscle fibers at late stages.[35,39] Hence, it appears that without Jeb/Alk signals to specify the visceral founders, all visceral mesoderm cells become fusion-competent cells by default and can be incorporated into somatic muscles by fusing with somatic founder myoblasts. Earlier defects in *jeb* mutants include the premature disappearance of Fas III and *bin* expression as well as defects in cell rearrangements and migration of the visceral mesoderm primordial.[40] Altogether, these data demonstrate that the cells receiving Jeb/Alk signals function indeed as founder cells, which provide critical roles in organizing the migration, fusion, and differentiation of the developing visceral mesoderm.

Patterning of the Trunk Visceral Mesoderm along the Anterior-Posterior Axis

As described above, the trunk visceral mesoderm is assembled from the dorsal portions of the P domains of each mesodermal parasegment. Although each these metameric units contains only part of the original mesodermal parasegment, historically they have also been called "parasegments" (of the visceral mesoderm), and herein we adhere to this convention. Based upon the expression patterns of various markers in the visceral mesoderm, it has been shown that the cells within each of these parasegments (PS) become regionally distinct along the antero-posterior axis. This program initiates in the founder cell rows, and upon myoblast fusion is imposed on all nuclei within the trunk visceral mesoderm. Preliminary data on this subject were obtained through the late-phase expression patterns of *tin*, *bap*, a *bap-lacZ* reporter[4] (which is driven by an enhancer that is distinct from the early-acting enhancer discussed above), as well as the cell adhesion molecule Connectin (Con).[45] However, a recent study by Hosono et al has provided much more detailed insight into this process.[46] These authors show that, in the founder cell rows, *tin* and *bap* expression is reactivated in identical metameric domains that also express the FGF-encoding gene *branchless* (*bnl*). In addition, *hedgehog* (*hh*, and likewise *hh-lacZ*) and Con are expressed in metameric domains with different anterior-posterior registers. Based upon the combinatorial, exclusive, and absent expression of these markers in different cells, each parasegment of the visceral mesoderm can be subdivided into six (in the thorax only five) distinct regions or subdomains (Fig. 6). Although it is not clear whether all of these subdomains are functionally significant, at least the metameric expression of *hh* and *bnl* appears to be important for the development of the midgut and associated tissues. Specifically, visceral mesodermal *hh* appears to play a role in regulating the formation of gastric caeca, whereas the expression domains of *bnl* are required to guide the proper migration of the segmental buds of the visceral branch of the trachea.[46] These buds express the FGF receptor Breathless (Btl) and visceral mesodermal Bnl may serve as a localized chemoattractant for these cells.

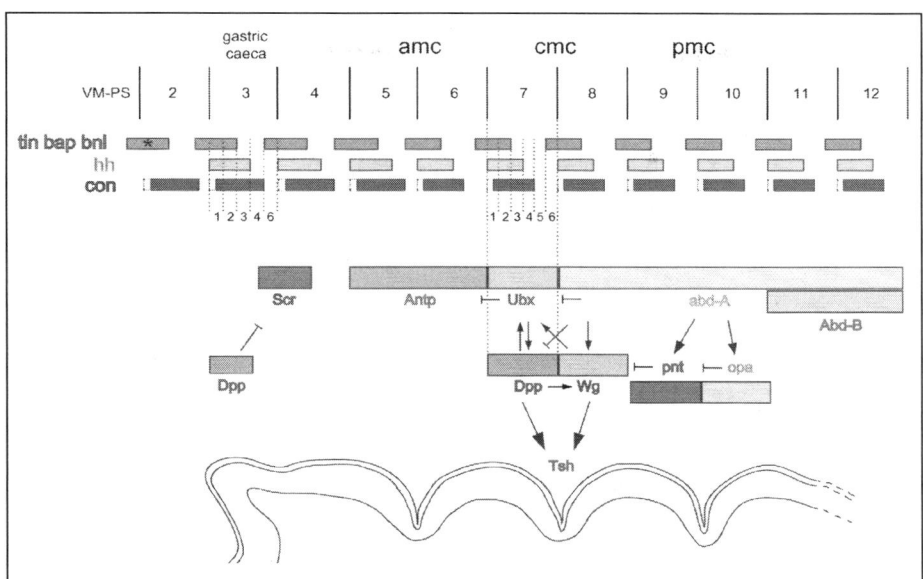

Figure 6. Summary of the anterior-posterior subdivisions of the visceral mesoderm and the genetic interactions patterning the midgut mesoderm. Top: Midgut extending from parasegments PS2-12 (limits are indicated by vertical bars); the position of the castric caeca and of midgut constrictions (amc, cmc and pmc; anterior, central and posterior midgut constrictions) are also shown. Middle: Visceral mesoderm PS subdivisions based upon metameric gene expression patterns. The anterior limit of each PS is defined by the expression in the visceral mesoderm of Con (weak: stippled, and strong: violet boxes). *tin*, *bap* and *bnl* are metamerically expressed in PS3-12 (green boxes) and *bnl* is additionally expressed in PS2 (asterisk). *hh* is expressed in the visceral mesoderm PS3-12 at stage 11 (orange boxes). Note the absence of the fifth subdomain from thoracic visceral mesoderm PSs. Bottom: Expression domains of the homeotic genes known to be expressed in the visceral mesoderm in relation to *dpp* and *wg* expression are represented by boxes. The regulatory network among midgut PS7-8 patterning genes is diagrammed. The homeotic-regulated signals Wg and Dpp, produced on either side of the future central constriction, play a critical role in central midgut constriction formation. The *pnt*- and *opa*-expressing domains define the posterior midgut constriction of the fourth chamber. Gastric caeca depend on the expression of Scr, *dpp* and *hh* in the visceral mesoderm PS3. A color version of this figure is available online at http://www.Eurekah.com.

The expression domains of *tin*, *bap*, *bnl*, *hh*, and Con arise near the anterior and/or posterior portions of the founder cell rows within each parasegment during stage 10-11, when the visceral mesoderm precursor clusters expand along the anterosterior axis (compare with Fig. 2A,B). During this process, the anterior founder cells of each cluster remain close to the ectodermal parasegment borders and, hence, the source of Wg and Hh signals. Likewise, the posterior founder cells of each cluster reach the borders of the next parasegment during the posterior expansion of the clusters, which brings them into the vicinity of the adjacent ectodermal Wg stripes during stage 11. These spatial relationships are in agreement with the finding that the metameric domains of *tin/ bap/ bnl*, *hh*, and Con, and hence the visceral mesodermal subdomains, are established through activating and repressing functions of Wg and Hh signals during these stages.[46] Thus, the visceral mesoderm develops in at least two steps under the control of ectodermal Wg and Hh signals. First, during stage 9-10 Wg is suppressing visceral mesoderm precursor formation (see above).[18,26] Second, during stages 10-11, Wg and Hh are responsible for establishing regional patterns within each visceral mesodermal parasegment, which are necessary for normal midgut and tracheal development.

In addition to its metameric pattern, the trunk visceral mesoderm becomes subdivided into broader domains along the antero-posterior axis, some of which span the width of several parasegments. During later stages of midgut development, this subdivision is reflected in the appearance of gastric caeca in the anterior portion and of three constrictions at specific positions within the main portion of the midgut, which subsequently develop into midgut loops. The formation and placement of these structures is dependent on the homeotic (Hox) genes *Sex combs reduced* (*Scr*), *Antennapedia* (*Antp*), *Ultrabithorax* (*Ubx*), and *abdominal-A* (*abd-A*) (reviewed in refs. 47, 48). The expression of each of these homeotic genes defines a specific domain along the antero-posterior axis of the visceral mesoderm (Fig. 6). Embryos lacking the function of a particular homeotic gene fail to develop the constriction in the region where the gene is normally expressed.[13] The formation of the midgut constrictions is dependent not only on the homeotic genes, but also on their downstream target genes such as *dpp*, *wg*, *teashirt* (*tsh*), *odd-skipped* (*odd*) and *pointed* (*pnt*) (reviewed in refs. 47, 48) (Fig. 6).

Herein, we exemplify this complex regulatory process by focusing on *dpp*. Spatially-restricted expression of *dpp* in PS7 of the visceral mesoderm (which again initiates specifically in the founder cells of PS7[9]) is crucial for normal midgut morphogenesis.[49-51] Previous studies identified Ubx and Extradenticle (Exd) as direct and positively acting upstream regulators of *dpp* in the visceral mesoderm, which bind to essential sites in a visceral mesoderm-specific *dpp* enhancer element.[49,52,53] Additionally,[49-51] proposed that the activation of *dpp* depends on tissue-specific cofactor(s) in the developing visceral mesoderm. More recent studies have shown that *dpp* activation requires specific Bin binding sites in this *dpp* enhancer element, suggesting that Bin serves as a tissue-specific cofactor of hometic gene products in the visceral mesoderm.[7] *dpp*, together with *wg*, is required to activate the transcription factor *teashirt* (*tsh*) in the central midgut constriction.[54] In addition, Dpp signals together with signals from the EGFR-ligand Vein, across germ layers induce the expression of the homeotic gene *labial* (*lab*) in the underlying endodermal midgut epithelium.[55-57] Taken together, these studies illustrate how regionally-restricted Hox factors, signaling molecules, and a visceral mesoderm-specific cofactor cooperate molecularly to promote morphogenetic changes during organogenesis of the midgut.

Visceral Mesoderm and Circular Muscles of the Foregut and Hindgut

Most of our efforts in the understanding of visceral mesoderm development have centered on the trunk visceral mesoderm. Several observations indicate that distinct mechanisms govern the development of the foregut and hindgut visceral mesoderm development. The mesodermal layers of foregut and hindgut are gradually assembled around the invaginating stomodeal and proctodeal tubes, which unlike the endodermal midgut rudiments are already patterned prior to this process (reviewed in refs. 47, 58). This situation would allow for the possibility that signaling from the ectodermal gut epithelia to the visceral mesoderm can occur. These mesodermal primordia, termed foregut and hindgut visceral mesoderm, respectively, express *bap* and *bin*, as well as high *twist* levels.[4,7,16,59] The expression of these genes persists and, in late stage embryos, covers the entire foregut and hindgut tubes. However, *bap* and *bin* do not seem to be required genetically for foregut and hindgut muscle development, although high Twist levels do appear to be important.[4,7,16] In the presumptive hindgut, which has been studied in more detail than the foregut, the *bap*-marked visceral mesodermal cells arise just posteriorly to the 15th ectodermal Wg stripe, in close contact with another, broader Wg-expressing domain within the proctodeal ectoderm.[16,59] Indeed, *wg* is required for normal *bap* induction and hindgut visceral mesoderm formation, which involves ectoderm-to-mesoderm signaling by Wg.[16,59] In addition to the induction of *bap* and the up-regulation of *twi*, *wg* also up-regulates the expression of the FGF-receptor encoding gene *htl* in these cells.[16] Temporally overlapping but mostly subsequent to this activity of Wg, Htl-mediated signaling is required to maintain the differentiation program of the hindgut visceral mesoderm.[16] In agreement with this observation, *thisbe* (*ths*), one of two genes encoding FGF8-related proteins that appear to serve as

Htl ligands, is highly expressed in the developing hindgut (and foregut) epithelium. Lack of *ths* together with the second gene, *pyramus* (*pyr*), leads to the absence of Mef2-positive hindgut visceral muscles.[60] Analogous mechanisms appear to operate during development of the foregut visceral mesoderm.[16] The positive regulation of *bap* by Wg in the hindgut and foregut visceral mesoderm contrasts with the negative regulation of *bap* in the trunk visceral mesoderm (see above) and is controlled by a separate enhancer element (Lee and Frasch, unpublished data).

Caudal Visceral Mesoderm and Longitudinal Midgut Muscles

The caudal visceral mesoderm (CVM) can be visualized via reporter gene expression driven by enhancers of *crocodile* (*croc*),[61] *mef-2*[62] and *couch-potato* (*cpo*).[6] The earliest specific marker for the CVM is *bHLH54F*, which encodes a *twist*-related basic helix-loop-helix transcription factor.[5] Initial *bHLH54F* expression during the late blastoderm stage occurs in a ventral/posterior domain, and subsequently at the posterior end of the ventral furrow, suggesting that the precursor cells of CVM may be determined before gastrulation.[5] This domain matches the mesodermal portion of the transverse stripe of expression of *brachyenteron* (*byn*), a Brachyury-related T-box gene.[63] The *byn* stripe in turn overlaps with the expression domain of the terminal gap genes *tailless* (*tll*) and *forkhead* (*fkh*), whereas it abuts the more posterior expression domain of another terminal gap gene, *huckebein* (*hkb*).

The genetic studies suggest that activation of the expression of *byn* by *tll* and its repression by *hkb* restrict *byn* expression to the striped domain that includes the presumptive CVM.[8,64] In addition, *tll* activates *fkh*, leading to coexpression of *byn* and *fkh* in the presumptive CVM.[8] It needs to be established whether *byn* and/or *fkh*, in combination with the mesoderm-determining genes snail (sna) and/or twist (twi), control the activation of *bHLH54F* specifically within the presumptive CVM. In parallel to *byn*, *hkb* also represses the mesoderm-determining gene *snail* (*sna*), which together with *tll* is required for upregulating the expression of *Zn finger homeodomain 1* (*zfh-1*) in the CVM during gastrulation.[8] As a result, *zfh-1*, *byn*, *fkh*, and *bHLH54F* become coexpressed in the CVM.

Mutant embryos for *byn*, *fkh* or *zfh-1* (data on *bHLH54F* are presently not available) show disruptions of caudal visceral mesoderm development.[8] Furthermore, based upon the complete absence of CVM in *byn*, *fkh* double mutant embryos (and likewise, in *tll*), compared to the less severe CMV phenotypes in the single mutants, *byn* and *fkh* appear to collaborate in specifying CVM.[8] This collaboration between *byn* and *fkh* is additionally dependent on the presence of high levels of expression of *zfh-1* in this posterior mesodermal domain. Accordingly, ectopic expression of both *byn* and *fkh* generates ectopic CVM in an anterior mesodermal area that also contains high endogenous *zfh-1* expression. The phenotypic analysis of loss- and gain-of-function with *byn* suggests a major role of *byn* in regulating the adhesive properties of CVM cells, which in turn may be a prerequisite for proper migration and survival of these cells.[8]

bin is starting to be expressed in the CVM at stage 10 and longitudinal midgut muscles are absent in *bin* mutants, suggesting that the CVM is either not formed or fails to migrate.[7] Several CVM-specific markers are expressed normally in *bin* mutants, which indicates that *bin* is not required for specifying CVM.[7] Together with similar data with *bap* mutants (Frasch, unpublished data) these observations suggest that the absence of longitudinal midgut muscles is an indirect effect caused by the absence of the trunk visceral mesoderm in these mutants. Hence, the longitudinal visceral muscle progenitors, which migrate in close contact with and along the trunk visceral mesoderm, require the trunk visceral mesoderm as a substratum for their normal migration.

During their migration, the CVM cells start expressing *duf* and become the founder cells of the longitudinal midgut muscles.[9,10] Upon reaching their destinations, these founder cells fuse with the remaining visceral fusion-competent cells that were not utilized for the previous fusions with circular visceral muscle founders. Perhaps, the observed defects in the circular midgut musculature in *byn* mutant embryos[8] are a result of the availability of supernumerary fusion competent cells, which may then be able to fuse with circular muscle fibers and contribute to abnormally large syncitia.

Conclusion

The past decade has brought enormous advances in our understanding of the mechanisms governing visceral muscle development in *Drosophila*. We have learned that the molecular patterning mechanisms generating the visceral muscle primordia in the early mesoderm are closely intertwined with the ones generating the primordial cells of heart and somatic muscles. In addition, we have gained information on genes and mechanisms that distinguish the development and differentiation of visceral muscles from that of the somatic and cardiac muscles. In the near future, additional information from genetic, genomic, and molecular studies will fill many of the remaining gaps in our knowledge. Furthermore, it has become apparent that many of the genes and mechanisms regulating early *Drosophila* trunk visceral mesoderm development are related to those acting in the development of the splanchnic mesoderm in vertebrates. In the vertebrates, this tissue not only forms visceral muscles but also the mesodermal portions of many other internal organs. Hence, we expect that the insights from *Drosophila* will also be useful for gaining a better understanding of the mechanisms of visceral muscle development and processes of organogenesis in vertebrate species.

References

1. Goldstein MA, Burdette WJ. Striated visceral muscle of Drosophila melanogaster. J Morphol 1971; 134:315-334.
2. Brown N, Gregory S, Martin-Bermudo M. Integrins as mediators of morphogenesis in Drosophila. Dev Biol 2001; 223:1-16.
3. Klapper R, Heuser S, Strasser T et al. A new approach reveals syncytia within the visceral musculature of Drosophila melanogaster. Development 2001; 128:2517-2524.
4. Azpiazu N, Frasch M. tinman and bagpipe: Two homeo box genes that determine cell fates in the dorsal mesoderm of Drosophila. Genes Dev 1993; 7:1325-1340.
5. Georgias C, Wasser M, Hinz U. A basic-helix-loop-helix protein expressed in precursors of Drosophila longitudinal visceral muscles. Mech Dev 1997; 69:115-124.
6. Campos-Ortega JA, Hartenstein V. The embryonic development of Drosophila melanogaster. 2nd ed. Berlin: Springer Verlag 1997.
7. Zaffran S, Küchler A, Lee HH et al. biniou (FoxF), a central component in a regulatory network controlling visceral mesoderm development and midgut morphogenesis in Drosophila. Genes Dev 2001; 15:2900-2915.
8. Kusch T, Reuter R. Functions for Drosophila brachyenteron and forkhead in mesoderm specification and cell signalling. Development 1999; 126:3991-4003.
9. San Martin B, Ruiz-Gomez M, Landgraf M et al. A distinct set of founders and fusion-competent myoblasts make visceral muscles in the Drosophila embryo. Development 2001; 128:3331-3338.
10. Klapper R, Stute C, Schomaker O et al. The formation of syncytia within the visceral musculature of the Drosophila midgut is dependent on duf, sns and mbc. Mech Dev 2002; 110:85-96.
11. Reuter R, Grunewald B, Leptin M. A role for the mesoderm in endodermal migration and morphogenesis in Drosophila. Development 1993; 119:1135-1145.
12. Tepass U, Hartenstein V. Epithelium formation in the Drosophila midgut depends on the interaction of endoderm and mesoderm. Development 1994; 120:579-590.
13. Reuter R, Scott MP. Expression and function of the homoeotic genes Antennapedia and Sex combs reduced in the embryonic midgut of Drosophila. Development 1990; 109:289-303.
14. Dettman R, Turner F, Raff E. Genetic analysis of the Drosophila β3-tubulin gene demonstrates that the microtubule cytoskeleton in the cells of the visceral mesoderm is required for morphogenesis of the midgut endoderm. Dev Biol 1996; 117-135.
15. Singer J, Harbecke R, Kusch T et al. Drosophila brachyenteron regulates gene activity and morphogenesis in the gut. Development 1996; 122:3707-3718.
16. San Martin B, Bate M. Hindgut visceral mesoderm requires an ectodermal template for normal development in Drosophila. Development 2001; 128:233-242.
17. Dunin-Borkowski O, Brown N, Bate M. Anterior-posterior subdivision and the diversification of the mesoderm in Drosophila. Development 1995; 121:4183-4193.

18. Azpiazu N, Lawrence P, Vincent J-P et al. Segmentation and specification of the Drosophila mesoderm. Genes Dev 1996; 10:3183-3194.
19. Riechmann V, Irion U, Wilson R et al. Control of cell fates and segmentation in the Drosophila mesoderm. Development 1997; 124:2915-2922.
20. Perez Sanchez C, Casas-Tinto S, Sanchez L et al. DmFoxF, a novel Drosophila fork head factor expressed in visceral mesoderm. Mech Dev 2002; 111:163-166.
21. St Johnston RD, Gelbart WM. Decapentaplegic transcripts are localized along the dorsal-ventral axis of the Drosophila embryo. EMBO J 1987; 6:2785-2791.
22. Staehling-Hampton K, Hoffmann FM, Baylies MK et al. dpp induces mesodermal gene expression in Drosophila. Nature 1994; 372:783-786.
23. Frasch M. Induction of visceral and cardiac mesoderm by ectodermal Dpp in the early Drosophila embryo. Nature 1995; 374:464-467.
24. Yin Z, Frasch M. Regulation and function of tinman during dorsal mesoderm induction and heart specification in Drosophila. Dev Genet 1998; 22:187-200.
25. Xu X, Yin Z, Hudson J et al. Smad proteins act in combination with synergistic and antagonistic regulators to target Dpp responses to the Drosophila mesoderm. Genes Dev 1998; 12:2354-2370.
26. Lee H, Frasch M. Wingless effects mesoderm patterning and ectoderm segmentation events via induction of its downstream target sloppy paired. Development 2000; 127:5497-5508.
27. Baylies M, Martinez Arias A, Bate M. Wingless is required for the formation of a subset of muscle founder cells during Drosophila embryogenesis. Development 1995; 121:3829-3837.
28. Lawrence P, Bodmer R, Vincent J. Segmental patterning of heart precursors in Drosophila. Development 1995; 121:4303-4308.
29. Wu X, Golden K, Bodmer R. Heart development in Drosophila requires the segment polarity gene wingless. Dev Biol 1995; 169:619-628.
30. Hinz U, Wolk A, Renkawitz-Pohl R. Ultrabithorax is a regulator of β3 tubulin expression in the Drosophila visceral mesoderm. Development 1992; 116:543-554.
31. Kremser T, Gajewski K, Schulz R et al. tinman regulates the transcription of the beta3 tubulin gene (betaTub60D) in the dorsal vessel of Drosophila. Dev Biol 1999; 216:327-339.
32. Zaffran S, Frasch M. The beta 3 tubulin gene is a direct target of bagpipe and biniou in the visceral mesoderm of Drosophila. Mech Dev 2002; 11485-93.
33. Bradley P, Myat M, Comeaux C et al. Posterior migration of the salivary gland requires an intact visceral mesoderm and integrin function. Dev Biol 2003; 257:249-262.
34. Gabay L, Seger R, Shilo B-Z. MAP kinase in situ activation atlas during Drosophila embryogenesis. Development 1997; 124:3535-3541.
35. Lee HH, Norris A, Weiss JB et al. Jelly belly protein activates the receptor tyrosine kinase Alk to specify visceral muscle pioneers. Nature 2003; 425:507-512.
36. Englund C, Loren CE, Grabbe C et al. Jeb signals through the Alk receptor tyrosine kinase to drive visceral muscle fusion. Nature 2003; 425:512-516.
37. Barad M, Erlebacher A, McGinnis W. Despite expression in embryonic visceral mesoderm, H2.0 is not essential for Drosophila visceral muscle morphogenesis. Dev Genet 1991; 12:206-211.
38. Kolsch V, Paululat A. The highly conserved cardiogenic bHLH factor hand is specifically expressed in circular visceral muscle progenitor cells and in all cell types of the dorsal vessel during Drosophila embryogenesis. Dev Genes Evol 2002; 212:473-485.
39. Stute C, Schimmelpfeng K, Renkawitz-Pohl R et al. Myoblast determination in the somatic and visceral mesoderm depends on notch signalling as well as on milliways (mili(Alk)) as receptor for jeb signalling. Development 2004; 131:743-754.
40. Weiss J, Suyama K, Lee H et al. A Drosophila LDL receptor repeat-containing signal required for mesoderm migration and differentiation. Cell 2001; 107:387-398.
41. Morris S, Kirstein M, Valentine M et al. Fusion of a kinase gene, ALK, to a nucleolar protein gene, NPM, in nonHodgkin's lymphoma. Science 1994; 263:1281-1284.
42. Loren C, Scully A, Grabbe C et al. Identification and characterization of DAlk: A novel Drosophila melanogaster RTK which drives ERK activation in vivo. Genes Cells 2001; 6:531-544.
43. Carmena A, Gisselbrecht S, Harrison J et al. Combinatorial signaling codes for the progressive determination of cell fates in the Drosophila embryonic mesoderm. Genes Dev 1998; 15:3910-3922.
44. Carmena A, Buff E, Halfon M et al. Reciprocal regulatory interactions between the Notch and Ras signaling pathways in the Drosophila embryonic mesoderm. Dev Biol 2002; 244:226-242.
45. Bilder D, Scott M. hedgehog and wingless induce metameric pattern in the Drosophila visceral mesoderm. Dev Biol 1998; 201:43-56.
46. Hosono C, Takaira K, Matsuda R et al. Functional subdivision of trunk visceral mesoderm parasegments in Drosophila is required for gut and trachea development. Development 2003; 130:439-449.

47. Skaer H. The alimentary canal. In: Bate M, Martinez-Arias A, eds. The Development of Drosophila melanogaster. New York: Cold Spring Harbor Laboratory Press, 1993:2:941-1012.
48. Bienz M. Homeotic genes and positional signalling in the Drosophila viscera. Trends Genet 1994; 10:22-26.
49. Reuter R, Panganiban GEF, Hoffmann FM et al. Homeotic genes regulate the spatial expression of putative growth factors in the visceral mesoderm of Drosophila embryos. Development 1990; 110:1031-1040.
50. Masucci JD, Hoffmann FM. Identification of two regions from the Drosophila decapentaplegic gene required for embryonic midgut development and larval viability. Dev Biol 1993; 159:276-287.
51. Hursh D, Padgett R, Gelbart W. Cross regulation of decapentaplegic and Ultrabithorax transcription in the embryonic visceral mesoderm of Drosophila. Development 1993; 117:1211-1222.
52. Sun B, Hursh D, Jackson D et al. Ultrabithorax protein is necessary but not sufficient for full activation of decapentaplegic expression in the visceral mesoderm. EMBO J 1995; 14:520-535.
53. Manak J, Mathies L, Scott M. Regulation of a decapentaplegic midgut enhancer by homeotic proteins. Development 1994; 120:3605-3619.
54. Mathies L, Kerridge S, Scott M. Role of the teashirt gene in Drosophila midgut morphogenesis: Secreted proteins mediate the action of homeotic genes. Development 1994; 120:2799-2809.
55. Immergluck K, Lawrence P, Bienz M. Induction across germ layers in Drosophila mediated by a genetic cascade. Cell 1990; 62:261-268.
56. Panganiban GEF, Reuter R, Scott MP et al. A Drosophila growth factor homolog, decapentaplegic, regulates homeotic gene expression within and across germlayers during midgut morphogenesis. Development 1990; 110:1041-1050.
57. Szuts D, Eresh S, Bienz M. Functional intertwining of Dpp and EGFR signaling during Drosophila endoderm induction. Genes Dev 1998; 12:2022-2035.
58. Lengyel J, Iwaki D. It takes guts: The Drosophila hindgut as a model system for organogenesis. Dev Biol 2002; 213:1-19.
59. Hoch M, Pankratz M. Control of gut development by fork head and cell signaling molecules in Drosophila. Mech Dev 1996; 58:3-14.
60. Stathopoulos A, Tam B, Ronshaugen M et al. pyramus and thisbe: FGF genes that pattern the mesoderm of Drosophila embryos. Genes Dev 2004; 18:687-699.
61. Häcker U, Kaufmann E, Hartmann C et al. The Drosophila fork head domain protein crocodile is required for the establishment of head structures. EMBO J 1995; 14:5306-5317.
62. Nguyen H. Drosophila mef2 expression during mesoderm development is controlled by a complex array of cis-acting regulatory modules. Dev Biol 1998; 204:550-566.
63. Kispert A, Herrmann B, Leptin M et al. Homologs of the mouse Brachyury gene are involved in the specification of posterior terminal structures in Drosophila, Tribolium, and Locusta. Genes Dev 1994; 15:2137-2150.
64. Murakami R, Shigenaga A, Kawakita M et al. Aproctus, a locus that is necessary for the development of the proctodeum in Drosophila embryos, encodes a homolog of the vertebrate Brachyury gene. Roux's Arch Dev Biol 1995; 205:89-96.
65. Riechmann V, Rehorn K, Reuter R et al. The genetic control of the distinction between fat body and gonadal mesoderm in Drosophila. Development 1998; 125:713-723.

CHAPTER 7

Development of the Larval Somatic Musculature

Ana Carmena and Mary Baylies*

Abstract

The larval somatic musculature of *Drosophila* is arranged in a highly stereotyped pattern of 30 muscle fibers per hemisegment. Each muscle possesses a distinctive set of properties: size, shape, orientation, attachments to the epidermis and specific innervation. These qualities make each myofiber a unique element within the pattern. In this chapter we review what is known to date about how a muscle acquires its particular identity. We discuss the mechanisms that underlie the specification of muscle founder cells and fusion-competent myoblasts and how both types of myoblasts contribute to the final pattern of each larval muscle. While a body of evidence supports a crucial role for founder cells in conferring particular muscle identities, the function of fusion-competent myoblasts during muscle morphogenesis remains an open question for future investigation in the field.

Introduction

During *Drosophila* embryogenesis, mesodermal cells undergo a stereotypical series of movements, cell fate decisions and morphological changes to create a complex pattern of thirty larval muscles per abdominal hemisegment (Fig. 1A). Each muscle consists of a single syncytium, which is created by the fusion of neighboring myoblasts. While all muscles share general properties, such as the contractile proteins and neurotransmitter receptors, each muscle in the pattern is a unique structure that can be identified by its orientation, size, shape, epidermal attachments and motoneuron innervation. The acquisition of these muscle specific properties during myogenesis depends upon the prior specification of a special class of myoblasts called founder cells (Fig. 1B). Muscle founder cells were originally identified morphologically,[1] and through the specific expression of the homeodomain protein Slouch (S59) in subsets of these cells.[2] Each larval muscle is prefigured by a single founder cell, which seeds muscle formation by fusing with surrounding fusion-competent myoblasts. The number of fusion events is variable (3-25), depending on the specific muscle.[3] Analysis of embryos carrying mutations in genes required for the fusion process, such as *myoblast city*, supports the idea that each founder cell contains the "identity" information for an individual muscle: when fusion is blocked, founder cells make mononucleated muscles that are correctly innervated and properly oriented. These muscles, however, are much smaller and often are unable to attach to tendon cells. Under these conditions, fusion-competent myoblasts remain undifferentiated, although they can express general muscle-markers such as the contractile protein Myosin.[4] Thus, it has been proposed that each founder cell contains the unique information required to direct the morphogenesis of a specific muscle.

*Corresponding Author: Mary Baylies—Program in Developmental Biology, Sloan Kettering Institute, Memorial Sloan Kettering Cancer Center, Box 310, 1275 York Avenue, New York, New York 10021, U.S.A. Email: m-baylies@ski.mkscc.org.

Muscle Development in Drosophila, edited by Helen Sink. ©2006 Eurekah.com and Springer Science+Business Media.

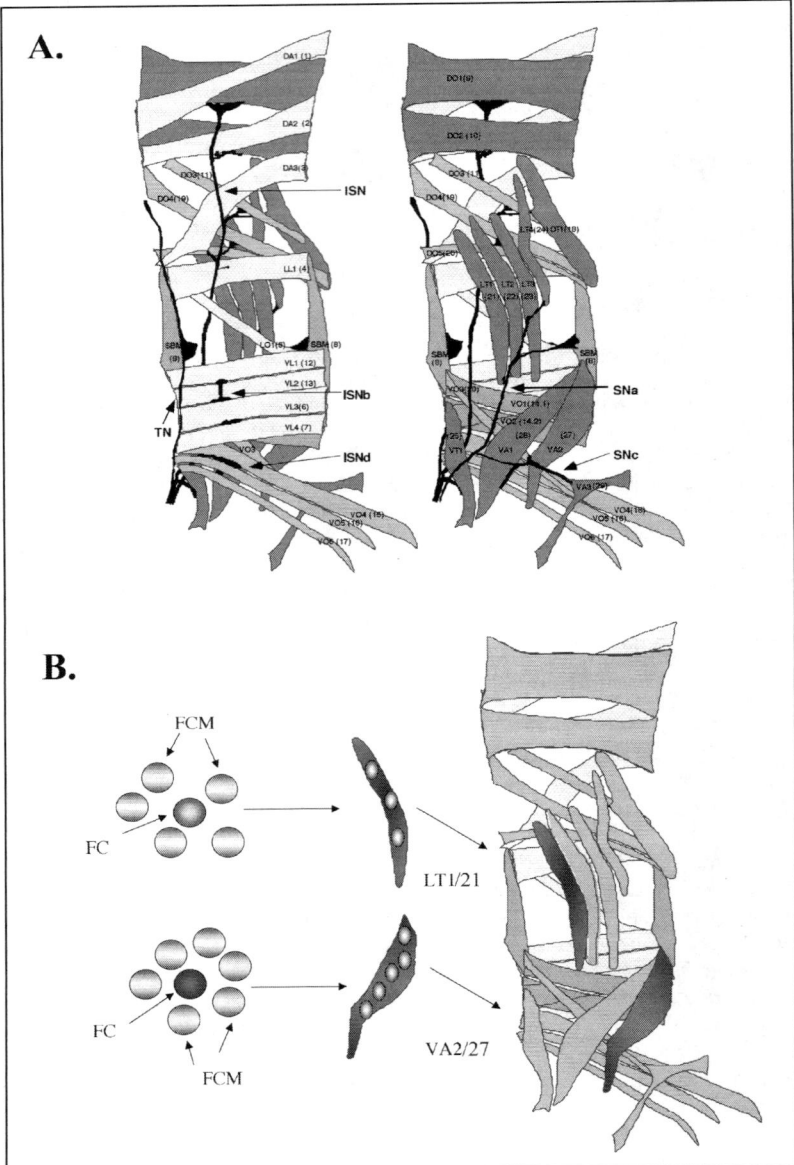

Figure 1. The *Drosophila* larval somatic muscles. A) Cartoon showing the internal (left) and external (right) views of abdominal segments A2-A7 muscle pattern (adapted from Landgraf, 1999[57]). Dorsal is at top, anterior is left. Identification of the muscles as in Bate, 1993[3] and in Crossley, 1978.[58] Muscle position (D, dorsal; L, lateral; V, ventral) followed by orientation (A, acute; L, longitudinal; O, oblique; T, transverse); SBM, segment border muscle. Motoneurons are shown for reference. ISN, intersegmental nerve (with tracks b and d); SN, segmental nerve (with tracks a and c); TN, transverse nerve. B) The body wall muscles arise from a specialized set of myoblasts called founder cells (FC). Founder cells fuse between 3-25 times to neighboring fusion-competent myoblasts (FCM) depending on the muscle. Lateral transverse muscle 1 (LT1/21) and ventral acute muscle (VA2/27) are shown.

The purpose of this chapter is to review the molecules and mechanisms underlying the "birth" of the founder cell and fusion-competent myoblasts. We start post-gastrulation after mesodermal cells have completed their dorsal migration over the ectoderm and are being allocated to specific mesodermal cell fates, such as the somatic muscles or heart, with a view to the particular signals and transcriptional regulators key for this process. We then focus on the actual "birthing" process of founder cells, which are specified as sibling pairs from the division of progenitor cells. We highlight the dynamic interplay between several signal transduction pathways and mesoderm-specific transcriptional regulators required for progenitor specification as well as the molecules known to be involved in conferring a unique founder cell fate. Lastly, we consider the information content of the muscle founder cell and suggest how this information is mobilized to form a specific muscle.

Allocation of Somatic Mesoderm-Mesodermal Subdivision

Transplantation studies indicate that, at gastrulation, mesodermal cells have not yet committed to a particular mesodermal fate.[5] These studies show that although these cells are incapable of forming structures other than mesoderm, the specific mesodermal fate chosen (i.e., somatic muscles, heart) depends on the final position that a cell occupies within the mesoderm and its location relative to other germ layers. Consistent with these studies, both intrinsic and extrinsic signals have been shown to affect the allocation of mesodermal cells into different fates (Fig. 2).

The coordinated activities of the segmentation genes *even-skipped* (*eve*) and *sloppy-paired* (*slp*) as well as the ectodermal signals Hedgehog (Hh) and Wingless (Wg) are required to divide each mesodermal segment into two domains across the anterior-posterior axis[6,7] (Fig. 2). Cells in the Eve/Hh domain differentiate into visceral muscle, fat body, heart and a subset of glia, while cells in the Slp/Wg domain differentiate into somatic muscles and heart. While Eve is detected transiently in the mesoderm at the onset of gastrulation, Slp is detected in the mesoderm both at gastrulation and during the subdivision period. Eve and Slp function as transcriptional regulators, which pattern the mesoderm both directly by regulating expression of mesodermal genes, and indirectly by activating the ectodermal signaling pathways Hh and Wg/

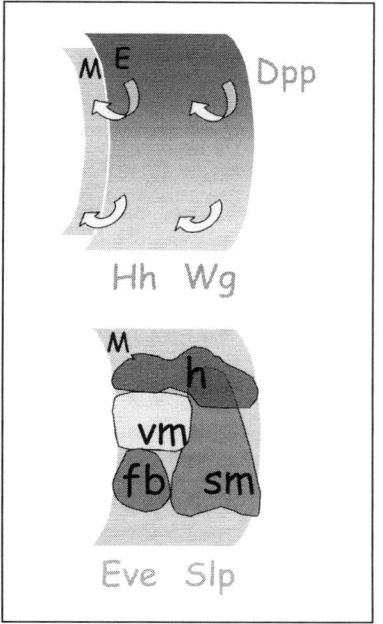

Figure 2. Intrinsic transcriptional regulators and extrinsic signals are required for mesodermal subdivision. The transcriptional regulators Eve and Slp work in concert with the ectodermally produced Dpp, Wg and Hh signals to pattern the mesoderm. M, mesoderm; E, ectoderm; h, heart; vm, visceral muscle; fb, fat body; sm, somatic muscle.

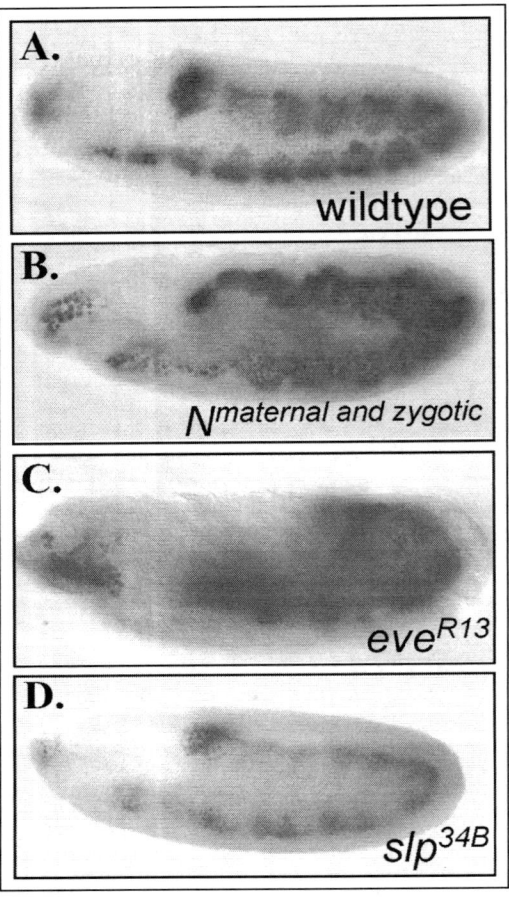

Figure 3. Modulation of Twist requires the concerted action of a number of regulators. Lateral views of stage 10 embryos stained with an antibody to Twist. Anterior, left. A) Wild-type embryo shows a modulated Twist expression pattern in which each segment contains high and low domains of Twist expression. B) Embryos missing both the maternal and zygotic contributions of Notch fail to modulate Twist, resulting in uniform Twist expression. The mesoderm appears undulated due to ectopic neuroblast production in this background. C) *Eve* mutant embryos also fail to modulate Twist, maintaining high Twist levels throughout the mesoderm (cf[7]). D) Embryos lacking both *slp* loci express uniform low levels of Twist throughout the mesoderm (cf[17]).

Wnt. These signaling pathways reinforce and further refine the patterning initiated by Eve and Slp. As the mesoderm is subdivided along the anterior-posterior axis, the activity of the TGF-β family member Decapentaplegic (Dpp) stimulates the subdivision along the dorsal-ventral axis. Dpp, secreted by the dorsal ectoderm, acts on the underlying mesoderm to regulate dorsal fate genes, such as *tinman*, and to repress ventrally expressed genes such as *pox meso*[8,9] (Fig. 2).

From the view of somatic myogenesis, a critical output of this subdivision process is the modulation of Twist expression. The characteristic uniform high Twist expression required to specify mesoderm at gastrulation becomes modulated into a segmentally repeated pattern of low and high Twist expression along the anterior-posterior axis of the embryo.[10-12] Cells expressing low Twist levels differentiate into visceral muscles and fat body, while cells expressing high Twist levels develop into somatic muscles.[13,14] Increasing Twist expression in mesodermal domains where Twist levels are usually low blocks formation of visceral mesoderm and fat body and leads to formation of ectopic somatic muscles in their place. Conversely, uniform low Twist levels interfere with somatic myogenesis but permit the development of other tissues.[15,16] These results indicate that the role of Twist during mesodermal subdivision is to promote formation of the somatic muscles.

During the subdivision period, the transcriptional regulators Slp and Eve as well as the Wg and Hh signaling pathways modulate Twist expression (Fig. 3). Specifically, prior to and during subdivision, the expression of Slp in the mesoderm overlaps with the high Twist domain primordial.[17] Genetic experiments revealed that Slp activates Twist expression; *slp* loss-of-function mutants express Twist uniformly at low levels throughout the mesoderm, while panmesodermal overexpression of Slp has the opposite effect: Twist is expressed at uniform high levels.[7,17] Since Slp and Twist colocalize in the mesoderm and analysis of the *twist* promoter revealed several conserved Slp-binding sites, it is likely that Slp directly regulates *twist*.[18] Slp activity is, in turn, regulated by Wg signaling and by Eve. Ectodermal Wg signaling directly induces mesodermal *slp* expression.[17] However, contrary to what happens in *slp* mutant embryos, *wg* loss- and gain-of-function mutants properly modulate Twist into low and high domains. Thus, the effect that Slp has on Twist appears to be only partially mediated by Wg.[7,18] Moreover, Eve and Hh signaling also affect Twist indirectly as Eve, which functions as both an ectodermal upstream regulator and a mesodermal target of Hh signaling, interferes with Slp function.[6,7]

Notch signaling also regulates Twist expression during the subdivision period.[19] Embryos that completely lack Notch fail to modulate Twist into low and high domains, resulting in the uniform maintenance of high Twist levels (Fig. 3). Conversely, global mesodermal overexpression of a constitutively active form of the Notch receptor represses Twist expression: compared to wild-type embryos, most cells express low levels of Twist. Analysis of the mechanism utilized by Notch to regulate *twist* revealed that Notch and its transcriptional effector Suppressor of Hairless[20] have multiple inputs into *twist*. Notch signaling regulates *twist* both directly, through a conserved Su(H) site on the *twist* promoter, and indirectly, by activating proteins that repress *twist*. Genetic data suggested that these "repressors of *twist*" are the bHLH transcriptional repressors encoded by the *Enhancer of split [E(spl)]* gene complex and the HLH protein Extramacrochaetae (Emc). E(spl) appears to repress Twist directly through sites in the *twist* promoter, whereas Emc inhibits Twist by sequestering the bHLH protein Daughterless, which is required to up-regulate Twist.[19]

Birth of a Muscle Founder Cell

Muscle founder cells are born as distinct sibling pairs by the division of progenitor cells. These progenitors, in turn, segregate from clusters of equivalent cells ("promuscle groups") that express the bHLH protein Lethal of Scute (L'sc).[21] L'sc is expressed in 19 clusters per hemisegment that appear sequentially in the embryonic mesoderm from late stage 10 (5h after egg laying) until stage 12 (7h 30 after egg laying). In each equivalence group, L'sc expression is progressively restricted to one cell, the progenitor. Expression of L'sc is lost when the progenitor divides, so that L'sc is no longer detected in founder cells. The number of L'sc-expressing clusters/progenitors can account for all the 30 founder cells required per hemisegment to seed each specific myofiber. The absence or excess of *l'sc* function leads to losses or duplications, respectively, of a subset of Slouch+-founder cells and somatic muscles. These phenotypes indicate a role of L'sc during early myogenesis. However, *l'sc* loss-of-function phenotype is weaker than that expected for its wide pattern of expression in all promuscle clusters/progenitors. This suggests that there are other factors that collaborate with L'sc to specify progenitor fate.

Different signaling pathways are required throughout the process of muscle progenitor specification. Particularly, for two-dorsal L'sc+ clusters/progenitors that also express the identity protein Eve (L'sc clusters 2 and 15), it has been shown that individual progenitors are progressively determined by unique combinations of intercellular signals.[22] First, the intersecting domains of Wg and Dpp, secreted from the ectoderm, define a broad region in the dorsal mesoderm that is competent to respond to a subsequent Ras-Mitogen Activated Protein Kinase (MAPK) signaling. This prepatterned region is revealed by the expression of L'sc (i.e., "precluster"). Then, Ras-MAPK is locally activated, within the competent region, by two receptor tyrosine kinases (RTKs), the *Drosophila* EGF receptor (DER) and the FGF receptor Heartless (Htl).[23,24] RTK local activation restricts L'sc expression to a subset of cells, within the precluster, which constitute a promuscle cluster or equivalence group. Subsequently, Ras activates Eve in all cells of the cluster. Finally, individual progenitors are singled out from each group of equivalent cells under the opposing influences of the positive Ras activity and the lateral inhibitory function of the Notch signaling pathway (Fig. 4A).

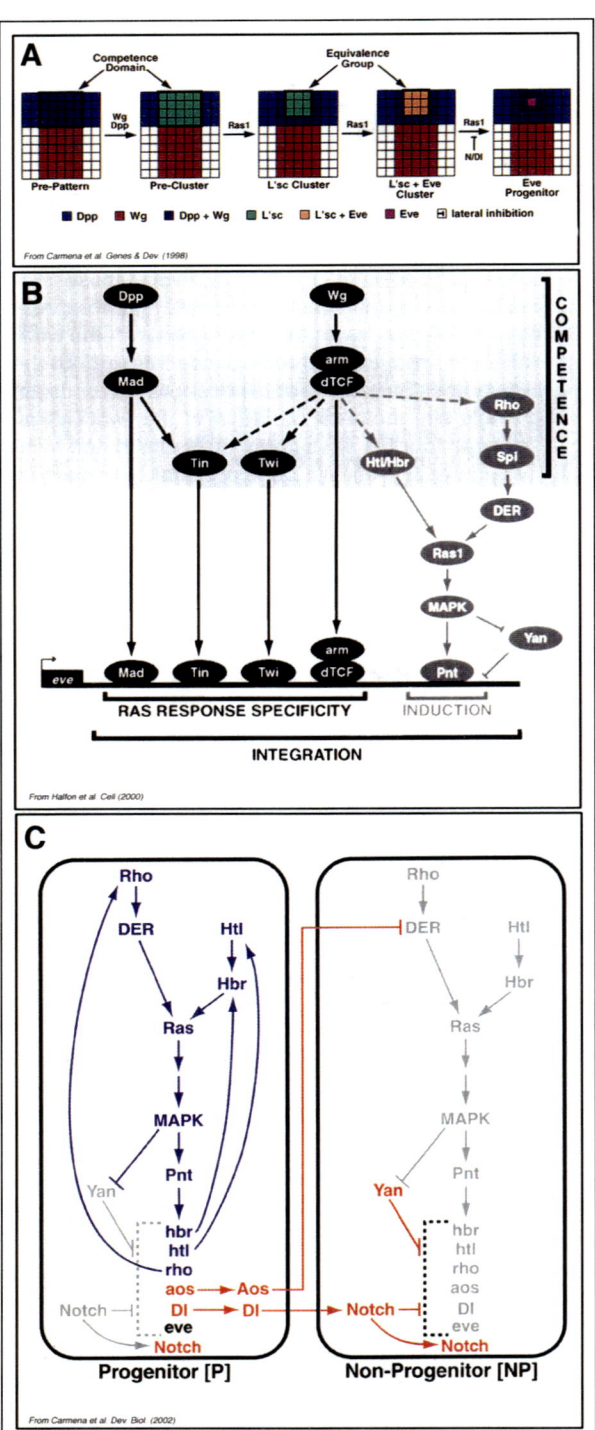

Figure 4. See legend on next page.

Figure 4, shown on previous page. Combinatorial code and integration of signals during progenitor specification. A) Model for the combinatorial and concerted action of Wg, Dpp, Ras1 and Dl/N signals throughout Eve progenitor specification. The intersecting domains of expression of Wg (red) and Dpp (blue), secreted from the overlying ectoderm, prepattern the dorsal mesoderm (purple). This "competence domain" is revealed by the expression of L'sc (green). Subsequently, Ras1 is locally activated in a subset of cells within the competence domain restricting L'sc to this subgroup of cells, which constitute an equivalence group. Ras1 also activates the identity protein Eve (red), which colocalizes with L'sc in this cluster of cells (yellow). Finally, a lateral inhibitory signal promoted by the neurogenic genes Delta (Dl) and Notch (N) restricts the potential to become a progenitor to a single cell of the equivalence group. B) A model for developmental competence, signal integration and RTK/Ras response specificity in the determination of mesodermal Eve progenitors. Wg and Dpp confer competence to mesodermal cells by activating proximal components of the Ras-MAPK signaling cascade (Rho, Htl and Hbr) and mesoderm-specific transcription factors (Tin and Twi). The integration of tissue-specific and signaling activated factors (Mad, dTCF, Pnt and Yan) at the Eve enhancer results in a specific response to Ras-MAPK signaling. C) Cross-talk between Ras and Notch signaling pathways during muscle and heart Eve+ progenitor specification. Notch competes with Ras-MAPK signaling pathway by inhibiting different components (hbr, htl, rho, aos) of the Ras signaling cascade. Ras also cooperates with Notch signaling by up-regulating the Notch ligand Delta (Dl) and the secreted DER inhibitor Argos (Aos), which synergizes with Dl to inhibit Ras-MAPK signaling in cells of the equivalence group that will adopt a "nonprogenitor" fate. In addition, multiple positive and negative feedback loops are established: Ras activates proximal components of its own pathway (Rho, Htl and Hbr) and Notch upregulates its own expression and downregulates its ligand Dl.

Once progenitors are specified, they move into close contact with the ectoderm and divide asymmetrically, giving rise to two founder cells.[25,26] Different proteins, known to have a crucial function during neural asymmetric cell divisions, have also been implicated in the asymmetric division of muscle progenitors.[25-29] Inscuteable (Insc), an adaptor cytoskeletal protein, and Numb, a membrane associated protein that contains a phosphotyrosine-binding domain, localize as cortical crescents on opposite sides of dividing progenitor cells. Insc provides positional information required for Numb asymmetric segregation into one of the sibling founder cells. This is critical for the specification of distinct sibling cell fates, as Numb binds and suppresses Notch signaling in the sibling cell into which it segregates.[30] Two different models, which are not mutually exclusive, have been proposed to explain the mechanism by which Numb inhibits Notch during neural asymmetric cell divisions. One model shows that Numb inhibits Notch through α-Adaptin, a protein involved in receptor-mediated endocytosis.[31] The authors show that Numb contributes to the endocytosis of Notch and the consequent repression of Notch signal by polarizing the distribution of α-Adaptin. The second model points to Sanpodo (Spdo), a four-pass transmembrane protein that interacts physically with Notch and is required for Notch signaling.[32,33] According to this model, Numb inhibits Notch by regulating Spdo membrane localization: Numb binds Spdo and inhibits Spdo localization at the membrane; consequently, Notch is inactive in this cell. This last mechanism of Notch inhibition by Numb may be also operating in the mesoderm, as Spdo has been shown to have an important role in this tissue during asymmetric cell divisions.[27,29] Another protein necessary during both neural and muscle asymmetric cell divisions is Partner of Numb (Pon).[28] Pon colocalizes and interacts physically with Numb. Loss- and gain-of-function mutant analysis of *pon* shows that Pon forms part of the molecular machinery crucial for Numb asymmetric localization. Finally, a new protein implicated in Insc regulation has been identified very recently.[34] This protein, called Abstrakt (Abs), belongs to a family of RNA-dependent ATPases or DEAD-box proteins. The authors show that Abs binds the *insc* mRNA and is essential in the posttranscriptional regulation of *insc*. In *abs* mutant embryos, the levels of Insc protein drop significantly with the consequent failure in the proper localization of cell fate determinants (i.e., Insc, Pon). Abs is also required for both neural and muscle progenitors' asymmetric divisions.

Thus, both intrinsic and extrinsic cues are required for the assignation of specific cell fates to muscle founders. Shortly after they are born, both sibling founder cells express a specific set of identity proteins, which usually are already expressed in the progenitor and occasionally at the cluster stage. However, as the siblings mature, they begin to display distinct characteristics. For example, one of the two siblings loses the expression of identity proteins that were expressed in the progenitor, whereas the other sibling keeps the expression of such proteins.[21,25,35] In other cases, only one sibling starts to express specific markers that were not previously expressed in the progenitor.[36] These observations reveal the asymmetric nature of the progenitor division and the unique fate of each founder cell. Founder cells will start to fuse with neighboring fusion-competent myoblasts, seeding the formation of individual and distinctive muscle fibers.

Integration of Signal Transduction Pathways during Muscle Progenitor Specification

Multiple signaling pathways and transcriptional regulators are involved in the determination of muscle progenitors. However, how mesodermal cells integrate this information is not completely understood. Transcriptional convergence has been shown to be an important mechanism for integrating signals and for achieving tissue-specific responses.[37,38] In these studies, the authors identify an enhancer from the muscle identity gene *eve* that recapitulates Eve expression during muscle and heart progenitor specification (MHE, Muscle and Heart Enhancer). This enhancer contains binding sites for two classes of transcriptional regulators: signal-activated and tissue-restricted factors. Among the first class are the nuclear effectors of Wg (dTCF), Dpp (Mad) and Ras (Pointed and Yan) signaling pathways; among the mesoderm-specific factors are Tinman and Twist. All these factors are integrated at the MHE where they function synergistically to promote *eve* transcription. This enhancer model explains how Wg and Dpp provide competence to mesodermal cells to respond to Ras-MAPK, by regulating proximal components of the Ras-MAPK pathway, and how the specificity of Ras response is achieved, since Wg and Dpp also regulate Twist and Tinman expression (Fig. 4B).

Additional networks of signal cross-regulation have been described during muscle progenitor specification. For example, extensive reciprocal interactions occur between opposing Ras/MAPK and Notch signaling pathways during the formation of Eve+ progenitors.[36] Notch competes with Ras by down-regulating different components of the Ras-MAPK cascade, including Htl, Heartbroken (Hbr, also known as Stumps and Dof)[39-41] and Rhomboid (Rho).[23,42] Co-operation between Ras and Notch signals was also observed: Ras upregulates the Notch ligand Delta and the epidermal growth factor receptor antagonist Argos.[43] Then, Delta and Argos synergize to inhibit nonautonomously the Ras-MAPK signaling cascade. In addition, multiple feedback loops, positive and negative, are established. For instance, Ras upregulates proximal components of its own pathway, including Htl, Hbr and Rho, in the prospective progenitor. In addition, Notch upregulates its own expression and represses Delta and Argos in nonprogenitor cells (Fig. 4C). All these reciprocal interactions combine to generate the signal thresholds that are essential for proper specification of progenitors and nonprogenitors from groups of initially equivalent cells.

Taken together, these findings show that signal transduction pathways do not act as isolated linear cascades during muscle progenitor specification. Rather, their effects are intertwined at multiple levels to form an integrated network of cross-talk nodes and feedback loops.

Muscle Identity Is Regulated by Muscle Founder Identity Genes

While recent work has given us insight to the specification of founder cells, we know little about the next step in muscle morphogenesis, when the information and processes within a founder cell coordinates and elaborates specific muscle traits. Some clues to this next step have come from the observation that the specific combinations of signaling inputs that a given founder cell receives results in the production of a unique set of molecular determinants, founder cell identity proteins, that gives each muscle fiber its shape, size and connection pattern. Eleven identity genes, including the homeobox genes *S59* (also named *slouch*),[2,21,44] *apterous* (*ap*),[45] *muscle segment homeobox (msh),*[46] *eve,*[22,47,48] *ladybird,*[49] the *myc*-related HLH encoding gene *collier,*[50] the bHLH encoding gene, *nautilus (nau),*[51,52] *vestigial*[53] and the zinc finger encoding gene *Krüppel (Kr)*[54] have been found to date to be expressed in different, sometimes overlapping, subsets of founder cells. These genes have been called "muscle or founder cell identity genes", initially by virtue of the fact that they identify, by their expression pattern, specific founder cells and muscles. Loss- and gain-of-function experiments have been performed on 7 of the 11 known founder cell identity genes.[3,55] These data indicated a role for these regulators in determining final muscle characteristics; for example, loss of *Kr* function leads not only to complete muscle transformations (i.e., specific muscle losses and duplications) but also incomplete transformations of specific muscles (the specific muscle forms in the correct position but has aberrant features such as changes in shape and attachments). Hence, a gene is called a muscle/founder cell identity gene if it fulfils the following: (1) it encodes a transcriptional regulator; (2) it is expressed in subsets of founder cells; and (3) it has a known or presumed role in specifying the developmental fates of individual muscles, leading to complete and/or partial transformations in muscle identity.

Both the overlapping pattern of founder cell identity gene expression in different sets of muscle founder cells and genetic analyses have led to the hypothesis that individual muscle identities are specified by combinatorial codes of identity genes. For example, the muscle VA2/27 (muscle letter/number hereon) founder cell expresses and maintains Kr, Slouch, Ap, and Nau and are thought to all contribute to VA2/27 identity.[44,54] However, loss- and gain-of-function experiments revealed added complexity to the combinatorial hypothesis, suggesting that these transcriptional regulators may target different aspects of the morphogenic process in different founder cells. It appears that some founder cell identity genes may control other founder cell transcriptional regulators whereas others appear to control directly specific attributes of the muscle morphogenesis. For example, Kr and Slouch are expressed in the muscle progenitor that divides to give rise to the founder cells for muscles VA2/27 and VA1/26 (Fig. 5). Both proteins are maintained in the VA2/27 muscle founder cell and fade in the VA1/26 muscle founder cell. In the absence of mesodermal Kr, however, Slouch expression is initiated in the muscle progenitor, but is lost in the VA2/27 founder. This VA2/27 muscle founder cell fails to give rise to a VA2/27 muscle; instead the VA2/27 founder cell has been transformed into its sibling VA1/26 founder, which subsequently develops into a second VA1/26 muscle. Thus the embryo has two VA1/26 muscles and no VA2/27 muscles. Ectopic Kr expression in both VA1/26 and VA2/27 muscle founder cells leads to expression of Slouch in both VA1/26 and VA2/27 muscle founders and the subsequent loss of the VA1/26 muscle and duplication of the VA2/27 muscle[54] (Fig. 5). Loss of Slouch function itself in VA2/27 founder cell does not lead to a complete VA2/27 -> VA1/26 transformation as would have been predicted from a simple combinational model in which all transcriptional regulators act together on the same target promoters. In this case, muscle VA2/27 is partially transformed: the muscle forms but with aberrant features (altered shape and attachment) and altered gene expression {maintained Kr and Nau (VA2/27 markers) but failed to express Ap}. Loss of Nau or Ap, which are also expressed in the VA2/27 founder cell, has no effect on VA2/27 development, possibly due to functional redundancy with Kr, Slouch or yet unknown proteins.[44] Taken together, these experiments indicated that there may be a hierarchy among identity genes in different founder cells: some identity genes regulate the expression of others (e.g., Kr can regulate Slouch and presumably other regulators for VA2/27) to dictate a complete muscle identity, others (Slouch for VA2/27) regulate only a subset of the specific program and still others (Ap, Nau for VA2/27) have no effect possibly due to redundancy.

Loss- and gain-of-function experiments of these muscle gene regulators also suggested that the specific founder cell context determines the role or place of muscle regulators within the hierarchy. For example, Slouch is required for the normal development of all abdominal muscles that are derived from Slouch founder cell. However loss of Slouch activity has different effects in individual muscle types. As stated above, *slouch* mutant embryos form muscle VA2/27 with

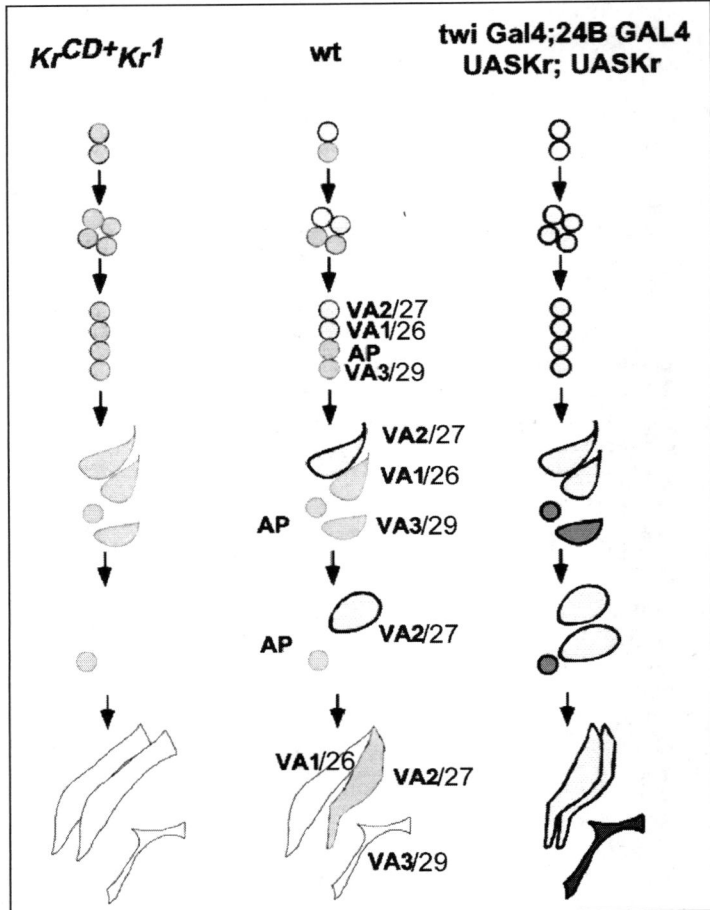

Figure 5. The founder cell identity gene, *Krüppel*, regulates VA2/27 and VA1/26 muscle identity in part by controlling expression of Slouch. Lineage of the VA1,2,3/26,27,29 muscles. Slouch (grey); Kr co-localization with Slouch noted by heavy black outlining of progenitor, founder cell and developing muscle. In wild-type embryos (middle panel) Kr and Slouch are coexpressed in a single progenitor that divides to give rise to VA2/27 and VA1/26 founder cells. Slouch and Kr are maintained in VA2/27 as it differentiates to form its muscle; Kr is not maintained in VA1/26; Slouch is also expressed in another progenitor that divides to give rise to VA3/29 and an adult progenitor (AP), but fades in these and VA1/26 as the muscles form. In *Kr* loss-of-function (left; Kr gap gene function rescued in the blastoderm), Slouch expression is initiated, but not maintained. As a result, two VA1/26 muscles form; VA3/29 forms normally. In *Kr* gain-of-function (right), Slouch is maintained in the founder cells: during differentiation, it fades from VA3/29. Maintained Kr and Slouch in VA1/2 founder cells leads to the duplication of VA2. (Adapted from ref. 54).

aberrant features. However, muscles VT1/25 and VA3/29, which also normally express Slouch, are completely transformed into muscles with different identities.[44] Hence the combination of transcriptional regulators, the place within the hierarchy and thus, the specific role of a transcriptional regulator within a specific founder cell appears to control the unique traits elaborated during muscle morphogenesis.

A Molecular Definition of Muscle Founder Cell and Fusion-Competent Myoblasts

Founder cell identity genes have given us a glimpse into the regulation of muscle identity. However, additional studies are needed to determine how these regulators translate the specific aspects of muscle morphogenesis. For example, we do not know the genes or the mechanisms that control how the final number of fusion events is regulated, how a particular tendon cell is selected or how final shape of the muscle is achieved. In addition, while many founder cells have had identity genes assigned to them, there are many muscle founder cells in which no such genes have been localized. Recently, a step forward has been taken to better understand the translation of founder cell identity into muscle morphogenesis. A large-scale gene expression analysis was performed to identify genes differentially expressed in founder cells versus fusion-competent myoblasts. To enrich for each cell type, embryos derived from *Toll*[10b] mutants to obtain primarily muscle-forming mesoderm, and expression of activated forms of Ras or Notch to induce founder cell or fusion-competent myoblast fate, respectively, were collected. The transcripts present in embryos of each genotype were compared using hybridization to cDNA microarrays encompassing 40% of the *Drosophila* genome. Among the 83 genes differentially expressed, genes known to be enriched in founder cells or fusion-competent myoblasts such as *htl* or *hibris*, previously characterized genes with unknown roles in muscle development, and predicted genes of unknown function were found. Additional studies of newly identified genes discovered new patterns of gene expression restricted to one of the two types of myoblasts and also remarkable muscle phenotypes. Some genes such as *phyllopod* play a critical role during specification of particular muscles, while others including *tartan* and *cadmus* are necessary for normal muscle morphogenesis.[56]

The simplest view of the "founder cell" hypothesis is that each founder cell contains all the information for the development of a particular muscle. Fusion-competent myoblasts, in contrast, have been seen as a naïve group of myoblasts, entrained to a particular muscle program upon fusion to the founder cell. The array data indicate that these two groups of myoblasts have distinct transcriptional profiles and raise the possibility of a greater role for fusion-competent myoblasts in determining the final muscle morphology. The mechanisms underlying the complex morphological changes that occur during migration and cell fusion, as well as changes in cell shape and cell physiology, require a rich and dynamic program of transcriptional activity. A major challenge for the future of in muscle biology lies in exploring the function of the new genes identified in this and similar screens during the period of muscle morphogenesis. The key to understanding this process will be to find the molecular interactions among these newly identified genes and the known molecular networks that build and coordinate the thirty unique muscle elements in each *Drosophila* embryonic hemisegment.

Acknowledgements

A. Carmena and M. Baylies were supported by a Human Frontiers Postdoctoral Fellowship and by National Institutes of Health Grant GM56989.

References

1. Bate M. The embryonic development of larval muscles in Drosophila. Development 1990; 110:791-804.
2. Dohrmann C, Azpiazu N, Frasch M. A new Drosophila homeo box gene is expressed in mesodermal precursor cells of distinct muscles during embryogenesis. Genes Dev 1990; 4:2098-2111.
3. Bate M. The mesoderm and its derivatives in the development of Drosophila melanogaster. In: Bate M. Martinez-Arias A, eds. New York: Cold Spring Harbor Laboratory Press, 1993:1013-1090.
4. Rushton E, Drysdale R, Abmayr SM et al. Mutations in a novel gene, myoblast city, provide evidence in support of the founder cell hypothesis for Drosophila muscle development. Development 1995; 121:1979-1988.

5. Beer J, Technau G, Campos-Ortega J. Lineage analysis of transplanted individual cells in embryos of Drosophila melanogaster: IV commitment and proliferative capabilities of individual mesodermal cells. Roux's Arch Dev Biol 1987; 196:222-230.

6. Azpiazu N, Lawrence PA, Vincent J-P et al. Segmentation and specification of the Drosophila mesoderm. Genes Dev 1996; 10:3183-3194.

7. Riechmann V, Irion U, Wilson R et al. Control of cell fates and segmentation in the Drosophila mesoderm. Development 1997; 124:2915-2922.

8. Frasch M. Induction of visceral and cardiac mesoderm by ectodermal Dpp in the early Drosophila embryo. Nature 1995; 374:464-467.

9. Staehling-Hampton K, Hoffman FM, Baylies MK et al. dpp induces mesodermal gene expression in Drosophila. Nature 1994; 372:783-786.

10. Simpson P. Maternal-zygotic gene interactions during the formation of the dorso-ventral pattern in Drososphila embryos. Genetics 1983; 105:615-632.

11. Thisse B, Stoetzel C, Gorosotiza-Thisse C et al. Sequence of the twist gene and nuclear localization of its protein in endomesodermal cells of early Drosophila embryos. EMBO J 1988; 7:2175-2183.

12. Leptin M. Twist and snail as positive and negative regulators during Drosophila mesoderm development. Genes & Development 1991; 5:1568-1576.

13. Baylies MK, Bate M. twist: A myogenic switch in Drosophila. Science 1996; 272:1481-1484.

14. Dunin-Borkowski O, Bate M, Brown N. Anterior-posterior subdivision and the diversification of the mesoderm in Drosophila. Development 1995; 121:4183-4193.

15. Bate M, Baylies MK. Intrinsic and extrinsic determinants of mesodermal differentiation in Drosophila. Semin Cell Dev Biol 1996; 7:103-111.

16. Castanon I, Von Stetina S, Kass J et al. Dimerization partners determine the activity of the Twist bHLH protein during Drosophila mesoderm development. Development 2001; 128(16):3145-3159.

17. Lee HH, Frasch M. Wingless effects mesoderm patterning and ectoderm segmentation events via induction of its downstream target sloppy paired. Development 2000; 127(24):5497-5508.

18. Cox V. Signal transport, levels and integration: New lessons from wingless regulation of Drosophila mesoderm development [Ph.D]. New York: Development Biology, Weill School of Medicine at Cornell University, 2004.

19. Tapanes-Castillo A, Baylies M. Notch signaling patterns Drosophila mesodermal segments by regulating the bHLH transcription factor twist. Development 2004; 131:2359-2372.

20. Bray S, Furriols M. Notch pathway: Making sense of Suppressor of Hairless. Curr Biol 2001; 11(6):R217-221.

21. Carmena A, Bate M, Jiménez F. lethal of scute, a proneural gene, participates in the specification of muscle progenitors during Drosophila embryogenesis. Genes Dev 1995; 9:2373-2383.

22. Carmena A, Gisselbrecht S, Harrison J et al. Combinatorial signaling codes for the progressive determination of cell fates in the Drosophila embryonic mesoderm. Genes & Development 1998a; 15:3910-3922.

23. Buff E, Carmena A, Gisselbrecht S et al. Signalling by the Drosophila epidermal growth factor receptor is required for the specification and diversification of embryonic muscle progenitors. Development 1998; 125:2075-2086.

24. Michelson AM, Gisselbrecht S, Zhou Y et al. Dual functions of the heartless fibroblast growth factor receptor in development of the Drosophila embryonic mesoderm. Dev Genet 1998b; 22(3):212-229.

25. Carmena A, Murugasu-Oei B, Menon D et al. Inscuteable and numb mediate asymmetric muscle progenitor cell divisions during Drosophila myogenesis. Genes Dev 1998b; 12:304-315.

26. Baylies MK, Bate M, Ruiz-Gomez M. The specification of muscle in Drosophila. Cold Spring Harb Symp Quant Biol 1997; LXII:385-393.

27. Park M, Yaich LE, Bodmer R. Mesodermal cell fate decisions in Drosophila are under the control of the lineage genes numb, Notch, and sanpodo. Mech Dev 1998; 75(1-2):117-126.

28. Lu B, Rothenberg M, Jan LY et al. Partner of numb colocalizes with Numb during mitosis and directs Numb asymmetric localization in Drosophila neural and muscle progenitors. Cell 1998; 95(2):225-235.

29. Ward EJ, Skeath JB. Characterization of a novel subset of cardiac cells and their progenitors in the Drosophila embryo. Development 2000; 127(22):4959-4969.

30. Guo M, Jan LY, Jan YN. Control of daughter cell fates during asymmetric division: Interaction of Numb and Notch. Neuron 1996; 17(1):27-41.

31. Berdnik D, Torok T, Gonzalez-Gaitan M et al. The endocytic protein alpha-Adaptin is required for Numb-mediated asymmetric cell division in Drosophila. Dev Cell 2002a; 3(2):221-231.

32. Dye CA, Lee JK, Atkinson RC et al. The Drosophila sanpodo gene controls sibling cell fate and encodes a tropomodulin homolog, an actin/tropomyosin-associated protein. Development 1998; 125(10):1845-1856.

33. O'Connor-Giles KM, Skeath JB. Numb inhibits membrane localization of Sanpodo, a four-pass transmembrane protein, to promote asymmetric divisions in Drosophila. Dev Cell 2003; 5(2):231-243.

34. Irion U, Leptin M. Developmental and cell biological functions of the Drosophila DEAD-box protein Abstrakt. Curr Biol 1999; 9(23):1373-1381.

35. Ruiz-Gomez M, Bate M. Segregation of myogenic lineages in Drosophila requires numb. Development 1997a; 124:4857-4866.

36. Carmena A, Buff E, Halfon M et al. Reciprocal regulatory interactions between the Notch and Ras signaling pathways in the Drosophila embryonic mesoderm. Dev Biol 2002; 244:226-242.

37. Halfon MS, Carmena A, Gisselbrecht S et al. Ras pathway specificity is determined by the integration of multiple signal-activated and tissue-restricted transcription factors. Cell 2000; 103:63-74.

38. Knirr S, Frasch M. Molecular integration of inductive and mesoderm-intrinsic inputs governs even-skipped enhancer activity in a subset of pericardial and dorsal muscle progenitors. Dev Biol 2001; 238(1):13-26.

39. Imam F, Sutherland D, Huang W et al. stumps, a Drosophila gene required for fibroblast growth factor (FGF)-directed migrations of tracheal and mesodermal cells. Genetics 1999; 152:307–318.

40. Michelson AM, Gisselbrecht S, Buff E et al. Heartbroken is a specific downstream mediator of FGF receptor signalling in Drosophila. Development 1998a; 125:4379–4389.

41. Nolan K, Barrett K, Lu Y et al. Myoblast city, the Drosophila homolog of DOCK180/CED-5, is required in a Rac signaling pathway utilized for multiple developmental processes. Genes & Development 1998; 12:3337-3342.

42. Freeman M. Feedback control of intercellular signalling in development. Nature 2000; 408(6810):313-319.

43. Freeman M, Klambt C, Goodman CS et al. The argos gene encodes a diffusible factor that regulates cell fate decisions in the Drosophila eye. Cell 1992; 69(6):963-975.

44. Knirr S, Azpiazu N, Frasch M. The role of the NK-homeobox gene slouch (S59) in somatic muscle patterning. Development 1999; 126(20):4525-4535.

45. Bourgouin C, Lundgren SE, Thomas JB. apterous is a Drosophila LIM domain gene required for the development of a subset of embryonic muscles. Neuron 1992; 9:549-561.

46. Nose A, Isshiki T, Takeichi M. Regional specification of muscle progenitors in Drosophila: The role of the msh homeobox gene. Development 1998; 125:215-223.

47. Frasch M, Hoey T, Rushlow C et al. Characterization and localization of the even-skipped protein of Drosophila. EMBO J 1987; 6(3):749-759.

48. Su M, Fujioka M, Goto T et al. The Drosophila homeobox genes zfh-1 and even-skipped are required for cardiac-specific differentiation of a numb-dependent lineage decision. Development 1999; 126:3241-3251.

49. Jagla T, Bellard F, Lutz Y et al. ladybird determines cell fate decisions during diversification of Drosophila somatic muscles. Development 1998; 125:3699-3708.

50. Crozatier M, Vincent A. Requirement for the Drosophila COE transcription factor collier in formation of an embryonic muscle: Transcriptional response to Notch signalling. Development 1999; 126(7):1495-1504.

51. Michelson A, Abmayr S, Bate M et al. Expression of a MyoD family member prefigures muscle pattern in Drosophila embryos. Genes Dev 1990; 4:2086-2097.

52. Paterson BM, Walldorf U, Eldridge J et al. The Drosophila homologue of vertebrate myogenic-determination genes encodes a transiently expressed nuclear protein marking primary myogenic cells. Proc Natl Acad Sci USA 1991; 88(9):3782-3786.

53. Bate M, Rushton E. Myogenesis and muscle patterning in Drosophila. C R Acad Sci III 1993; 316(9):1047-1061.

54. Ruiz-Gómez M, Romani S, Hartmann C et al. Specific muscle identities are regulated by Krüppel during Drosophila embryogenesis. Development 1997; 124(17): 3407-14.

55. Frasch M. Controls in patterning and diversification of somatic muscles during Drosophila embryogenesis. Current opinion in Genetics & Development. 1999; 9:522-529.

56. Artero R, Furlong, E., Beckett, K et al. Notch and Ras signaling pathway effector genes expressed in fusioncompetent and founder vells during Drosophila myogenesis. Development 2003; 130(25):6257-72.

57. Landgraf M, Baylies M, Bate M. Muscle founder cells regulate defasciculation and targeting of motor axons in the Drosophila embryo. Curr Biol 1999; 9(11):589-592.

58. Crossley AC. The morphology and development of the Drosophila muscular system. In: Ashburner M, Wright T, eds. The Genetics and Biology of Drosophila, Vol. 2b. New York: Academic Press, 1978:499-560.

CHAPTER 8

Muscle Morphogenesis:
The Process of Embryonic Myoblast Fusion

Susan M. Abmayr* and Kiranmai S. Kocherlakota

Abstract

One important aspect of myogenic differentiation in *Drosophila melanogaster*, as in many other organisms, is the generation of multinucleate muscle fibers through the fusion of myoblasts. This process cannot be initiated until the myoblasts have differentiated to a point at which they become competent to fuse and express genes associated with the fusion process. The myoblasts must then identify and adhere to their fusion partners, a process that is seeded by the founder cell and involves its recruitment of fusion-competent myoblasts. Cell adhesion molecules that are members of the Immunoglobulin Superfamily mediate recognition between these two populations of myoblasts. In subsequent events, intracellular proteins that are essential for the fusion process are recruited to the plasma membrane, where they likely contribute to reorganization of the cytoskeleton through activation of small GTPases. Morphological changes associated with fusion include recruitment of electron dense vesicles and formation of fusion plaques at points of cell-cell contact. Molecules such as Antisocial/Rolling Pebbles, Myoblast City and Loner, among others, function at these stages. While little is known at present about the actual molecules or process by which the lipid bilayers break down, it seems likely that known proteins essential for fusion will provide an entry into their identification. This chapter provides a review of the literature describing these genes and a discussion of their proposed roles in the pathway.

Introduction

In organisms ranging from *Drosophila* to mammals, the musculature is comprised of an elaborate array of distinct fibers that are generated by the fusion of committed myoblasts. As discussed elsewhere in this volume, the muscle fibers differ from each other in features that include location, pattern of innervation, site of attachment, and size. These unique features are controlled by information contained within a muscle specific myoblast—the founder cell. Founder cells initiate fusion with surrounding "naïve" fusion-competent myoblasts, which in turn, take on the identity and features of the original founder cell. The final size of the muscle fiber is then attained through multiple rounds of fusion of the developing syncitia with additional fusion-competent myoblasts. The smallest muscles of the embryo will be formed by fusion of as few as 3-5 cells, whereas larger muscles require approximately 30-35 cells.

Several distinct events occur in the developing myoblasts to achieve this ultimate end. After the myoblasts become specified as either founder cells or fusion-competent myoblasts, these cells must identify and adhere to their fusion partners. Recognition is specific and directional, in that fusion does not appear to occur between two equivalent myoblasts.[1] Instead, a founder

*Corresponding Author: Susan M. Abmayr—The Stowers Institute for Medical Research, 1000 E. 50th Street, Kansas City, Missouri 64110, U.S.A. Email: sma@stowers-institute.org

Muscle Development in Drosophila, edited by Helen Sink. ©2006 Eurekah.com and Springer Science+Business Media.

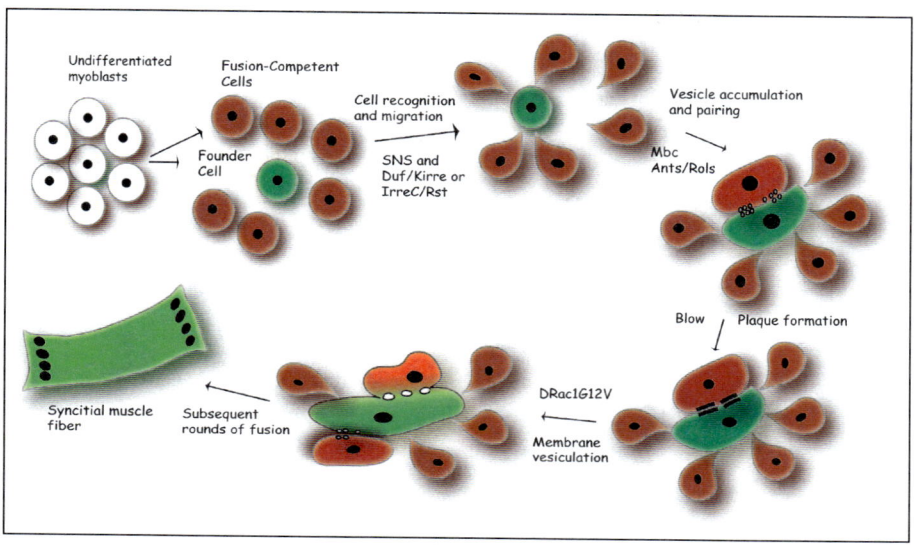

Figure 1. A schematic representation of the steps involved in myoblast fusion. Following specification of the founder cell and the fusion-competent myoblast, the latter extend filopodia toward the founder cell. The two cells align with each other and elongate. Paired electron-dense vesicles appear on either side of the associated plasma membranes and later resolve into plaques. Regions of membrane discontinuity are observed as the membrane breaks down near these plaques. Further rounds of fusion are required for each muscle fiber to achieve its full size and nuclei number. The placement of specific genes in the pathway has been inferred from the step that is blocked in the corresponding mutants.

cell initiates the process by recruitment of a fusion-competent myoblast. Both cell specific and nonmuscle specific proteins contribute to this process, and genetic studies in *Drosophila* embryos have identified a number of key players in recent years. These studies begin to define the genes and steps necessary for the recognition and adhesion between myoblasts and, perhaps, the fusion of two lipid bilayers into one.

Ultrastructure of the Adherent Myoblast Interface

In a detailed morphological analysis of fusing myoblasts at the level of the electron microscope, Doberstein et al[2] described a series of intracellular events that accompany fusion and are shown diagrammatically in Figure 1. The first obvious change is the accumulation of clusters of electron dense vesicles on the cytoplasmic sides of opposed plasma membranes of two associated myoblasts. These vesicles, which have not been observed in other cell types, will eventually align with one another across the intervening membranes to form "paired vesicles".[2] While founder cells and fusion-competent myoblasts were not identified in this ultrastructural analysis, we infer that the vesicles appear in both cell types as well as in the developing myotube and associated fusion-competent cells.

After alignment, the vesicles are thought to resolve into electron dense plaques, termed "prefusion complexes", that are observed on the cytoplasmic sides of the corresponding plasma membranes. These plaques are reminiscent of structures identified in fusing vertebrate myoblasts that are also thought to result from the fusion of electron dense vesicles to the plasma membrane.[3] As vesicles are being recruited and plaques form at sites of cell-cell contact, the myoblasts elongate to maximize contact points. Multiple pores are then observed in the fusing membranes, adjacent to the electron dense fusion plaques. This morphology suggests that fusion occurs at multiple sites,[2] similar to that seen previously in the Lepidopteran *Antheraea*

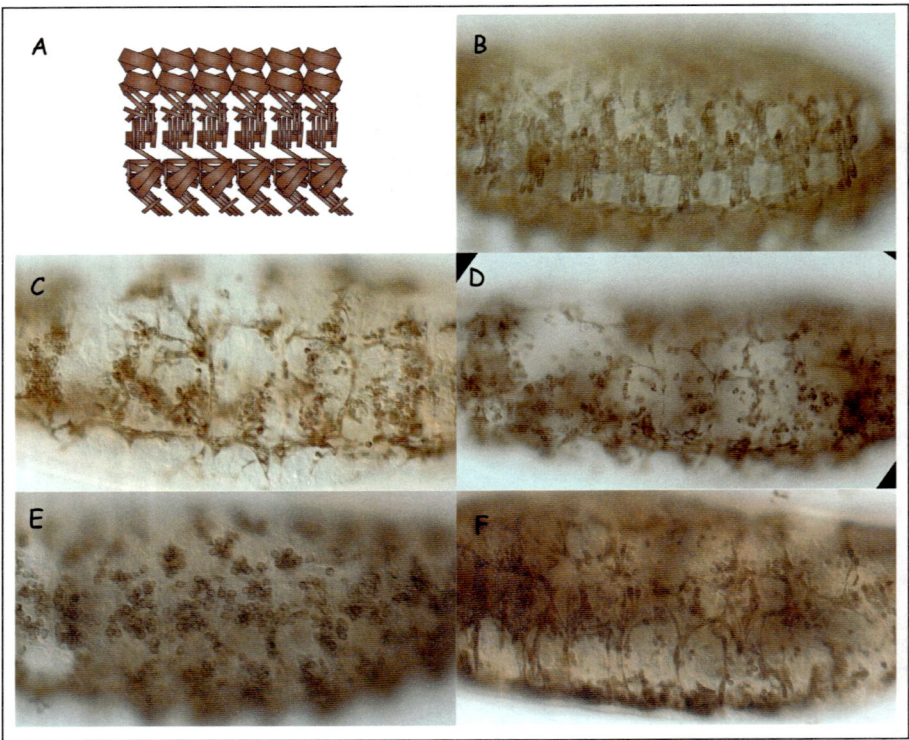

Figure 2. Visualization of the muscle pattern in embryos of various genetic backgrounds. A) Schematic of the pattern of abdominal muscles. B) wild-type embryo: C) embryo homozygous for a mutation in *sns*; D) embryo homozygous for a deficiency that removes both *duf/kirre* and *irreC/rst*; E) embryo homozygous for a mutation in *mbc*; F) embryo homozygous for a mutation in *blow*.

polyphemus.[4] Hence it does not appear that myoblast fusion in *Drosophila* utilizes a simple zipper mechanism in which it originates from a single site. The fusion process is completed as the membrane vesiculates and is removed. In an effort to advance our understanding of genes associated with these morphological changes, Doberstein et al[2] carried out similar studies on mutant embryos in which myoblast fusion does not occur to determine the point at which fusion was arrested. The results of these studies are discussed at appropriate places below and indicated in the schematic in Figure 1.

Myoblast Recognition and Adhesion

Prior to myoblast fusion in the *Drosophila* embryo, fusion-competent myoblasts must identify, migrate to, and adhere to the founder cell with which they will eventually fuse. Genetic studies in *Drosophila* have identified three members of the Immunoglobulin Superfamily (IgSF) that are essential for myoblast recognition and adhesion (reviewed in refs. 5-8; shown in Fig. 2 and/or Fig. 3). These include *sticks and stones* (*sns*),[9] *dumfounded/kin of irreC* (*duf/kirre*)[10] and *irregular-chiasm-C/roughest* (*irreC/rst*).[6] The *hibris* (*hbs*) gene, which encodes a fourth IgSF member in *Drosophila* myoblasts, appears to regulate the fusion process.[5,11,12]

Sticks-and-Stones (sns)

The *sns* locus was identified on the basis of its mutant phenotype in embryos, in which there is a complete absence of multinucleate muscle fibers and a correspondingly large number

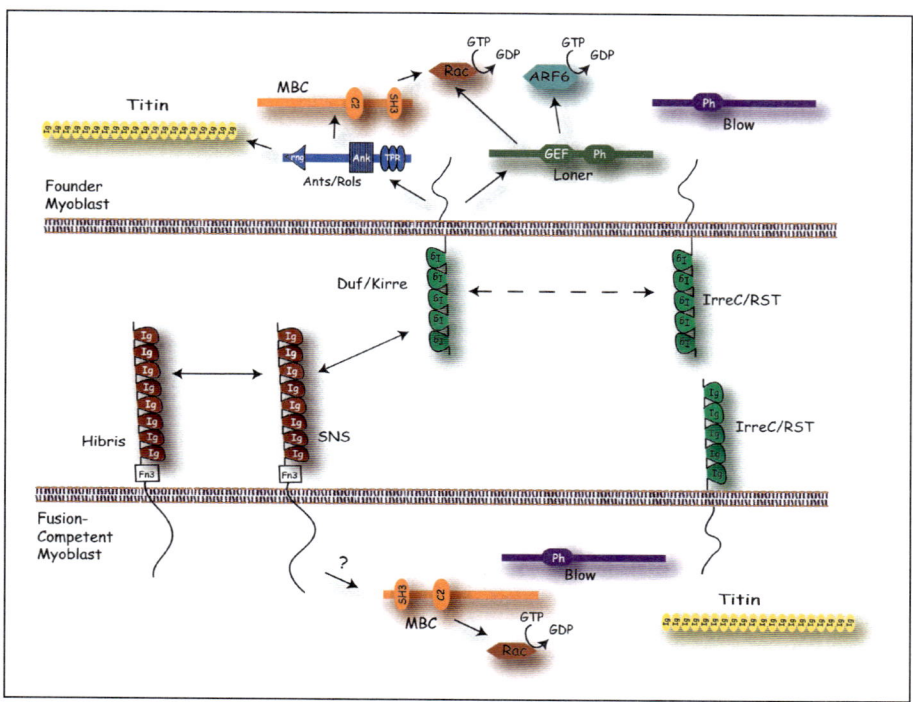

Figure 3. A signal transduction cascade inferred from known proteins involved in myoblast fusion. Candidate functional domains of the proteins involved have been indicated. The model includes only molecules whose functions and/or interactions have been demonstrated or can be inferred from their structure. Bold single ended arrows indicate the proposed direction of signal transmission. The double sided dashed arrow between Duf/Kirre and IrreC/RST indicates functional redundancy, and that the molecules are interchangeable. The double sided bold arrow between SNS and Hibris indicates a repressive relationship demonstrated by genetic interactions. ADP Ribosylation Factor (ARF6), Ankyrin repeats (Ank), Antisocial/Rolling Pebbles (Ants/Rols), Blown Fuse (Blow), Dumbfounded/Kin of IrreC (Duf/Kirre), Fibronectin III (FN3), Guanine nucleotide diphosphate (GDP), Guanine nucleotide exchange factor (GEF), Guanine nucleotide triphosphate (GTP), Immunoglobulin-like domain (Ig), Irregular chiasm C/Roughest (IrreC/RST), Myoblast City (MBC), Pleckstrin homology (PH), Src homology 3 (SH3), Sticks and Stones (SNS), Tetratricopeptide repeats (TPR).

of unfused Myosin-expressing myoblasts (Fig. 2).[9] The unfused fusion-competent myoblasts do not extend filopodia towards the founder cells, suggesting that recognition of these cells does not occur in the absence of *sns*. Consistent with a role at the cell surface, *sns* is predicted to encode a cell adhesion molecule with 8 extracellular immunoglobulin-like (Ig-like) domains, a single fibronectin type-III (FN-III) domain, a transmembrane region, and a cytoplasmic region. The *sns* transcript and protein are expressed exclusively in the fusion-competent myoblasts of the somatic (Fig. 3)[9] and visceral[13,14] musculature, and no expression is observed in the founder cells. SNS protein is evident just prior to fusion and decreases rapidly as fusion is completed. Both biochemical and wholemount embryo analyses have indicated that it localizes to the plasma membrane of myoblasts prior to fusion.[9] SNS shares homology with human Nephrin, which has been implicated in Congenital Nephrotic Syndrome,[15,16] *Drosophila* Hbs[11,12] and *C. elegans* SYG-2.[61] Of note, genetic and molecular studies have revealed that mutations in the *sns* locus[9] and EMS-induced mutants in the *rolling stone* (*rost*),[17,18] locus are actually allelic, and map to the genetic position of *sns*.[9,19] This finding is particularly relevant because the

EMS-induced *rost*[15] mutation, which was examined at the ultrastructural level by Doberstein,[2] actually represents a mutation in *sns*. Interestingly, it appears that fusion is arrested at a fairly late stage in these mutant embryos, since electron dense plaques accumulate but the plasma membrane does not breakdown. However, since the sequence lesion in the *rost*[15] has not yet been identified, it may be a hypomorphic allele in which the SNS protein is partially functional. Support for this possibility is provided by the mutant phenotype of *rost*[15] alleles, in which myoblast fusion does not appear to be completely blocked.[18]

The membrane localization of SNS in combination with its identification as a cell adhesion molecule suggest the possibility that SNS may mediate recognition and/or adhesion between founder cells and fusion-competent myoblasts. On closer examination of the *sns* mutant phenotype, it is also apparent that the fusion-competent myoblasts are unable to migrate toward founder myoblasts. This myoblast behavior is in contrast to that observed in other fusion mutant backgrounds, such as *myoblast city (mbc)* and *blown fuse (blow)*,[20] in which myoblasts have clearly associated with each other but fusion does not proceed (Fig. 2). Thus, SNS seems to act as a receptor on the surface of fusion-competent myoblasts that responds to an attractant generated by the founder cells, inducing migration via activation of an intracellular signal transduction cascade. Consistent with this possibility, the cytoplasmic region of SNS contains potential phosphorylation sites, proline rich regions, and stretches of homology with related proteins that may mediate interaction with cytoplasmic proteins. The presence of these motifs may indicate that it directs intracellular events that regulate myoblast migration, recognition and/or adhesion. Consistent with this hypothesis, the transmembrane and cytodomains are essential for SNS function (Banerji and Abmayr, unpublished). Moreover, candidate extracellular ligands for SNS that serve as attractants for fusion-competent myoblasts have been identified,[6,10] and are discussed below.

Dumbfounded/Kin of Irregular Chiasm C (Duf/Kirre) and Kin of Irregular Chiasm C/Roughest (IrreC/Rst)

Complementing expression of SNS in the fusion-competent myoblasts is the expression of two other IgSF members in the founder cells. One gene, *dumbfounded/kin of irreC (duf/kirre)*, was identified through its association with the founder cell specific enhancer trap line *rP298*,[10] which directs expression of β-galactosidase specifically in the founder cells.[21] It is predicted to encode a single pass membrane spanning protein with 5 extracellular Ig-like domains and a cytoplasmic tail.[10] Early in development, expression of the *duf/kirre* transcript is observed at low levels in the developing mesoderm. This broad expression becomes restricted to a limited number of cells in the embryo from which the founder cells arise (as depicted in Fig. 3). It is not expressed in the fusion-competent myoblasts. Like *sns*, the *duf/kirre* transcript remains detectable in muscle precursors as long as they are fusing, but its level drops quickly after fusion is completed.

A second member of the IgSF, *irreC/rst*, is located 127 kilobases (kb) away from *duf/kirre*. The *irreC/rst* gene was identified on the basis of defects in axonal projections in the adult brain.[22] The encoded protein is 45% similar to Duf/Kirre, and includes 5 extracellular Ig-like domains, a transmembrane region and a highly homologous cytoplasmic tail.[6,22,23] Interestingly, both Duf/Kirre and IrreC/RST share homology with Neph1 which interacts with Nephrin in the kidney,[62] and SYG-1 which is the binding partner of SYG-2 in *C.elegans*.[61,63]

Like the *duf/kirre* transcript, IrreC/RST protein is detected in regions of the embryo where the founder cells and fusion-competent myoblasts arise. In contrast to Duf/Kirre, IrreC/RST is not restricted to the founder cells and is expressed in at least some fusion-competent myoblasts (as depicted in Fig. 3).[6] While its role in these fusion-competent myoblasts remains unclear, IrreC/RST clearly serves a function redundant with that of Duf/Kirre in the founder cells. Specifically, examination of embryos bearing a deficiency that removes both genes revealed a complete absence of differentiated muscle fibers and large number of unfused Myosin-expressing myoblasts (Fig. 2).[10] However, expression of either *duf/kirre* or *irreC/rst* in the mesoderm was sufficient to rescue this defect.[6,10]

Duf/Kirre and IrreC/RST both appear to act as attractants for fusion-competent myoblasts, since expression of either protein in the embryonic ectoderm is sufficient to target migration of these cells.[6,10] Of much interest is the question of whether Duf/Kirre functions, in part, as a secreted form that sets up a concentration gradient recognized by the migrating fusion-competent myoblasts. Consistent with this possibility, a cleaved form of Duf/Kirre is detected in trans-fected S2 cells,[24] and a truncated ectodomain is detected in the culture media (Galletta and Abmayr, unpublished data). However, the fusion-competent myoblasts may also be capable of extending long processes that scan for, and contact, the founder cell.

Based on their mutant phenotypes, Duf/Kirre, IrreC/RST and SNS all play roles in the recognition of founder cells by fusion-competent myoblasts. In the absence of *sns*, the fusion-competent cells remain round and do not appear to extend filopodia. In contrast, the fusion-competent cells of embryos lacking *duf/kirre* and *irreC/rst* extend filopodia, but these projections are randomly oriented rather than directed toward the founder cells.[10] This behav-ior is consistent with a model in which the Duf/Kirre and IrreC/RST attractants may be ligands for the SNS receptor on the surface of fusion-competent myoblasts, as depicted in Figure 3. Consistent with a role in adhesion of founder cell: fusion-competent myoblast recognition, these proteins are capable of mediating adhesive events in cultured cells. Though the biological relevance remains unclear, aggregation experiments in S2 cells indicate that Duf/Kirre and IrreC/RST mediate homotypic aggregation.[12,64,65] More importantly, cells expressing SNS adhere to cells expressing Duf/Kirre or IrreC/RST, while expression of SNS alone is not sufficient to induce cell aggregation.[12,65] It remains to be determined whether the interaction of SNS-expressing cells with cells expressing either Duf/Kirre or IrreC/RST reflects a direct mo-lecular interaction between these molecules. However, the above cell interactions in conjunc-tion with the restricted patterns of expression of these molecules appears to satisfy the need for directional fusion machinery in the embryo.

Hibris (hbs)

A fourth IgSF member that is expressed in the developing musculature, and plays a role in myoblast fusion, is encoded by *hibris (hbs)*.[11,12] Hbs was identified in a screen for transcripts that were differentially expressed in founder cells versus fusion-competent myoblasts[11] and, simultaneously, through database searches for members of the IgSF in *Drosophila*.[12] Hbs is predicted to have 8 or 9 Ig-like domains, a FN-III domain, a transmembrane spanning region and a cytoplasmic region. Like SNS, the cytoplasmic region contains potential phosphoryla-tion sites, tyrosine residues that are conserved with SNS and/or Nephrin, and a candidate PEST sequence. Hbs is 48% identical and 63% similar to SNS and 29% identical and 44% similar to human Nephrin.[11] It is expressed earlier in development than SNS, and can be seen in precursors to tissues other than muscle that include the trachea and nervous system. Like SNS, Hbs expression becomes restricted to myoblasts by embryonic Stage 12. This expression includes a large subset of the fusion-competent myoblasts but does not include the founder cells (Fig. 3).[11,12] In cells that express both proteins, SNS and Hbs co-localize at discrete points on the cell surface.[11]

These features suggest that Hbs and SNS, like Duf/Kirre and IrreC/RST,[23] might have arisen from a single gene and serve redundant functions. However, loss-of-function alleles have revealed that *hbs* plays a quite different role in myoblast fusion. Specifically, it is not essential for viability, and mutants survive to become semi-fertile adults. These flies do display a rough-eyed phenotype, consistent with *hbs* expression in imaginal discs[11,12] *hbs* mutant embryos do, how-ever, exhibit a modest increase in the number of unfused myoblasts, and may have missing or smaller muscles. However, these defects are not sufficient to impair survival. Like the embry-onic loss-of-function phenotype, targeted expression of Hbs in the developing mesoderm causes muscle loss and an increase in the number of unfused myoblasts. This effect is mediated through the cytoplasmic domain, since expression of this domain alone mimics the overexpression phe-notype[11,12] and is not observed upon ectopic expression of either a secreted or membrane bound extracellular domain. Interestingly, the mild myoblast fusion defect of *hbs* mutant embryos is dominantly suppressed by loss of one copy of *sns*. Conversely, *sns* mutations enhance the phenotype seen by overexpression of Hbs in the mesoderm. Thus, an antagonistic interaction occurs between *hbs* and *sns* mutants, as depicted in Figure 3.[11] Artero et al have proposed three models by which *hbs* and *sns* could antagonize one another during myoblast fusion. (1) SNS and Hbs could compete for the same extracellular ligand. This model is reasonable if Duf/Kirre is the ligand for SNS, since Hbs-expressing cells interact heterotypically with Duf/Kirre-expressing cells in culture.[12] However, the lack of myoblast fusion defects in embryos expressing the extracellular domain of Hbs argues against this hypothesis. (2) Hbs and SNS may combine to form a "negative" receptor. In this scenario the Hbs/SNS coreceptor may respond differently to ligand than the "positive" SNS receptor. (3) Hbs and SNS may converge on an intracellular downstream target that plays a role in regulating fusion. It must be noted that all of these models accommodate Hbs as a nonessential regulator of SNS function. However, further study will be necessary to understand the exact role of Hbs in myoblast fusion.

Intracellular Events Associated with Myoblast Fusion

Following recognition and adhesion between myoblasts, cytoplasmic machinery to direct migration and membrane fusion must be in place. One enticing possibility is that the cell surface molecules involved in recognition and adhesion may themselves initiate intracellular signaling events via their cytoplasmic tails, and play a role in recruiting muscle specific machinery to the sites of fusion. A second possibility is that myoblast fusion will require more broadly-expressed proteins for reorganization of the cytoskeleton that underlies the fusing membranes. Lastly, cytoplasmic events may bring about morphological changes associated with migration, perhaps also requiring cytoskeletal rearrangement. It remains to be determined whether the intracellular events directing migration and fusion involve the same molecules or converge on a common pathway. Proteins that may function in these processes and/or interact with IgSF members have been identified and are discussed below.

The Antisocial/Rolling Pebbles (Ants/Rols)-Myoblast City(MBC)-Drosophila Rac1(Drac1) Associated Pathway

Mutant *myoblast city (mbc)* embryos are characterized by the absence of multinucleate muscle fibers and presence of large numbers of unfused myoblasts (Fig. 2).[25] The morphology of these mutant embryos differs from that seen in embryos lacking either *sns* or *duf/kirre* and *irreC/rst* in that the unfused fusion-competent myoblasts migrate to, and cluster around, the founder cells.[20] Thus, MBC appears to be required for fusion but not necessarily migration. At the ultrastructural level, the number of prefusion complexes seen in these embryos was significantly reduced, suggesting that it might play a role in vesicle accumulation.[2] The *mbc* locus encodes a cytoplasmic protein with extensive homology to *C. elegans* Ced-5[26] and human *Dock180*[27] All three of these proteins contain src homology 3 (SH3) domains at their N-terminus, multiple proline-rich, putative SH3 binding domains at their C-terminus, and additional blocks of homology that may be associated with guanine nucleotide exchange activity (see below). MBC is expressed in a wide variety of tissues in the developing embryo.[28] Consequently, the defects seen in *mbc* mutant embryos are not limited to myoblast fusion, and include incomplete dorsal closure of the epidermis, abnormal fasciculation of the ventral nerve cord neurons, and severely impaired migration of border cells in the adult ovary.[28-30] Orthologs of *mbc* are involved in diverse processes that include cell engulfment, cell migration, epithelial morphogenesis and oncogenic transformation.[26,27] While these processes appear diverse at first glance, they have in common the potential reorganization of the cytoskeleton. In fact, *mbc* mutant embryos exhibit perturbations in the actin cytoskeleton along the leading edge of the migrating epidermis during dorsal closure.[28]

Numerous studies have suggested that the small GTPase Rac1 is a major target of the MBC/Dock-180/Ced-5 family (Fig. 3),[30-35] and is associated with these cytoskeletal changes. Rac1 is a member of the Rho family of small GTPases, which are often associated with regulation of changes in the cytoskeleton. In vertebrate cell culture systems, two groups have recently demonstrated that a conserved region of Dock 180, termed the "Docker" domain or "Dock Homology Region-2"(DHR-2),[36,37] functions as an unconventional guanine nucleotide exchange factor (GEF) for Rac1. Expression of the DHR-2 region increases the GTP loading of Rac1 in 293-T cells[36] and appears to be dependent on coexpression of the ELMO1 protein. ELMO1 is the vertebrate ortholog of *C. elegans* Ced-12 which, like Ced-5, was identified on the basis of a defect in the engulfment of cell corpses.[26,31,38-40] These data provide compelling evidence that Dock180 and Ced-5 function through Rac1. In the *Drosophila* musculature, expression of dominant negative or constitutively active forms of DRac1 in the mesoderm causes defects in myoblast fusion,[1] presumably by interfering with critical cytoskeletal events. Ultrastructural analysis of fusing myoblasts in embryos expressing constitutively active DRac1 revealed the presence of prefusion complexes, electron dense vesicles and isolated fusion pores. The plasma membranes of the myoblasts are in close apposition, but do not appear to vesiculate.[2] For these reasons, constitutively active Rac1 is thought to block myoblast fusion at a relatively late step. Recently, double mutant loss-of-function alleles of Drac1 and Drac2 have been shown to exhibit significant defects in myoblast fusion,[42] as anticipated from analysis of embryos expressing the above constructs. In addition, Drac1 and *mbc* have been linked genetically in the eye, in which loss of one copy of *mbc* suppresses the overexpression phenotype of Drac1.[30]

A biochemical link between MBC, and associated DRac1, and the IgSF members discussed earlier is provided by the Antisocial (Ants) /Rolling pebbles (Rols) protein (Fig. 3). Ants/Rols was identified in three independent screens for genes affecting embryonic muscle development.[24,43,44] Myoblast fusion does not occur in *ants/rols* mutant embryos. Ultrastructural analysis of these mutant embryos revealed small syncitia containing 2-4 nuclei.[44] However fusion did not proceed further, and neither prefusion complexes nor electron dense plaques were observed. The predicted Ants/Rols protein contains several domains with the potential to mediate protein-protein interaction, including a RING finger, 9 ankyrin repeats, 3 tetratricopeptide repeats (TPRs) and a coiled-coil region.[24,43,44] In the embryo, Ants/Rols is expressed in the founder myoblasts at a time consistent with myoblast fusion. Moreover, Duf/Kirre and Ants/Rols are present on the myoblast membrane, and colocalize at points of cell:cell contact.[43] Immunoprecipitations from cotransfected S2 cells has demonstrated that it can interact biochemically with the N-terminal region of MBC and with Duf/Kirre.[24] In addition, the *mbc* locus was identified as a dominant enhancer of the *irreC/rst* eye phenotype, suggesting that MBC interacts with IrreC/RST.[45] Together, these data support a model in which MBC is recruited by Duf/Kirre to discrete points on the membrane via Ants/Rols, leading to localized changes in the cytoskeleton through DRac1 (Fig. 3).

*The Loner-*Drosophila *ADP Ribosylation Factor 6 Associated Pathway*

The *loner* locus was identified in a genetic screen for embryos defective in myoblast fusion.[46] In *loner* mutant embryos, the fusion-competent myoblasts appear to recognize and extend filopodia toward the founder cells but do not fuse into syncitia. Expression of Loner protein is restricted to founder cells (Fig. 3), where it is localized in punctate foci that, in some cases, include Ants/Rols.[46] Interestingly, it loses this characteristic localization and becomes more diffuse within the cytoplasm of embryos lacking *duf/kirre* and *irreC/rst*. It is localized to discrete points in the cytoplasm of transfected S2 cells, but gets recruited to the membrane at sites of cell contact in the presence of Duf/Kirre. Loner localization is not affected in *ants/rols* mutant embryos and vice versa, suggesting that they function independently even though they can colocalize.

The Loner protein appears to function as a GEF, and is present in three different isoforms. All of these contain a Sec7 domain, a pleckstrin homology (PH) domain, a coiled coil domain and an IQ-motif. Sec7 domains are often directly associated with GEF activity toward ADP ribosylation factors (ARF), while PH domains enhance the activity of GEFs. Analysis of Loner deletion mutants, assayed by their ability to rescue the mutant phenotype, revealed that the Sec7 and PH domains are essential.[46] In light of its homology to Sec7, Chen et al hypothesized that Loner functions as a GEF for the small GTPase dARF6, and demonstrated that the Sec7 domain was sufficient to cause GDP/GTP exchange in GDP release assays. Consistent with this association between Loner and dARF6, expression of a dominant negative form of dARF6^{T27N} in founder cells disrupts the fusion process.[46] ARF6 is known to act through membrane lipid modifications,[47] and enhances Rac mediated actin cytoskeleton remodeling.[48] Interestingly, Loner may also associate with DRac1, since DRac1 localization to discrete points in founder cells is lost in *loner* mutant embryos.[46] Thus, Loner may provide an alternative to the MBC associated pathway for regulating DRac1. It may also function in an independent pathway to regulate cytoskeletal rearrangements and/or lipid modifications that are critical in the fusion process.

Other Genes with Roles in Myoblast Fusion

Several genes have been identified that appear to be involved in myoblast fusion on the basis of a mutant or overexpression phenotype, though their specific biochemical role remains unclear. The *blown fuse (blow)* locus was identified in a screen for lethal mutations defective in guidance of motoneurons.[2] However *blow* is essential for myoblast fusion (Fig. 2), and the neuronal defect is an indirect consequence of the lack of somatic muscles. The *blow* transcript is restricted to myoblasts, and appears just prior to fusion. The predicted Blow protein, which contains a putative PH domain, appears to reside in the cytoplasm (Fig. 3).[2] In *blow* mutant embryos, electron dense vesicles accumulate in prefusion complexes, but electron dense plaques are not observed. Assuming that plaques form from an association between vesicles and the plasma membrane, it is possible that Blow is involved in this step.

Two genes that have not been fully characterized also contribute to myoblast fusion. The *singles bar* gene appears to encode a hydrophobic protein that is essential for myoblast fusion,[49] though the fusion-competent myoblasts migrate to the founder cells.[10] This mutant phenotype suggests that *singles bar* functions after cell migration and adhesion, possibly in membrane associated events critical during the fusion process. Embryos mutant for the bona fide *rolling stone (rost)* locus, originally identified through a P-element line with insertions in both the *rost* and *sns* loci,[9,19] have defects in myoblast fusion.[17,18] The transcript associated with this locus is expressed in the mesoderm, and a promoter-lacZ reporter that includes 400 bp upstream of the ATG is expressed in mature muscles.[17] The predicted Rost protein is hydrophobic and may span the membrane up to 7 times. It is enriched in biochemical preparations of partially-purified membranes,[17] supporting its association with the membrane. While the true *rost* loss-of-function phenotype remains unclear, embryos expressing antisense *rost* do exhibit a block in myoblast fusion.[17]

D-Titin, a protein of approximately 2.0 MDa, that is composed primarily of Ig-like, FN-III and PEVK domains,[50,51] serves as an elastic scaffold for both muscle sarcomeres and for chromosomes.[51-53] It is expressed at a high level in myoblasts prior to fusion and appears to accumulate on the myoblast surface, often at points of contact between myoblast and myotube.[50,51] Interestingly, recruitment of D-Titin to points of cell-cell contact in founder myoblasts is dependent on the Ants/Rols protein discussed above (Fig. 3).[43] Analysis of D-Titin mutant embryos has shown that it is essential for the integrity of the sarcomere, but also plays a role in myoblast fusion.[50,51] Thus, it is possible that D-Titin is playing a role in either directly or indirectly organizing the actin cytoskeleton. The involvement of a muscle structural protein in the myoblast fusion process is not unique to Titin. Recent studies from the Bernstein lab[54] have shown that Paramyosin plays a role in myoblast fusion and accumulates at discrete points between fusing myoblasts. Lastly, a biochemical screen for proteins that interact with MBC led to the identification of dCrk, a cytoplasmic adapter protein with one SH2 domain and two SH3 domains.[60] Mutants in the dCrk locus have not been described. However, expression of a membrane-targeted myristylated form of dCrk in the musculature severely perturbs myoblast fusion (Galletta and Abmayr, unpublished data).

The genes described above likely do not represent the entire complement of molecules in the founder cells and fusion-competent myoblasts that are required for fusion. Efforts to identify additional genes have included traditional genetic approaches as well as targeted molecular approaches, with a variety of results. For example, the utilization of green fluorescent protein (GFP) to mark muscle cells dramatically increased the feasibility of phenotypic screens, and Chen et al are likely to have identified loss-of-function alleles of other critical loci. A microarray approach was carried out by Furlong et al,[55] and utilized *twist* deficient embryos to identify loci preferentially expressed in the mesoderm[55] This screen resulted in the isolation of *gleeful* (*gfl*), independently identified by two other groups and termed *lameduck* (*lmd*) and *myoblasts incompetent* (*minc*).[56,57] The *lmd/minc/gfl* gene, is involved in specification of the fusion-competent myoblasts. Many other genes identified in this screen remain excellent candidates, and simply await functional analysis. These include kinases, actin binding proteins, Calcium binding molecules and small GTPases. Identification of transcripts differentially expressed in the founder cells versus the fusion-competent myoblasts has been mentioned earlier in the context of the discovery of the *hbs* locus. This screen has yielded other potential players in the fusion process,[58] including cytoplasmic factors such as kinases or proteins that can mediate protein:protein interactions. The general usefulness of this approach in identifying functionally important genes will require further characterization of these candidate proteins. Finally, a genome-wide protein-interaction map was generated, and has identified proteins that could be components of the myoblast fusion pathway on the basis of their ability to interact with molecules described earlier in this chapter.[59] The relevance of the proteins identified in this screen also await functional analysis. Nevertheless, the large number of candidates identified in these screens has the potential for significant impact on the field of myoblast fusion.

Conclusion

In the last decade, remarkable advances have been made in our understanding of *Drosophila* myogenesis in general. Similarly, knowledge of the molecules through which the embryonic myoblasts migrate, recognize and fuse to each other did not exist a decade ago. Many advances have been made possible by new technologies that pair molecular biology with genetics or cell biology, and include in situ hybridization, confocal microscopy, large-scale genome wide genetic screens, and the availability of the *D. melanogaster* genome sequence to name a few. Nevertheless, there is much more to be learned. Among the issues to be investigated are the downstream pathways associated with binding of adhesion molecules at the cell surface, the molecular components of the fusion associated vesicles, details of the fusion process itself, and the mechanism through which muscle size is determined. Hopefully the rate of progress will lead to significant advances in our understanding of these processes in another decade.

References

1. Baylies MK, Bate M, Ruiz-Gomez M. Myogenesis: A view from Drosophila. Cell 1998; 93(6):921-927.
2. Doberstein SK, Fetter RD, Mehta AY, Goodman CS. Genetic analysis of myoblast fusion: Blown fuse is required for progression beyond the prefusion complex. J Cell Biol 1997; 136(6):1249-1261.
3. Rash JE, Fambrough D. Ultrastructural and electrophysiological correlates of cell coupling and cytoplasmic fusion during myogenesis in vitro. Dev Biol 1973; 30:166-186.
4. Stocker RF. Electron microscopic observations on the fusion of myoblasts in Antheraea polyphemus (Lepidoptera). Experientia 1974; 30(8):896-898.
5. Dworak HA, Sink H. Myoblast fusion in Drosophila. Bioessays 2002; 24(7):591-601.
6. Strunkelnberg M, Bonengel B, Moda LM et al. rst and its paralogue kirre act redundantly during embryonic muscle development in Drosophila. Development 2001; 128(21):4229-4239.
7. Taylor MV. Muscle development: Molecules of myoblast fusion. Curr Biol 2000; 10(17):R646-648.
8. Taylor MV. Muscle differentiation: How two cells become one. Curr Biol 2002; 12(6):R224-228.
9. Bour BA, Chakravarti M, West JM et al. Drosophila SNS, a member of the immunoglobulin superfamily that is essential for myoblast fusion. Genes Dev 2000; 14(12):1498-1511.
10. Ruiz-Gomez M, Coutts N, Price A et al. Drosophila dumbfounded: A myoblast attractant essential for fusion. Cell 2000; 102(2):189-198.

11. Artero RD, Castanon I, Baylies MK. The immunoglobulin-like protein hibris functions as a dose-dependent regulator of myoblast fusion and is differentially controlled by ras and notch signaling. Development 2001; 128(21):4251-4264.
12. Dworak HA, Charles MA, Pellerano LB et al. Characterization of Drosophila hibris, a gene related to human nephrin. Development 2001; 128(21):4265-4276.
13. San Martin B, Bate M. Hindgut visceral mesoderm requires an ectodermal template for normal development in Drosophila. Development 2001; 128(2):233-242.
14. Klapper R, Stute C, Schomaker O et al. The formation of syncytia within the visceral musculature of the Drosophila midgut is dependent on duf, sns and mbc. Mech Dev 2002; 110(1-2):85-96.
15. Kestila M, Lenkkeri U, Mannikko M et al. Positionally cloned gene for a novel glomerular protein-Nephrin-is mutated in congenital nephrotic syndrome. Mol Cell 1998;1:572-582.
16. Lenkkeri U, Mannikko M, McCready P et al. Structure of the gene for congenital nephrotic syndrome of the finnish type (NPHS1) and characterization of mutations. Am J Hum Genet 1999; 64:51-61.
17. Paululat A, Goubeaud A, Damm C et al. The mesodermal expression of rolling stone (rost) is essential for myoblast fusion in Drosophila and encodes a potential transmembrane protein. J Cell Biol 1997; 138:337-348.
18. Paululat A, Burchard S, Renkawitz-Pohl R. Fusion from myoblasts to myotubes is dependent on the rolling stone gene (rost) of Drosophila. Development 1995; 121:2611-2620.
19. Paululat A, Holz A, Renkawitz-Pohl R. Essential genes for myoblast fusion in Drosophila embryogenesis. Mech Dev 1999; 83(1-2):17-26.
20. Abmayr SM, Balagopalan L, Galletta BJ et al. Cell and molecular biology of myoblast fusion. Int Rev Cytol 2003; 225:33-89.
21. Nose A, Isshiki T, Takeichi M. Regional specification of muscle progenitors in Drosophila: The role of the msh homeobox gene. Development 1998; 125:215-223.
22. Ramos RG, Igloi GL, Lichte B et al. The irregular chiasm C-roughest locus of Drosophila, which affects axonal projections and programmed cell death, encodes a novel immunoglobulin-like protein. Genes Dev 1993; 7(12B):2533-2547.
23. Strunkelnberg M, de Couet HG, Hertenstein A et al. Interspecies comparison of a gene pair with partially redundant function: The rst and kirre genes in D. virilis and D. melanogaster. J Mol Evol 2003; 56(2):187-197.
24. Chen EH, Olson EN. Antisocial, an intracellular adaptor protein, is required for myoblast fusion in Drosophila. Dev Cell 2001; 1(5):705-715.
25. Rushton E, Drysdale R, Abmayr SM et al. Mutations in a novel gene, myoblast city, provide evidence in support of the founder cell hypothesis for Drosophila muscle development. Development 1995; 121:1979-1988.
26. Wu YC, Horvitz HR. C. elegans phagocytosis and cell-migration protein CED-5 is similar to human DOCK180. Nature 1998; 392(6675):501-504.
27. Hasegawa H, Kiyokawa E, Tanaka S et al. DOCK180, a major CRK-binding protein, alters cell morphology upon translocation to the cell membrane. Mol Cell Biol 1996; 16(4):1770-1776.
28. Erickson MRS, Galletta BJ, Abmayr SM. Drosophila myoblast city encodes a conserved protein that is essential for myoblast fusion, dorsal closure and cytoskeletal organization. J Cell Biol 1997; 138(3):589-603.
29. Duchek P, Somogyi K, Jekely G et al. Guidance of cell migration by the Drosophila PDGF/VEGF receptor. Cell 2001; 107(1):17-26.
30. Nolan KM, Barrett K, Lu Y et al. Myoblast city, the Drosophila homolog of DOCK180/CED-5, is required in a rac signaling pathway utilized for multiple developmental processes. Genes Dev 1998; 12:3337-3342.
31. Gumienny TL, Brugnera E, Tosello-Trampont AC et al. CED-12/ELMO, a novel member of the CrkII/Dock180/Rac pathway, is required for phagocytosis and cell migration. Cell 2001; 107(1):27-41.
32. Albert ML, Kim JI, Birge RB. alphavbeta5 integrin recruits the CrkII-Dock180-rac1 complex for phagocytosis of apoptotic cells. Nat Cell Biol 2000; 2(12):899-905.
33. Reddien PW, Horvitz HR. CED-2/CrkII and CED-10/Rac control phagocytosis and cell migration in C. elegans. Nat Cell Biol 2000; 2(3):131-136.
34. Kiyokawa E, Hashimoto Y, Kobayashi S et al. Activation of Rac1 by a Crk SH3-binding protein, DOCK180. Genes Dev 1998; 12(21):3331-3336.
35. Kiyokawa E, Hashimoto Y, Kurata T et al. Evidence that DOCK180 up-regulates signals from the CrkII-p130Cas complex. J Biol Chem 1998; 273:24479-24484.
36. Cote JF, Vuori K. Identification of an evolutionarily conserved superfamily of DOCK180-related proteins with guanine nucleotide exchange activity. J Cell Sci 2002; 115(Pt 24):4901-4913.

37. Brugnera E, Haney L, Grimsley C et al. Unconventional Rac-GEF activity is mediated through the dock180-ELMO complex. Nat Cell Biol 2002; 4(8):574-582.

38. Zhou Z, Caron E, Hartwieg E et al. The C. elegans PH domain protein CED-12 regulates cytoskeletal reorganization via a Rho/Rac GTPase signaling pathway. Dev Cell 2001; 1(4):477-489.

39. Wu YC, Tsai MC, Cheng LC et al. C. elegans CED-12 acts in the conserved crkII/DOCK180/ Rac pathway to control cell migration and cell corpse engulfment. Dev Cell 2001; 1(4):491-502.

40. Chung S, Gumienny TL, Hengartner MO et al. A common set of engulfment genes mediates removal of both apoptotic and necrotic cell corpses in C. elegans. Nat Cell Biol 2000; 2(12):931-937.

41. Luo L, Liao YJ, Jan LY et al. Distinct morphogenetic functions of similar small GTPases: Drosophila Drac1 is involved in axonal outgrowth and myoblast fusion. Genes Dev 1994; 8:1787-1802.

42. Hakeda-Suzuki S, Ng J, Tzu J et al. Rac function and regulation during Drosophila development. Nature 2002; 416(6879):438-442.

43. Menon SD, Chia W. Drosophila rolling pebbles: A multidomain protein required for myoblast fusion that recruits D-Titin in response to the myoblast attractant dumbfounded. Dev Cell 2001; 1(5):691-703.

44. Rau A, Buttgereit D, Holz A et al. rolling pebbles (rols) is required in Drosophila muscle precursors for recruitment of myoblasts for fusion. Development 2001; 128(24):5061-5073.

45. Tanenbaum SB, Gorski SM, Rusconi JC et al. A screen for dominant modifiers of the irreC-rst cell death phenotype in the developing Drosophila retina. Genetics 2000; 156(1):205-217.

46. Chen EH, Pryce BA, Tzeng JA et al. Control of myoblast fusion by a guanine nucleotide exchange factor, loner, and its effector ARF6. Cell 2003; 114(6):751-762.

47. Donaldson JG. Multiple roles for Arf6: Sorting, structuring, and signaling at the plasma membrane. J Biol Chem 2003; 278(43):41573-41576.

48. Donaldson JG. Myoblasts fuse when loner meets ARF6. Dev Cell 2003; 5(4):527-528.

49. Maeland AD, Bloor JW, Brown NH. Characterization of singles bar a new mutant in Drosophila melanogaster required for muscle development. Mol Cell Biol 1996; 7:39a.

50. Zhang Y, Featherstone D, Davis W et al. Drosophila D-titin is required for myoblast fusion and skeletal muscle striation. J Cell Sci 2000; 113(Pt 17):3103-3115.

51. Machado C, Andrew DJ. D-Titin: A giant protein with dual roles in chromosomes and muscles. J Cell Biol 2000; 151(3):639-652.

52. Gregorio CC, Granzier H, Sorimachi H et al. Muscle assembly: A titanic achievement? Curr Opin Cell Biol 1999; 11(1):18-25.

53. Trinick J, Tskhovrebova L. Titin: Aa molecular control freak. Trends Cell Biol 1999; 9(10):377-380.

54. Liu H, Mardahl-Dumesnil M, Sweeney ST et al. Drosophila paramyosin is important for myoblast fusion and essential for myofibril formation. J Cell Biol 2003; 160(6):899-908.

55. Furlong EE, Andersen EC, Null B et al. Patterns of gene expression during Drosophila mesoderm development. Science 2001; 293(5535):1629-1633.

56. Duan H, Skeath JB, Nguyen HT. Drosophila Lame duck, a novel member of the gli superfamily, acts as a key regulator of myogenesis by controlling fusion-competent myoblast development. Development 2001; 128(22):4489-4500.

57. Ruiz-Gomez M, Coutts N, Suster ML et al. myoblasts incompetent encodes a zinc finger transcription factor required to specify fusion-competent myoblasts in Drosophila. Development 2002; 129(1):133-141.

58. Artero R, Furlong EE, Beckett K et al. Notch and Ras signaling pathway effector genes expressed in fusion-competent and founder cells during Drosophila myogenesis. Development 2003; 130(25):6257-6272.

59. Giot L, Bader JS, Brouwer C et al. A protein interaction map of Drosophila melanogaster. Science 2003.

60. Galletta BJ, Niu XP, Erickson MR et al. Identification of a Drosophila homologue to vertebrate Crk by interaction with MBC. Gene 1999; 228(1-2):243-52.

61. Shen K, Fetter RD, Bargmann CI. Synaptic specificity is generated by the synaptic guidepost protein SYG-2 and its receptor, SYG-1. Cell 2004; 116(6):869-81.

62. Donoviel DB, Freed DD, Vogel H et al. Proteinuria and perinatal lethality in mice lacking NEPH1, a novel protein with homology to NEPHRIN. Mol Cell Biol 2001; 21(14):4829-36.

63. Shen K, Bargmann CI. The immunoglobulin superfamily protein SYG-1 determines the location of specific synapses in C. elegans. Cell 2003; 112(5):619-30.

64. Schneider T, Reiter C, Eule E et al. Restricted expression of the irreC-rst protein is required for normal axonal projections of columnar visual neurons. Neuron 1995; 15(2):259-71.

65. Galletta BJ. Chakravarti M, Banerjee R et al. SNS: Adhesive properties, localization requirements and ectodomain dependence in S2 cells and embryonic myoblasts. Mech Dev 121:1455-1468.

CHAPTER 9

Muscle Attachment Sites:
Where Migrating Muscles Meet Their Match

Talila Volk*

Abstract

The precise match between somatic muscles and their epidermal muscle attachment cells is achieved through a continuous dialogue between these two cell types. Initially, an intricate pattern of tendon precursors is produced within the epidermis. The tendon precursor cells guide the somatic myotubes to migrate and associate with their targets, the tendon cells. Following their association, the myotube provides a differentiation signal to the tendon cell driving its maturation. The establishment of hemi-adherence junctions between the muscle and the tendon cell represents a final differentiation step, triggering the typical cytoskeletal organization in each of these cell types, and ultimately allowing the muscle to contract and the tendon cell to withstand this contractive force.

Introduction

Muscle and tendon assembly in *Drosophila* provides an attractive system in which to study how cells coalesce into a complex tissue during embryonic development. The specific recognition by somatic muscles of their epidermal attachment cells[1] is critical to enable coordinated larval movements following embryonic development. Muscles and tendon cells develop from distinct germ layers, the mesoderm and ectoderm, respectively. Thus, the initial patterning and determination of each cell type takes place independently as a result of germ layer-specific patterning events. The spatial correspondence between individual muscles and tendons follows their initial determination, and is based on bi-directional signaling between these two cell types, which is critical for their terminal differentiation. Below I will discuss what is known to date regarding these major steps of tissue organization and differentiation.

Determination of Epidermal Muscle-Attachment Precursors

Transcriptional Regulation of Stripe

A key factor in tendon cell differentiation is the early growth response-like triple zinc finger transcription factor, Stripe (Sr).[1,2] Sr activity is necessary and sufficient to induce the fate of muscle attachment cells.[1,3] Embryos mutant for *sr* lack epidermal muscle attachment cells, resulting in the complete loss of organization of the somatic musculature. Furthermore, ectopic expression of Sr leads to the corresponding ectopic attraction of the muscles towards these sites, and to the transformation of Sr-expressing cells into tendon-like cells. This suggests that Sr is the key inducer of tendon cell fate.

*Talila Volk—Department of Molecular Genetics, Weizmann Institute of Science, Rehovot 76100, Israel. Email: lgvolk@wicc.weizmann.ac.il

Muscle Development in Drosophila, edited by Helen Sink. ©2006 Eurekah.com and Springer Science+Business Media.

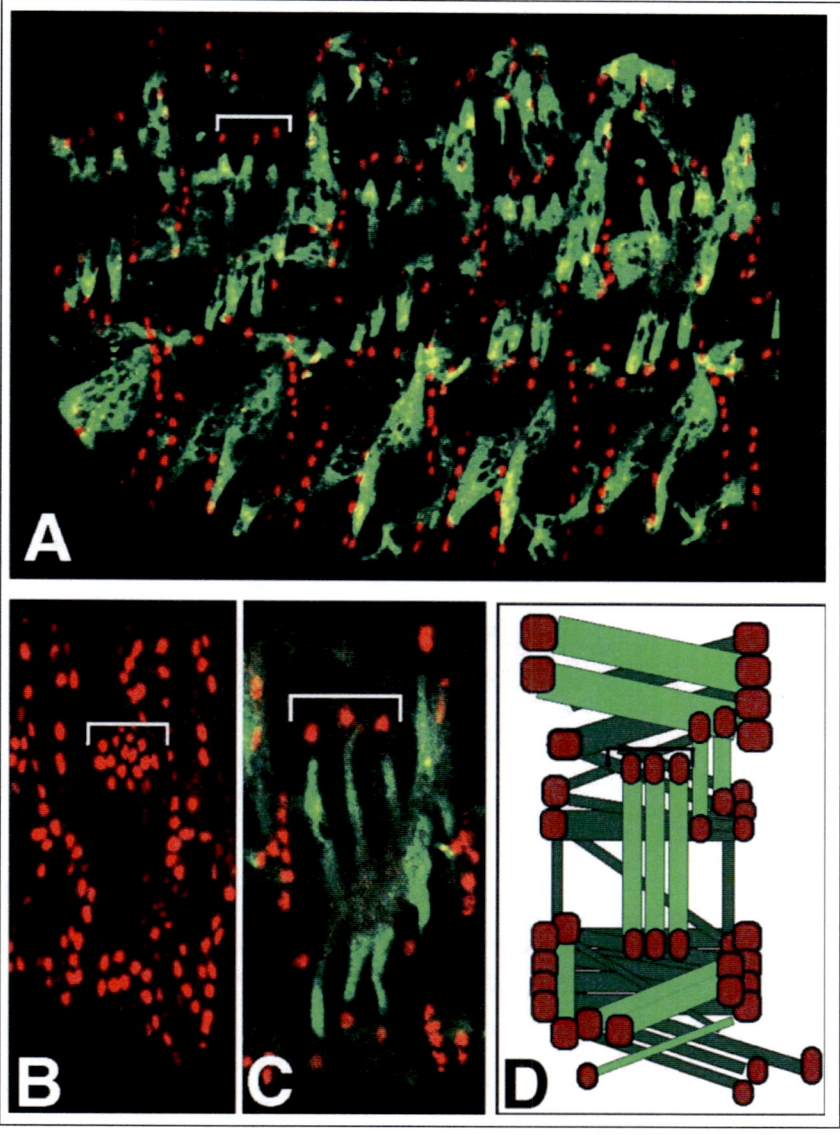

Figure 1. The pattern of embryonic *Drosophila* tendon cells. A) The pattern of tendon cells (visualized by an antibody against Stripe), and their precise match with the somatic muscles (detected by an antibody against myosin heavy chain (MHC), green), revealed in an embryo at stage 16. B) At an earlier developmental stage (stage 14-15), the initial number of tendon precursors (visualized by an antibody against b-galactosidase staining of a stripe-enhancer trap strain, red) is higher that the number of mature tendon cells (C, stage 16-17), where each tendon cell (labeled with an antibody against Stripe, red) is connected to a specific muscle (detected by an antibody against MHC, green). Notice the reduction in the number of lateral transverse tendon cells (marked by brackets), from ~14 cells at stage 14-15 to 3 cells at stages 16-17. D) Schematic view of the somatic muscles (green) and tendon cells (red) of a single abdominal segment (adapted from: Volk T. Singling out Drosophila tendon cells: A dialogue between two distinct cell types. Trends Genet 15:448-53, ©1999 Elsevier, with permission).

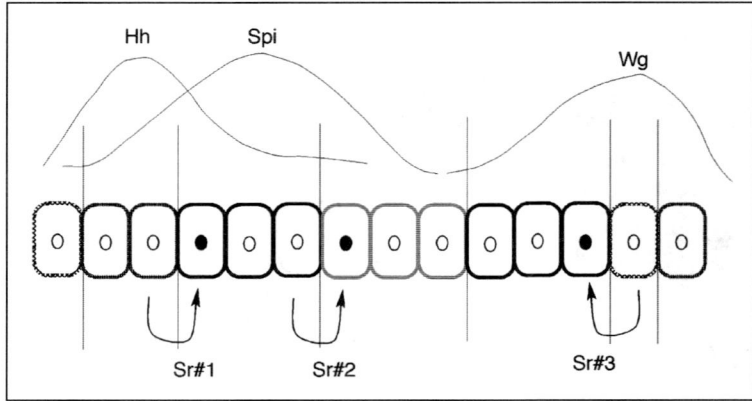

Figure 2. Transcriptional regulation of *stripe* by ectodermal patterning genes. The repeating pattern of tendon precursors is established through the regulation of expression of the *stripe* gene. Three signals, each emanating from their respective territory, induce Sr expression in an adjacent territory. Hh signaling induces Sr row 1 in the first row of Rho-expressing cells. Spi, through activation of the EGFR signaling, induces Sr row 2 in the first row of Ser-expressing cells. Wg signaling induces Sr row 3 in the adjacent anterior cells. A hypothetical activity landscape for each signal is diagrammed based on the affect of each signal on Sr expression (adapted from: Hatini V, DiNardo S. Divide and conquer: Pattern formation in Drosophila embryonic epidermis. Trends Genet 17:574-9, ©2001 Elsevier, with permission).

The pattern of Sr expression in the ectoderm is rather complex, both along the anterior-posterior and dorsal-ventral axes (Fig. 1).[4] Consistent with the intricate expression pattern of *sr* is a composite promoter, which responds to an array of patterning genes. For example, along the ventral ectoderm, three rows of Sr-expressing cells are detected and function as the binding sites for the ventral muscles (Fig. 2). The first row of Sr is induced by Hedgehog (Hh) activity in the adjacent territory in the first row of Rhomboid-expressing cells. The second Stripe row is induced by Spitz activation of the Epidermal Growth Factor Receptor (EGFR) pathway in the first row of Serrate-expressing cells. Sr row number 3 results from Wingless (Wg) activity in the adjacent anterior cells (see Fig. 2).[5,7] A detailed analysis of a *sr* enhancer that drives expression at the segment border cells provided the first direct evidence for functional Cubitus interruptus (Ci) and Pangolin binding sites corresponding to the Hh and Wg signaling pathways.[8] While Hh positively regulates *sr* transcription through Ci activity, Wg restricts Hh regulation to a single row of *sr*-expressing cells through the activity of Pangolin.[8] Although Hh, Wg and the EGFR may be responsible for the regulation of *sr* expression along the anterior-posterior axis, the genes controlling *sr* expression along the dorsal-ventral axis are yet to be elucidated.

Post Transcriptional Regulation of Stripe Expression

Sr is also regulated post transcriptionally by the RNA-binding protein, Held Out Wing (HOW). Embryos mutant for *how* maintain high Sr expression in late stages of embryonic development in all their tendon precursor cells.[10] *sr* mRNA stability was found to be directly regulated by HOW activity. *how* mRNA is spliced to produce two protein isoforms which function in opposing directions; HOW(L) represses *sr* mRNA levels by inducing their fast degradation, and HOW(S) elevates *sr* mRNA levels by inducing their stabilization.[12] In tendon precursors where *sr* is expressed at low levels, HOW(L) is highly expressed and moreover, its expression is further induced by Sr activity. Sr together with HOW(L) form a negative feedback loop maintaining the low levels of Stripe in these precursor cells. HOW(S) is detected in tendon cells in late developmental stages. HOW(S) activity leads to elevation of *sr* mRNA levels. Therefore, the switch between HOW(L) and HOW(S) may serve as a switch between the tendon precursor stage and the mature tendon cell. The signal responsible for the switch between HOW(L) and HOW(S) is yet to be discovered, but some evidence suggests that it occurs through activation of the EGFR pathway. HOW(S) is elevated only in cells where EGFR signaling is high, and Vein, the neuregulin-like ligand is required for this EGFR activation.[10]

In summary, since Sr is a potent transcription factor which, when highly expressed, is capable of altering the program of cell differentiation, its expression must be tightly controlled. At least three levels of Sr regulation were described: (1) transcriptional control by the segment polarity patterning gene products; (2) alternate post-transcriptional control by the opposing activities of HOW(L) or HOW(S), and (3) positive regulation in response to the muscle-dependent signal, Vein (see below).

The Guidance of the Muscle Towards Its Tendon Attachment Site

Each hemisegment of a *Drosophila* embryo contains 30 different types of somatic muscles. Each of these muscles acquires a specific fate due to the unique combinatorial expression of transcription factors inherited by the founder cell of a given muscle.[13] The specificity of a single muscle type is determined by its orientation, the number of fusion-competent myoblasts to be incorporated, and the morphology of the specific muscle.

In parallel to myotube formation, the myotube leading edges extend towards the epidermal attachment cells.[11] These immature tendon cells were suggested to guide the muscle leading edge towards the correct insertion site. Thus, Sr expression may induce genes whose products attract specific muscle cells. Experiments in which Sr was ectopically expressed in the ectoderm or in the salivary glands indicated that cells overexpressing Sr attract the myotubes towards ectopic sites, suggesting that Sr activates transcription of specific muscle-guiding cues.[1,3] Although the nature of these guiding cues is yet to be elucidated, these experiments suggest that the cues provided to guide the muscles act at short range, as only muscles that are in close proximity to the Sr-overexpressing cells are affected.

The ubiquitous expression of Sr in all muscle attachment cells raises the question of how individual attachment cells specifically guide distinct muscle types. Most of the tendon-specific genes identified so far do not exhibit differential expression pattern in specific attachment cells. However, there are some interesting exceptions. Slit is a leucine-rich repeat protein that mediates neuronal and myotube guidance. Slit is expressed by segment border tendon cells but not by the lateral tendon cells and, correspondingly, the lateral transverse muscles do not express its receptor, Roundabout (Robo) (see below).[14] It is possible that a combination of Sr and secreted, stage-specific factors present in the ectoderm like Slit, attract or repulse a specific muscle. Two examples of such a combination of Sr and local guiding factors have been described.

The first example is the ventral oblique muscles that are attracted by the ventral attachment cells expressing Sr, but are repelled by Slit activity secreted from the ventral midline cells. This results in a typical space detected between the edges of the ventral muscles on each hemisegment which overlaps the range of Slit activity at the ventral domain. The ventral muscles themselves express both the Robo and Robo2 receptors, enabling their response to Slit. Indirect evidence suggests that Slit may not only function to repulse certain muscle groups but also may be required to attract others. First, Slit is expressed in the segment border attachment cells to which Robo-positive longitudinal muscles attach. Secondly, when Robo is ectopically expressed in the lateral transverse muscles these muscles become attracted to the Slit-positive segment border attachment cells.[14]

The second example is combination between Stripe and Derailed activities, which guides the lateral transverse muscles. While these cells lack Robo expression, they express the receptor tyrosine kinase, Derailed. In mutants lacking Derailed, these muscles ignore the Stripe-expressing attachment sites located at the ventral and dorsal regions in the middle of each hemisegment and continue to elongate dorsally and ventrally.[35] Derailed has been shown to be a receptor for Wnt5 in the Drosophila CNS.[36] Similarly, the lateral transverse muscle could be influenced by altered Derailed activity due to a putative Wnt family member protein that is distributed in the ectoderm in the middle of each segment.

The Essential Role of Muscle Signaling in Tendon Cell Maturation

The precise match between a specific muscle and its attachment cell is achieved by a muscle-dependent signal. The initial number of ectodermal tendon precursors expressing Stripe is higher than the final number of muscle-bound tendon cells (see Fig. 1). For example, there are about 13 tendon precursors that correspond to the lateral transverse muscles in the middle of each segment at the dorsal and ventral positions. Of these, only three cells will maintain Stripe expression and develop into mature tendon cells connected to the three lateral transverse muscles.[4] What is the mechanism that determines whether a muscle attachment precursor develops into a mature cell? The exact correlation between the number of muscles and their corresponding mature tendon cells suggests a direct interaction between the two cell types. Also, in mutants where some of the muscle groups are missing, the corresponding attachment cells do not maintain Stripe expression and loose their potential to develop into tendon cells.[1] These experiments suggest that a differentiation factor provided by the muscles induces tendon cell maturation. In a screen for mutations affecting tendon cell differentiation, a mutation in Vein, a ligand of the EGFR was identified. In *vn* mutant embryos, Stripe expression is absent from all tendon cells, and when Vein is ectopically expressed in the ectoderm, it induces broad ectopic expression of Stripe.[37] Vein is a neuregulin-like factor produced by the muscle cell and apparently secreted at the leading myotube edge.[38,37] Interestingly, Vein protein is accumulated only at the space between the muscle-tendon junctions, suggesting a mechanism that presumably protects it from degradation.[15] At the junction site, Vein binds to the EGFR at the tendon cell membrane and activates the EGFR pathway within that cell. In contrast to other embryonic sites in which Vein serves to back up the potent EGFR Spitz ligand, in the tendon/muscle system Vein is the only activating ligand. Staining with di-phospho-ERK antibody, which identifies cells with activated EGFR signaling, indicates the pronounced and continuous activation of this pathway in tendon cells.[16] This continuous activation is unique to tendon cells, as normally EGFR activation is transient, and the continuous activation may reflect pronounced accumulation of the Vein ligand at muscle-attachment sites.

The Requirement for Muscle-Tendon Junctions for Proper Muscle Function

The formation of stable adherence junctions between the muscle and its corresponding tendon cell is essential to anchor the muscles to the epidermis and to withstand the force of muscle contraction. Hemi-adherence junctions are formed between the muscle and the extracellular matrix (ECM) secreted into the space between the muscle and the tendon cell, and similar junctions form between the tendon and the same ECM material. The formation of hemi-adherence junctions on both the tendon and the muscle sides is associated with alterations in muscle and tendon morphology. In the muscle, junction formation results in the arrest of filopodia formation at the leading edge and establishment of all the components required to build up the basal-adherence junction. In the tendon cell, junction formation is associated with rearrangement of the cytoskeletal network and the change of its morphology into a "half moon shape" cell (see Fig. 3). The hemi-adherence junctions formed on both the muscle and tendon sides are mediated by the *Drosophila* integrin adhesion receptors, the αPS2βPS heterodimer at the muscle side, and the αPS1βPS heterodimer at the tendon cell. These receptors bind Tiggrin and possibly Laminin, the major ECM components of the muscle-tendon junction.[9,17] The ECM appears as electron-dense material accumulated at the space between the muscle and tendon cell and contains, in addition to Tiggrin and Laminin, putative signaling proteins such as F-Spondin,[18] Masquarade[19] and Vein.[15] It is not clear what triggers the initial deposition of Tiggrin and Laminin at the muscle-tendon junction; however the muscle, the tendon, and hemocytes positioned in close proximity to the junction site were

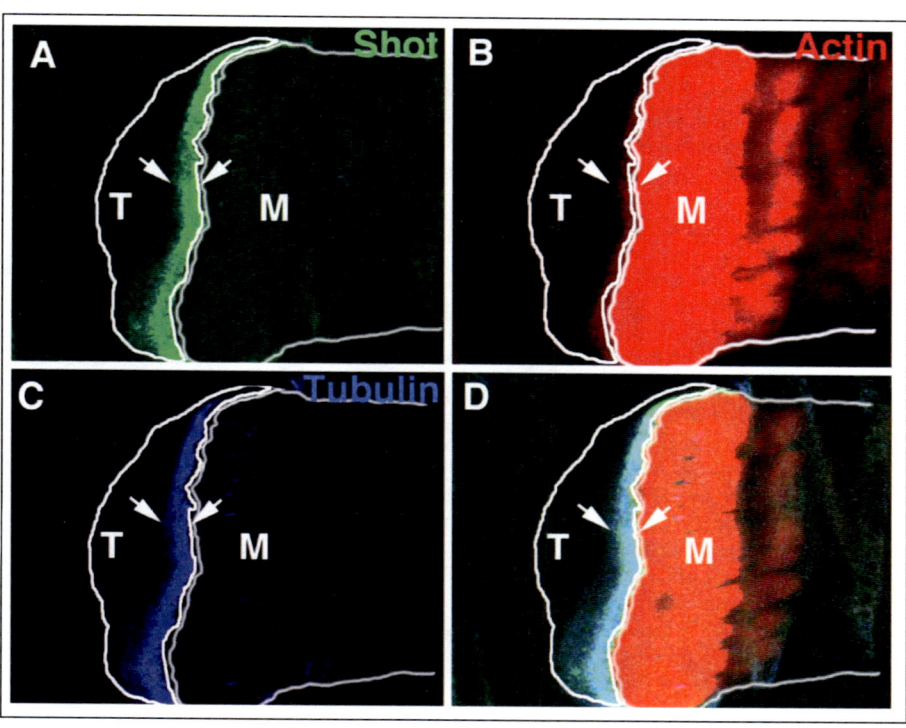

Figure 3. Shot is detected at a unique microtubule-rich domain that links the muscle-tendon hemiadherence junction and the cuticle. A–D) Lateral transverse muscle (M) and its specific tendon cell (T) are shown in a flat preparation of third instar larvae labeled with (A, green) Shot, (B, red) Phalloidin, and (C, Blue) Tubulin. D is the merged images of A–C. A white line indicates the outlines of the tendon and muscle cells. Arrows indicate the domain of muscle-tendon adherence junction. From: Subramanian A et al. Shortstop recruits EB1/APC1 and promotes microtubule assembly at the muscle-tendon junction. Curr Biol 13:1086-95, ©2003 Elsevier, with permission.

shown to secrete Tiggrin and Laminin.[20-22] In the absence of functional Integrins, such as in the *myospheroid* homozygous mutants, the muscles approach the tendon cells and form an initial contact (presumably mediated by cell-cell adhesion molecules such as Cadherins); however following the initial muscle contractions (around embryonic stage 16), the muscles round up and detach.[23-25]

The Cytoplasmic Composition of the Muscle and Tendon Junctional Complexes

A major function of Integrin-mediated adhesion is to recruit specific components to the cytoplasmic aspects of the cell at the junctional site, which reorganize the cytoskeletal networks within the cell, and transmit specific signals. During the past years, several cytoplasmic components in both the muscle and tendon adherence junctions, proven to be functionally important, were identified by genetic screens or by reverse genetics. Talin (encoded by the *rhea* gene)[17] is a multi-domain protein that, as in vertebrate cells, presumably binds the cytoplasmic tail of the Integrin β subunit upon ligand binding, and associates with actin filaments at its C-terminal domain.[26] Integrin-mediated adhesion is not functional in *rhea* mutant embryos and the muscles round up upon muscle contraction, similar to embryos lacking functional Integrins. Paxillin, an adaptor protein, containing 4 LIM domains, is localized at the cytoplasmic aspects of the junctional complex, and is thought to be involved in signal transduction due to its high content of phospho-tyrosine residues.[27] ILK is an integrin-linked kinase involved in the functional link between Integrins and actin microfilaments.[28] Finally, PINCH encodes for a multiple LIM-domain protein and is localized at the hemi-adherence muscle-tendon junctions.[29] Lack of functional PINCH in *steamer duck* mutants also results in muscle detachment and loss of actin microfilament organization. PINCH and ILK are thought to form a protein complex and may be involved in Integrin-mediated signaling.[30] All these gene products are functionally connected to the process by which Integrin-mediated adhesion organizes the actin cytoskeleton within the muscle and the tendon cell.

Recent data show that in the tendon cell, integrin-mediated adhesion does not lead to massive recruitment of the actin cytoskeleton, and actin levels in the mature tendon cell are extremely low. Rather, high levels of microtubules are detected in the tendon cells, organized as linear arrays stretched between the basal junction domain and the apical cuticle site. A key protein involved in linking these microtubule arrays to the basal hemi-adherence junction in the tendon cells is Shortstop (Shot) (also called Kakapo).[15,31,32] In contrast to the proteins described above Shot is expressed only in tendon cells and not in muscle cells. Shot is a multi-domain large protein containing an actin-binding domain at the N-terminal region, followed by a plakin domain, a rod-like domain consisting of spectrin-like repeats, and a C' terminal region (the Gas2 domain) which binds microtubules.[33] Shot is highly enriched at the basal hemi-adherence junctions of the tendon cell (Fig. 3). It recruits the microtubule arrays to this site by both direct binding to microtubule and by association with EB1, a microtubule-plus end binding protein.[34] The accumulation and organization of microtubules within the tendon cell is functionally essential for the ability of these cells to resist muscle contraction. When the levels of Shot are compromised (e.g., by the expression of *shot*-dsRNAi), microtubules detach from the basal junction, and the tendon cell elongates following muscle contraction, eventually breaking.

In summary, the establishment of hemi-adherence junctions in both the muscle and the tendon cell is considered to represent the final stage of maturation of both these cell types. The similar junctional composition leads to distinct cytoskeletal rearrangements in the two cell types critical for the correct function of the muscle and the tendon cell.

Concluding Remarks

Tissue differentiation is a multi-step process, which must be regulated continuously. Tendon cell differentiation represents a differentiation process in which the transitions from phase to phase are regulated by extrinsic signals. The details of the various differentiation steps and the signaling mechanisms controlling their processing has now been partially explained in molecular terms. Future characterization of similar processes in other tissues may indicate if parallel principles are implicated during tissue differentiation in other organisms.

Acknowledgments

I thank S. Shwarzbaum for critical reading of the manuscript. This work was supported by a grant from the Israeli Science Foundation.

References

1. Becker S, Pasca G, Strumpf D et al. Reciprocal signaling between Drosophila epidermal muscle attachment cells and their corresponding muscles. Development 1997; 124:2615-22.
2. Frommer G, Vorbruggen G, Pasca G et al. Epidermal egr-like zinc finger protein of Drosophila participates in myotube guidance. EMBO J 1996; 15:1642-9.
3. Vorbruggen G, Jackle H. Epidermal muscle attachment site-specific target gene expression and interference with myotube guidance in response to ectopic stripe expression in the developing Drosophila epidermis. Proc Natl Acad Sci USA 1997; 94:8606-11.

4. Volk T. Singling out Drosophila tendon cells: A dialogue between two distinct cell types. Trends Genet 1999; 15:448-53.
5. Hatini V, DiNardo S. Distinct signals generate repeating striped pattern in the embryonic parasegment. Mol Cell 2001; 7:151-60.
6. Schnepp B, Grumbling G, Donaldson T et al. Vein is a novel component in the Drosophila epidermal growth factor receptor pathway with similarity to the neuregulins. Genes Dev 1996; 10:2302-13.
7. Hatini V, DiNardo S. Divide and conquer: Pattern formation in Drosophila embryonic epidermis. Trends Genet 2001; 17:574-9.
8. Piepenburg O, Vorbruggen G, Jackle H. Drosophila segment borders result from unilateral repression of hedgehog activity by wingless signaling. Mol Cell 2000; 6:203-9.
9. Brown NH. Cell-cell adhesion via the ECM: Integrin genetics in fly and worm. Matrix Biol 2000; 19:191-201.
10. Nabel-Rosen H, Dorevitch N, Reuveny A et al. The balance between two isoforms of the Drosophila RNA-binding protein how controls tendon cell differentiation. Mol Cell 1999; 4:573-84.
11. Bate M, Rushton E. Myogenesis and muscle patterning in Drosophila. CR Acad Sci III 1993; 316:1047-61.
12. Nabel-Rosen H, Volohonsky G, Reuveny A et al. Two isoforms of the Drosophila RNA binding protein, how, act in opposing directions to regulate tendon cell differentiation. Dev Cell 2002; 2:183-93.
13. Baylies MK, Bate M, Ruiz Gomez M. Myogenesis: A view from Drosophila. Cell 1998; 93:921-7.
14. Kramer SG, Kidd T, Simpson JH et al. Switching repulsion to attraction: Changing responses to slit during transition in mesoderm migration. Science 2001; 292:737-40.
15. Strumpf D, Volk T. Kakapo, a novel cytoskeletal-associated protein is essential for the restricted localization of the neuregulin-like factor, vein, at the muscle-tendon junction site. J Cell Biol 1998; 143:1259-70.
16. Gabay L, Seger R, Shilo BZ. MAP kinase in situ activation atlas during Drosophila embryogenesis. Development 1997; 124:3535-41.
17. Brown NH et al. Talin is essential for integrin function in Drosophila. Dev Cell 2002; 3:569-79.
18. Umemiya T, Takeichi M, Nose A. M-spondin, a novel ECM protein highly homologous to vertebrate F-spondin, is localized at the muscle attachment sites in the Drosophila embryo. Dev Biol 1997; 186:165-76.
19. Murugasu-Oei B, Rodrigues V, Yang X et al. Masquerade: A novel secreted serine protease-like molecule is required for somatic muscle attachment in the Drosophila embryo. Genes Dev 1995; 9:139-54.
20. Fogerty FJ et al. Tiggrin, a novel Drosophila extracellular matrix protein that functions as a ligand for Drosophila alpha PS2 beta PS integrins. Development 1994; 120:1747-58.
21. Bunch TA et al. The PS2 integrin ligand tiggrin is required for proper muscle function in Drosophila. Development 1998; 125:1679-89.
22. Martin-Bermudo MD, Brown NH. The localized assembly of extracellular matrix integrin ligands requires cell-cell contact. J Cell Sci 2000; 113(Pt 21):3715-23.
23. MacKrell AJ, Blumberg B, Haynes SR et al. The lethal myospheroid gene of Drosophila encodes a membrane protein homologous to vertebrate integrin beta subunits. Proc Natl Acad Sci USA 1988; 85:2633-7.
24. Leptin M, Bogaert T, Lehmann R et al. The function of PS integrins during Drosophila embryogenesis. Cell 1989; 56:401-8.
25. Brown NH. Integrins hold Drosophila together. Bioessays 1993; 15:383-90.
26. Liu S, Calderwood DA, Ginsberg MH. Integrin cytoplasmic domain-binding proteins. J Cell Sci 2000; 113(Pt 20):3563-71.
27. Yagi R et al. A novel muscle LIM-only protein is generated from the paxillin gene locus in Drosophila. EMBO Rep 2001; 2:814-20.
28. Zervas CG, Gregory SL, Brown NH. Drosophila integrin-linked kinase is required at sites of integrin adhesion to link the cytoskeleton to the plasma membrane. J Cell Biol 2001; 152:1007-18.
29. Clark KA, McGrail M, Beckerle MC. Analysis of PINCH function in Drosophila demonstrates its requirement in integrin-dependent cellular processes. Development 2003; 130:2611-21.
30. Zhang Y et al. Assembly of the PINCH-ILK-CH-ILKBP complex precedes and is essential for localization of each component to cell-matrix adhesion sites. J Cell Sci 2002; 115:4777-86.
31. Gregory SL, Brown NH. kakapo, a gene required for adhesion between and within cell layers in Drosophila, encodes a large cytoskeletal linker protein related to plectin and dystrophin. J Cell Biol 1998; 143:1271-82.

32. Prokop A, Uhler J, Roote J et al. The kakapo mutation affects terminal arborization and central dendritic sprouting of Drosophila motorneurons. J Cell Biol 1998; 143:1283-94.
33. Lee S, Kolodziej PA. Short stop provides an essential link between F-actin and microtubules during axon extension. Development 2002; 129:1195-204.
34. Subramanian A et al. Shortstop recruits EB1/APC1 and promotes microtubule assembly at the muscle-tendon junction. Curr Biol 2003; 13:1086-95.
35. Callahan CA, Bonkovsky JL, Scully AL et al. derailed is required for muscle attachment site selection in Drosophila. Development 1996; 122:2761-7.
36. Yoshikawa S, McKinnon RD, Kokel M et al. Wnt-mediated axon guidance via the Drosophila Derailed receptor. Nature 2003; 422:583-8.
37. Yarnitzky T, Min L, Volk T. The Drosophila neuregulin homolog Vein mediates inductive interactions between myotubes and their epidermal attachment cells. Genes Dev 1997; 11:2691-700.

CHAPTER 10

Neuromuscular Development:
Connectivity and Plasticity

Louise Nicholson and Haig Keshishian*

Abstract

The *Drosophila* neuromuscular junction provides an excellent model system in which to study synaptic development. Axon outgrowth, target selection, and synaptogenesis have been extremely well characterized and occur with remarkable precision. Coupled with the powerful molecular genetic approaches available in *Drosophila*, this has lead to the identification of many genes involved in these processes. In this review, we examine the cellular and molecular mechanisms of guidance, target selection and synaptogenesis at the neuromuscular junction. We will also discuss how these synaptic contacts are refined and modified during periods of synaptic plasticity.

Introduction

The embryonic and larval neuromuscular junction (NMJ) of *Drosophila* is one of the leading model systems for examining synaptic development. It has been used successfully to investigate the cellular and molecular mechanisms that govern axonal guidance, target selection, synaptogenesis, and synaptic plasticity.[1,2] Several features have made the *Drosophila* NMJ especially suitable for the analysis of synaptic development. All of the motoneurons and muscle fibers are individually identifiable cells that establish specific and precisely wired connections.[2,3] As a result of this high degree of motoneuron target specificity, the NMJ provides a low noise environment for examining target selection and synaptic development. In addition, *Drosophila* is one of the preeminent genetic model organisms.[4] The genome is fully sequenced, and importantly there are numerous mutations available for genes that are involved in nervous system development. Sophisticated methods exist for the directed expression of transgenes to either side of the synapse and for visualizing individual neurons and muscles.[5,6] Finally, the muscle fibers are sufficiently large to allow the use of advanced electrophysiological and optical methods to study synaptic function and plasticity.[7-10]

Motoneuron Organization

Each abdominal body wall hemisegment from A2 through A7 is comprised of 30 muscle fibers that are innervated by about 35 individually identifiable motoneurons in a highly stereotyped and segmentally repeated fashion, shown in Figure 1.[3,11-13] These motoneurons fall into three classes based on synaptic morphology and physiological criteria.[14-17] Motoneurons with type I endings are characterized by their relatively short terminal branches and large boutons, and by the subsynaptic reticulum that the muscle fiber forms around the synapse. These can be

*Corresponding Author: Haig Keshishian—Molecular, Cellular and Developmental Biology Department, Yale University, P.O. Box 208103, New Haven, Connecticutt 06520, U.S.A. Email: haig.keshishian@yale.edu

Muscle Development in Drosophila, edited by Helen Sink. ©2006 Eurekah.com and Springer Science+Business Media.

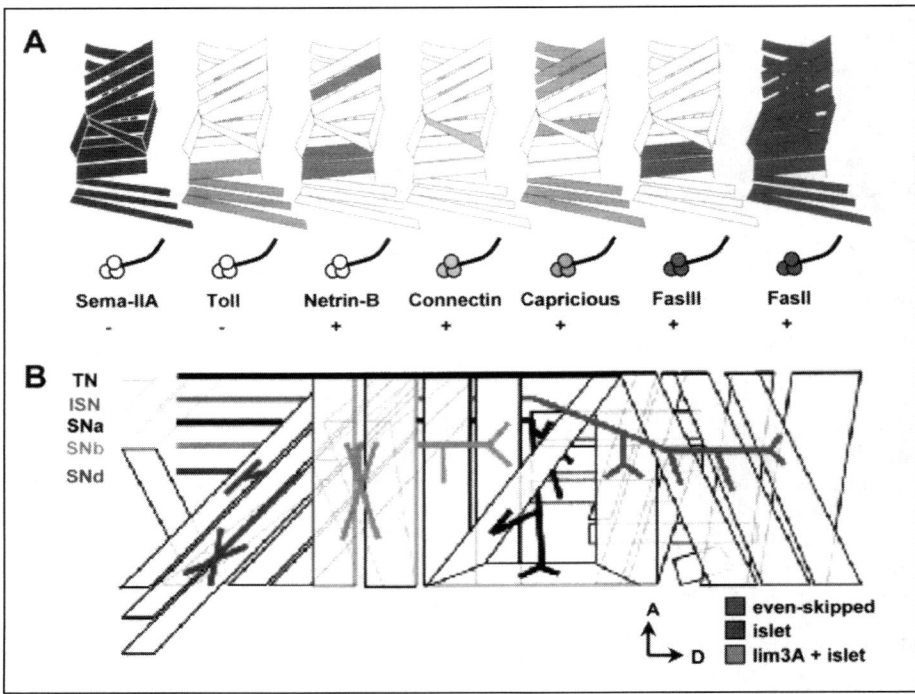

Figure 1. Molecular mechanisms governing neurite outgrowth and target selection at the *Drosophila* NMJ. A) Somatic muscle fibers of the embryonic bodywall express a diverse array of molecules that are either attractive (+) or repulsive (-) for motoneuron growth cones. The muscle fibers expressing the respective molecules, at the time the motoneuron growth cones arrive are indicated in color. Where an attractant is coexpressed by the innervating motoneurons it is also shown in color. Note that some muscle fibers express multiple attractive and repulsive molecules. It is thought that the relative attractiveness of a muscle for a motoneuron is due to a balance between attractive and repulsive signals. For each drawing the muscle fibers of a single abdominal hemisegment are shown, with dorsal to the top and anterior to the left. B) A single hemisegment of the *Drosophila* bodywall, showing five of the six major nerve branches that innervate the musculature. The nerves are color coded to indicate the transcriptional regulators that have been shown to play a role in determining motoneuronal pathway selection. A color version of this figure is available online at http://www.Eurekah.com.

further subdivided into Ib and Is. Type Ib endings are larger and produce smaller excitatory junction potentials than type Is. Type II endings are derived from just two motoneurons per hemisegment, and have long thin branches with numerous small boutons that are not associated with a subsynaptic reticulum. The majority of muscles are poly-innervated and receive both type I and type II innervation with most, if not all, motoneurons being glutamatergic.[18,19] Subsets may also express additional co-transmitters such as octopamine,[20] proctolin,[21] and PACAP.[22] Only muscle 12 in segments A2-A5 receives a type III ending, which is similar to type II in morphology, although the boutons are more oval-shaped and show a unique immunoreactivity for an insulin-like peptide.[17]

The muscle fiber target, position of the cell body within the central nervous system (CNS), and the neuroblast lineage that gives rise to each embryonic motoneuron has been determined, with many motoneurons being identifiable solely by their cell body position within the CNS.[12,13,23,24] In addition, a variety of molecular markers have been identified that can be used to distinguish specific motoneuron subsets. Each abdominal hemisegment is innervated by both ispilateral and contralateral motoneurons from both its own and from the next anterior CNS segment. Some synapse exclusively on individual muscles, whereas others target muscle fiber pairs or larger subsets of fibers. This suggests that muscles may be controlled both as groups and as individual fibers.

While there is no obvious topographic relationship between muscle location and the cell body position of the innervating motoneuron, the motoneuron dendrites are anatomically segregated in a way that reflects the position and orientation of the muscles they innervate.[13,25] This myotopic organization of the dendritic fields provides a central representation of the peripheral muscle field to presynaptic neurons and is probably matched by a corresponding functional organization of both sensory and interneuronal inputs to the motoneurons. This is a common feature of insect neuropil.[26,27]

Most motor axons exit the CNS via a common lateral exit point and then split into two main nerves, the intersegmental nerve (ISN), which innervates the dorsal muscle field, and the segmental nerve (SN), which innervates the ventral muscle field. A third smaller peripheral nerve, the transverse nerve (TN), exits the CNS dorsally and innervates a few muscle fibers in the mid-body wall.[2] Upon exiting the CNS, groups of axons defasciculate from the common motor axon pathways at specific choice points to form the five major nerve branches of the embryonic motor projection (Fig. 1). These axons remain fasciculated within each nerve branch until the target muscle domain is reached and then begin to defasciculate from one another as they contact their individual target muscle(s).

Axonal Outgrowth

Axons begin to extend into the periphery after the muscle fibers have formed, with the first ISN and SN motoneurons exiting the CNS at embryonic stage 13 and late stage 14, respectively.[16] Ablation experiments have shown that the presence of a specific muscle fiber is not required for motoneurons to reach their target site, indicating that the initial process of axon outgrowth is not dependent on unique guidance cues provided by the target muscle.[28,29] The pattern of outgrowth is highly stereotyped, suggesting that pioneer neurons contact specific intermediate targets to establish their axonal trajectory. In other systems, the intermediate targets that provide guidance cues have been shown to be pre-existing cellular landmarks, such as the segment boundary cells in the grasshopper limb, that have been coopted by the neuron to facilitate axon outgrowth.[30] This opportunistic behavior may underlie the ability of the growth cone to modify its response in the face of loss, duplication, or molecular manipulation of target, despite the apparent molecular stereotypy of the process of target selection (see below).

The pattern of motoneuron outgrowth is molecularly predetermined, as each motoneuron expresses a complement of proteins that enables it to interact and interpret guidance cues encountered in the environment to select the appropriate pattern of outgrowth. This begins with the specification of peripheral pathway preferences by the transcription factors Islet, Lim3, and Even-skipped (Eve), shown in Figure 1.[31,32] ISN motoneurons that innervate the dorsal muscle field express *eve*.[32] Loss of Eve function prevents axons in the ISN from extending dorsally, while ectopic *eve* expression can misdirect axons into the ISN and towards dorsal muscle targets. The ventral muscle field is innervated by SNb and SNd motoneurons (also referred to as ISNb and ISNd respectively). SNb motoneurons coexpress the LIM homeodomain genes *lim3* and *islet*, whereas SNd motoneurons express only *islet*.[31] In *lim3* mutants, SNb motoneurons fail to innervate their normal target muscles, and instead project into the SNd target area. Misexpression of *lim3* results in ectopic innervation of the SNb target area, suggesting that the expression of *islet* alone, or in combination with *lim3*, subdivides motoneurons into the SNd or SNb pathways respectively, providing a combinatorial code for pathway selection.

While these genes may control the choice of peripheral pathway, as transcription factors they can only regulate axon guidance indirectly and so far, the downstream transcriptional targets of these genes are unknown. The identity of the peripheral cues required for axon outgrowth is largely unknown although, given the ability of motoneurons to navigate to the correct target area in the absence of a specific target,[28,29] it seems possible that the periphery could be molecularly divided into broad target domains that attract each nerve branch. The cues for the formation of nerve branches appear to be partly derived from the founder myoblasts.[33] In the absence of founder myoblasts, motoneurons cannot defasciculate at branch points from the main nerve bundles but the presence of even a single founder myoblast can induce the formation of the appropriate nerve branch. However, the molecular nature of the guidance cues emanating from myoblasts remains to be elucidated.

Separation of motoneurons into specific nerve branches is mediated in part by the regulation of axonal adhesion between motoneurons. Fasciclin II (FasII), an immunoglobulin superfamily cell adhesion molecule is expressed by all motoneurons, while other adhesion molecules, such as Connectin (Con), are expressed by motoneuron subsets.[34-36] While in *fasII* mutants axon pathfinding by motoneurons appears largely unaffected, FasII overexpression results in an inability to defasciculate, causing axons to extend beyond their normal targets.[37,38] This suggests that, in order for axons to defasciculate from the common axonal pathway at the correct branch points, the adhesive forces provided by FasII and other neuronal cell adhesion molecules must be overcome. The observation that overexpression of cell adhesion molecules is sufficient to prevent defasciculation and overcome attractive guidance signals from the target muscle field implies that the balance between axon-axon attraction and axon-target attraction needs to be finely regulated.

FasII-mediated axonal adhesion is opposed by the transmembrane axonal repellent Semaphorin Ia (SemaIa), which is co-expressed with its receptor Plexin A (PlexA) on all motor axons.[39,40] In embryos lacking *semaIa* or *plexA*, SNb and SNa axons stall or fail to defasciculate at specific choice points. Loss of *plexA* also affects axons in the TN, causing ectopic projections onto ventral muscles. These defasciculation defects can be suppressed by reducing levels of FasII, although for SNa, where the cell adhesion molecule Con is also expressed, simultaneous reduction of both FasII and Con is required to suppress the *semaIa* loss-of-function defect.[41,42] Presumably during much of axon outgrowth, axonal attraction mediated by FasII and other adhesion molecules is sufficient to balance PlexA-SemaIa mediated repulsion. The activity of these cell adhesion molecules must therefore be modulated by other proteins at growth cone choice points.

The *beaten path (beat1a)* gene encodes a novel secreted protein expressed by motoneurons and required for motoneurons to defasciculate at branch points.[43] Dosage-sensitive interactions with both *fasII* and *conn* suggests that Beat1a promotes defasciculation by opposing FasII- and Con-mediated adhesion.[43] If Beat1a opposes FasII adhesion at growth cone choice points, resulting in a decrease of axon-axon attraction, this may shift the balance of attractive versus repulsive signals towards repulsion, enabling axons to defasciculate. However, the mechanism that enables Beat1a to act only at growth cone choice points remains unclear.

Attractive cues emanating from the musculature simultaneously attract motoneuron growth cones to the target muscle field. Sidestep, a novel transmembrane immunoglobulin superfamily member, is a key muscle-derived attractant for all motoneurons.[44] In *sidestep* mutants, motoneurons fail to defasciculate and innervate the target area, continuing to grow along the nerves instead. This loss-of-function phenotype, unlike the *beat* loss-of-function phenotype, cannot be overcome by reducing the levels of FasII, suggesting that target-derived attractive cues are needed in addition to a reduction in neuron-neuron adhesion for motoneurons to defasciculate and enter the muscle field.[44]

Axonal fasciculation and defasciculation during motor axon outgrowth is also regulated by receptor protein tyrosine phosphatases (RPTPs). Most RPTPs appear to be widely expressed in motoneurons, yet loss-of-function mutations in *dptp69D*, *dptp52F* and *DLAR* result in defasciculation or target field recognition defects that affect specific nerve branches, while *dptp99A* and *dptp10D* mutants exhibit no obvious defects.[45-48] Many guidance choices are unaffected until multiple RPTPs are eliminated.[46,48,49] These genetic interactions between RPTPs are dependent on context. DPTP99A, for example, which is partially redundant with DLAR for ISN outgrowth, appears to function in opposition to DLAR for the guidance of SNb axons. DPTP10D, which opposes other RPTPs during ISN guidance, seems to cooperate with DPTP99A during SNa outgrowth and bifurcation. The presence of both these phosphatases is sufficient to facilitate normal SNa outgrowth, while either one alone is insufficient.[48] Cooperative, competitive and partially redundant genetic interactions between specific combinations of RPTPs in different motoneurons therefore facilitate the outgrowth of each motoneuron pathway and suggest that guidance of specific axons likely involves the integration of information from multiple RPTP signaling pathways.

No ligands have yet been identified for *Drosophila* RPTPs but studies in vertebrates implicate secreted cytokines, neuronal CAMs and the laminin-nidogen complex as possible ligands (reviewed in ref. 50). While the intracellular substrates of most *Drosophila* RPTPs are unknown, potential substrates for DLAR include the cytoplasmic tyrosine kinase Abelson (Abl) and its phosphoprotein target Enabled (Ena), which function in actin assembly.[51,52] Abl shows antagonistic genetic interactions with DLAR during SNb guidance and both Abl and Ena bind to and are substrates for DLAR in vitro, suggesting that DLAR and Abl may mediate a phosphorylation-dependent switch that controls axon guidance by influencing actin assembly.[51]

Target Selection

Correct axon pathfinding brings motoneurons to the appropriate target region, but once there each motoneuron must still identify and synapse with its specific target muscle(s). Dye fill experiments have shown that motoneurons select their preferred target muscle fibers with remarkably few errors.[11,12] During target selection growth cones probe the surface of nearby muscle fibers, withdraw inappropriate contacts, and form stable synapses with correct targets. This suggests the growth cone samples many potential muscle partners but that specific mechanisms exist to enable each motoneuron to select its appropriate target. Muscle ablation and duplication experiments indicate that motoneuron target selection is based on cellular recognition.[28,29,53] Duplication of the muscle target results in appropriate innervation of both muscles.[53] If a motoneuron's normal target muscle is absent, the motoneuron grows to the correct target area but fails to form synapses on alternative muscle fibers.[28] However, ectopic synapses will eventually form on neighboring muscles.[28,29] This implies that specialized molecular recognition events occur between individual growth cones and their muscle targets, yet the motoneuron retains some degree of flexibility and is able to form connections with alternative partners in the absence of its preferred target(s).

While each motoneuron synapses with one, or at most a few, specific muscle fibers, target identity is likely to be specified by multiple factors, rather than by a single molecular label. For example, the candidate target recognition molecules Fasciclin III (FasIII), Con, and Capricious are all expressed by subsets of muscle fibers and also by the respective motoneurons that innervate them, as shown in Figure 1.[36,54,55] In each case, loss-of-function mutations result in largely normal motor axon guidance and synaptogenesis, whereas ectopic muscle expression can cause ectopic synapse formation, suggesting that multiple cues exist within muscle targets with partially overlapping functions that can compensate for the loss of one target recognition signal.[54-58] Both FasIII and Con appear to regulate target recognition through homophilic adhesion interactions, as motoneurons must also express FasIII or Con in order to recognize alternate FasIII- or Con-positive targets, and both can mediate in vitro cell adhesion.[56,58]

Further target recognition cues are provided by the *netrin* genes, *netrin-A* and *netrin-B*, which show highly restricted expression patterns in the embryonic musculature (Fig. 1).[59] Loss of both *netrin* genes results in partially penetrant defects in ISN and SNb, the nerves that innervate the dorsal and ventral Netrin-expressing muscle fields respectively. Motoneurons reach but fail to innervate their muscle targets, indicating that the main defect lies in target recognition.[40] Pan-muscular expression of either *netrin* gene also causes ISN projection defects. This suggests that differential netrin expression is necessary for these motoneurons to distinguish their targets from other muscles and thus promotes synaptogenesis.[40] Similar phenotypes are observed in embryos mutant for the Netrin receptor, *frazzled*, which is required for the attractive roles of *netrins* during motoneuron guidance.[40,60]

Toll, on the other hand, appears to be a negative regulator of synaptogenesis. Toll is a transmembrane receptor that is dynamically expressed in a subset of ventral muscles. The initial expression of Toll in these muscle fibers is subsequently down-regulated to coincide with the temporal order in which the muscles are innervated.[61,62] Motoneurons in *Toll* embryos often form alternate ectopic synapses on more proximal muscles and fail to innervate their correct targets, although this analysis is complicated by the loss of motoneurons that occurs in the *Toll* mutant.[61,62] Overexpression of Toll throughout the musculature results in innervation failure, despite motoneurons reaching the correct target muscle, suggesting that Toll inhibits synaptogenesis and that its initial expression may prevent motoneurons from innervating inappropriate muscle fibers prior to reaching their target(s).[62]

Motoneuron growth cones, especially those bound for more distal muscle targets, pass over many muscles during their progress. Transient expression of repellants could be a general strategy to enable axons to bypass alternative muscle targets. Intriguingly, Commissureless, which down-regulates Roundabout receptor expression during midline axon guidance by targeting it to the endosome,[63] is initially expressed on all muscle surfaces then subsequently internalized around the onset of synaptogenesis, and thus could potentially function to remove synaptic inhibitors from muscle surfaces.[64]

All muscles also express a low level of Semaphorin IIa (SemaIIa), a secreted repellant.[40,65] In embryos lacking *semaIIa*, the motor projection appears normal but motoneurons frequently form additional ectopic contacts on neighboring muscles. While motoneuron growth cones do transiently contact nontarget muscles during normal development, these inappropriate contacts are subsequently withdrawn.[11,12,66] The expression of SemaIIa on muscles therefore appears to provide a basal level of growth cone repulsion that normally prevents these ectopic contacts from being stabilized.[40]

FasII is also expressed by all muscles. Increasing the level of pan-muscular expression of FasII causes a significant increase in ectopic innervation of ventral muscles, suggesting that FasII promotes synaptogenesis.[67] The ectopic innervation phenotype caused by FasII overexpression can be enhanced if SemaIIa is eliminated at the same time and can be suppressed by SemaIIa overexpression.[40] FasII and SemaIIa therefore provide a balance of attractive and repulsive signaling from all muscles, with additional molecular cues being necessary for motoneurons to select a specific muscle.

The target recognition strategies outlined above can be illustrated for the RP3 motoneuron, perhaps the best-studied motoneuron in *Drosophila*. RP3 innervates the cleft between muscle fibers 6 and 7. These muscles express Netrin-B and FasIII, positive target recognition signals for RP3, and transiently express the putative repellant Toll. Analysis of these cues suggests that it is the balance of attractive and repulsive signals received by the growth cone that determines target selection. In *netrin* mutants, RP3 enters the appropriate ventral muscle area but fails to innervate the muscle 6/7 cleft.[40] In embryos that additionally lack SemaIIa, relatively normal RP3 innervation is restored, suggesting that NetrinB-mediated attraction enables RP3 to overcome SemaIIa-mediated repulsion.[40] Pan-muscular FasIII overexpression promotes ectopic synapse formation[54] and Toll overexpression cause RP3 innervation failure.[62] Simultaneous overexpression of FasIII and Toll however results in virtually wild-type RP3 innervation.[68] This suggests that when both are overexpressed, each molecule negates the other's ectopic expression. This implies that other cues can allow the growth cone to select its correct targets.[68] The relative levels of attractive and repulsive cues expressed on the muscle surface, rather than the absolute presence or absence of a specific protein, therefore appear to determine the outcome of each situation.

Establishing the Synapse

Synapse formation is completed in less than 4 hours, and the morphological changes that occur during synaptic differentiation have been described in detail. Motoneuron growth cones are initially thin, flat, highly labile structures extending many long filopodial processes, up to 15 μm in length, that explore the surface of target and nearby nontarget muscle fibers.[11,12,16,69-71] Confocal time lapse studies of vitally-labeled cells have shown that these growth cones are surprisingly fast moving, with filopodial projections that extend and retract on a 5-10 minute timescale.[66,72,73] The growth cone sampling is accompanied by corresponding "myopodial" extrusions from the muscle fibers. The myopodia are dynamic actin-based filopodia, of 10 μm or longer, which become progressively clustered to the presumptive site of synapse formation coincident with and dependent upon the arrival of the motoneuron, and may facilitate synaptogenesis by increasing the area of contact between muscles and motoneurons.[74,75]

Neuronal filopodia and myopodia frequently surround one another during this period, with junctional structures characterized by the accumulation of low levels of electron-dense material forming at sites of close membrane association.[74] Such contacts appear to form only with target muscles and seem typical of growth cone-muscle interactions during the period of synaptic target recognition. These may provide the chance for membrane-bound target recognition molecules to participate in motoneuron-muscle interactions. Ectopic expression of Toll can prevent such contacts being established, whereas FasIII misexpression causes inappropriate formation of junctional structures on nontarget muscles, suggesting that target recognition molecules can influence the formation of ultrastructural associations.[74]

Soon after, inappropriate contacts are withdrawn and the growth cone thickens and consolidates into a few enlarged, rounded branches, or "prevaricosities", which are restricted to the nascent synapse.[71] Filopodia are shorter and presynaptic specializations begin to form. An influx of synaptic vesicles to the developing active zone occurs at this time and an increase in synaptotagmin expression is observed. Subsequently the prevaricosities constrict to form the first synaptic boutons, and individual bouton types can be distinguished at the mature nerve terminal. Glutamate is first detected in neuronal filopodia soon after motoneuron contact, and synaptic transmission begins within approximately 30 minutes of initial contact.[16,76]

These presynaptic changes are accompanied by the differentiation of the post-synaptic receptor field. Around the time of motoneuron contact, low levels of Glutamate receptor (GluR) expression begins to be detected across the whole muscle surface.[70,77] The subsequent localization of the GluRs to the burgeoning synaptic zone is one of the first signs of post-synaptic development. A large increase in GluR density at the NMJ follows.

While initial expression of GluRs occurs in the absence of innervation, the clustering of GluRs and their ensuing accumulation at the post-synaptic site appears to be dependent upon neuronal activity, as these processes do not occur in aneural muscles or in embryos in which neuronal activity is virtually abolished through mutation or by pharmacological blockade.[70,77,78] This is in contrast to the vertebrate NMJ where the clustering of Acetylcholine (Ach) receptors occurs without action potentials.[79-81]

Although the cellular events that underlie the process of synapse formation have been well studied, the means by which neuron-specific target recognition mechanisms initiate a common pathway of cytoskeletal reorganization and synaptic differentiation is poorly understood. Tetraspanins, membrane spanning proteins that function in cell signaling, adhesion and motility as part of a receptor complex, could potentially mediate such a transition.[82,83] Embryos lacking *latebloomer*, a *Drosophila* tetraspanin that is transiently expressed in embryonic motoneurons, show no apparent guidance or target recognition defects, but fail to initiate synaptic contact. Innervation appears normal in larvae however, indicating that loss of Latebloomer delays but does not prevent synapse formation.[82] Two other *Drosophila* tetraspanin genes are also expressed in embryonic motoneurons. Elimination of all three tetraspanins significantly increases the number of absent RP3 synaptic termini in embryos but synaptogenesis still eventually occurs, suggesting that tetraspanins facilitate but are not required for synapse formation.[83] Other molecules that mediate the transition from growth cone to synapse are yet to be identified.

Neuromuscular Activity and Connectivity

The preceding discussion of the cellular and molecular mechanisms governing target selection might lead one to conclude that *Drosophila* NMJs are largely hard-wired, with little room for experience-dependent remodeling or functional plasticity. However, one of the unexpected features of this system has been the degree to which neuromuscular activity actually shapes both normal connectivity and synaptic development.

As previously discussed, embryonic motoneuron growth cones make contacts with multiple muscle fibers during the early stages of synaptic development and subsequently undergo synaptic refinement.[11,76,84,85] Inappropriate motoneuronal contacts are withdrawn over a period of 3-4 hours during late embryogenesis, so that synapses are restricted to the target muscle fiber by the time the animal hatches. Jarecki and Keshishian[86] noted that synaptic refinement and connectivity in *Drosophila* is strongly influenced by electrical activity. Suppression of activity during late embryogenesis, using temperature sensitive mutations that affect Na$^+$ channel function, results in extensive ectopically placed motoneuronal contacts throughout the embryonic musculature. If activity is suppressed past the time of hatching the ectopic contacts become permanent functional synapses. However, the miswiring observed in the embryo is reversible, provided that activity is restored within a critical period that extends into the first larval instar. The observation that neuromuscular wiring errors can be corrected as a function of restored electrical activity strongly suggests that activity is normally involved in the refinement of connections during development, with a critical period that lasts about 12 hrs.

Ectopic synapse formation also occurs if a muscle is lacking innervation. When the RP motoneurons are ablated, other motoneurons can establish persistent functional ectopic contacts on the denervated muscle fibers.[87] These ectopic synapses originate from motoneurons innervating neighboring muscle targets or as collaterals from the axons of nearby nerves. Similarly, in *latebloomer* mutant embryos where synaptogenesis is delayed, there is an increase in ectopic innervation from transverse nerve axons, which do not express *latebloomer*.[82] These inappropriate contacts are eliminated once innervation occurs, which is accomplished during the critical period prior to hatching.

An important question raised by these observations is whether synaptic refinement requires presynaptic (i.e., motoneuronal) or postsynaptic (muscular) electrical excitability. This issue has been resolved using the directed expression of modified potassium channels that suppress electrical excitability by shunting K$^+$ current, so called "electrical knock out" or EKO channels.[88] Using GeneSwitch GAL4, an inducible variant of the bipartite GAL4/UAS expression system,[5,6] it has been possible to direct the expression of EKO channels to either muscle fibers or neurons to suppress excitability at specific times in development.[88] Presynaptic expression of EKO channels results in erroneous ectopic contacts similar to those observed using activity mutations, as well as in a higher frequency of long-lived elongated filopodial processes. By contrast, postsynaptic suppression of excitability has no effect on motoneuronal connectivity, even when the dosages used result in complete muscle paralysis. This shows that the ability of motoneurons to refine their synaptic contacts depends at a minimum on their presynaptic excitability, and that refinement can occur even when the muscle fiber is effectively silenced and experimentally forced to remain near its resting voltage.

Yoon and Keshishian[89] have further tested whether the refinement of synaptic contacts involved the function of either attractive or repulsive molecular signals acting on the growth cones. Using transheterozygote genetic interaction tests, they examined whether a connectivity phenotype emerged through the synergistic interaction of electrical inactivity and loss-of-function of several candidate genes. No synergizing effects of electrical activity were observed for the diffusible chemotropic molecule Netrin, the Netrin receptor Frazzled, the Slit receptor Roundabout, or the transmembrane repellant molecule SemaIa. By contrast, a partial loss-of-function of the muscle-secreted chemorepellant SemaIIa resulted in a significant increase in the frequency of ectopic synapses when combined with a partial suppression of presynaptic electrical excitability. This result shows that the refinement of synaptic connections during late embryogenesis involves in some way the diffusible repellant SemaIIa, and significantly, that the ability of a motoneuron to respond to SemaIIa is activity dependent.

This observation is intriguing, as it provides the first example where the in vivo responsiveness of growth cones to neurotropic signals is shown to be under the influence of electrical activity. A similar result has been recently observed in vitro, where the responsiveness of the vertebrate growth cones in turning assays to the secreted repellant SemaIII has been found to depend on electrical activity.[90]

The identity of the SemaIIa receptor in *Drosophila* remains unknown, although Plexin B is a good candidate.[91] As discussed above, SemaIIa is normally expressed at low levels by all muscle fibers, providing a repulsive signal that reduces the affinity of an exploring motoneuron to muscle fibers throughout the bodywall.[40] When the molecule is overexpressed in embryonic muscle the result is extensive motoneuron growth cone withdrawal and muscle denervation.[92] This indicates that all motoneurons likely possess one or more receptors for SemaIIa. Intriguingly, when the candidate receptor Plexin B is overexpressed in motoneurons a similar denervation effect is observed,[91] suggesting that motoneurons over-responded to a native repellant from the musculature, most likely SemaIIa. The results of Yoon and Keshishian[89] suggest that the ability of the motoneuron receptor (possibly Plexin B) to respond to SemaIIa-mediated repulsive signaling is modulated by presynaptic electrical excitability. In the situations where activity is blocked experimentally, sensitivity to the repulsive signal is reduced, and inappropriate, ectopic synapses become established throughout the bodywall.

Conclusions

Over the past fifteen years significant advances have been made in understanding how the remarkable fidelity with which *Drosophila* motoneurons form synaptic connections is achieved. The initial process of pathway and target selection appears to be governed largely by molecular cues. While a growing number of genes involved have been identified many more, such as the peripheral signals that trigger outgrowth to different target muscle fields, remain to be discovered. When this process is disrupted, the inherent degree of flexibility that exists in the system is revealed. It is apparent that motoneurons are capable of being opportunistic, exploiting vacated muscles or adopting alternate targets in the absence of their own. Motoneurons also undergo a period of synaptic refinement and there is a critical period during which electrical activity is required for the elimination of inappropriate synapses. We are now beginning to understand how electrical activity interacts with molecular signals during synaptic refinement. There is every reason to believe that the *Drosophila* NMJ will continue to provide rich insights into synaptic development and function.

Acknowledgments

We thank members of the Keshishian lab for comments on the manuscript and William Leiserson in particular for helpful discussions. Supported by grants from the NIH and NSF.

References

1. Gramates LS, Budnik V. Assembly and maturation of the Drosophila larval neuromuscular junction. Int Rev Neurobiol 1999; 43:93-117.
2. Keshishian H, Broadie K, Chiba A et al. The Drosophila neuromuscular junction: A model system for studying synaptic development and function. Annu Rev Neurosci 1996; 19:545-575.
3. Hoang B, Chiba A. Single-cell analysis of Drosophila larval neuromuscular synapses. Dev Biol 2001; 229(1):55-70.
4. Rubin GM, Lewis EB. A brief history of Drosophila's contributions to genome research. Science 2000; 287(5461):2216-2218.
5. Brand AH, Perrimon N. Targeted gene expression as a means of altering cell fates and generating dominant phenotypes. Development 1993; 118(2):401-415.

6. Osterwalder T, Yoon KS, White BH et al. A conditional tissue-specific transgene expression system using inducible GAL4. Proc Natl Acad Sci USA 2001; 98(22):12596-12601.
7. Singh S, Wu CF. Ionic currents in larval muscles of Drosophila. Int Rev Neurobiol 1999; 43:191-220.
8. Ng M, Roorda RD, Lima SQ et al. Transmission of olfactory information between three populations of neurons in the antennal lobe of the fly. Neuron 2002; 36(3):463-474.
9. Fiala A, Spall T, Diegelmann S et al. Genetically expressed cameleon in Drosophila melanogaster is used to visualize olfactory information in projection neurons. Curr Biol 2002; 12(21):1877-1884.
10. Yu D, Baird GS, Tsien RY et al. Detection of calcium transients in Drosophila mushroom body neurons with camgaroo reporters. J Neurosci 2003; 23(1):64-72.
11. Halpern ME, Chiba A, Johansen J et al. Growth cone behavior underlying the development of stereotypic synaptic connections in Drosophila embryos. J Neurosci 1991; 11(10):3227-3238.
12. Sink H, Whitington PM. Location and connectivity of abdominal motoneurons in the embryo and larva of Drosophila melanogaster. J Neurobiol 1991; 22(3):298-311.
13. Landgraf M, Bossing T, Technau GM et al. The origin, location, and projections of the embryonic abdominal motorneurons of Drosophila. J Neurosci 1997; 17(24):9642-9655.
14. Atwood HL, Govind CK, Wu CF. Differential ultrastructure of synaptic terminals on ventral longitudinal abdominal muscles in Drosophila larvae. J Neurobiol 1993; 24(8):1008-1024.
15. Jia XX, Gorczyca M, Budnik V. Ultrastructure of neuromuscular junctions in Drosophila: Comparison of wild type and mutants with increased excitability. J Neurobiol 1993; 24(8):1025-1044.
16. Johansen J, Halpern ME, Keshishian H. Axonal guidance and the development of muscle fiber-specific innervation in Drosophila embryos. Journal of Neuroscience 1989; 9(12):4318-4332.
17. Gorczyca M, Augart C, Budnik V. Insulin-like receptor and insulin-like peptide are localized at neuromuscular junctions in Drosophila. J Neurosci 1993; 13(9):3692-3704.
18. Jan LY, Jan YN. L-glutamate as an excitatory transmitter at the Drosophila larval neuromuscular junction. J Physiol 1976; 262(1):215-236.
19. Johansen J, Halpern ME, Keshishian H. Axonal guidance and the development of muscle fiber-specific innervation in Drosophila embryos. J Neurosci 1989; 9(12):4318-4332.
20. Monastirioti M, Gorczyca M, Rapus J et al. Octopamine immunoreactivity in the fruit fly Drosophila melanogaster. J Comp Neurol 1995; 356(2):275-287.
21. Anderson MS, Halpern ME, Keshishian H. Identification of the neuropeptide transmitter proctolin in Drosophila larvae: Characterization of muscle fiber-specific neuromuscular endings. J Neurosci 1988; 8(1):242-255.
22. Zhong Y, Pena LA. A novel synaptic transmission mediated by a PACAP-like neuropeptide in Drosophila. Neuron 1995; 14(3):527-536.
23. Bossing T, Udolph G, Doe CQ et al. The embryonic central nervous system lineages of Drosophila melanogaster. I. Neuroblast lineages derived from the ventral half of the neuroectoderm. Dev Biol 1996; 179(1):41-64.
24. Schmid A, Chiba A, Doe CQ. Clonal analysis of Drosophila embryonic neuroblasts: neural cell types, axon projections and muscle targets. Development 1999; 126(21):4653-4689.
25. Landgraf M, Jeffrey V, Fujioka M et al. Embryonic origins of a motor system: Motor dendrites form a myotopic map in Drosophila. PLoS Biol 2003; 1(2):E41.
26. Burrows M. *The Neurobiology of an Insect Brain.* Oxford: Oxford University Press, 1996.
27. Bacon JP, Murphey RK. Receptive fields of cricket giant interneurones are related to their dendritic structure. J Physiol 1984; 352:601-623.
28. Sink H, Whitington PM. Early ablation of target muscles modulates the arborisation pattern of an identified embryonic Drosophila motor axon. Development 1991; 113(2):701-707.
29. Cash S, Chiba A, Keshishian H. Alternate neuromuscular target selection following the loss of single muscle fibers in Drosophila. J Neurosci 1992; 12(6):2051-2064.
30. O'Connor TP, Duerr JS, Bentley D. Pioneer growth cone steering decisions mediated by single filopodial contacts in situ. J Neurosci 1990; 10(12):3935-3946.
31. Thor S, Andersson SG, Tomlinson A et al. A LIM-homeodomain combinatorial code for motor-neuron pathway selection. Nature 1999; 397(6714):76-80.
32. Landgraf M, Roy S, Prokop A et al. Even-skipped determines the dorsal growth of motor axons in Drosophila. Neuron 1999; 22(1):43-52.
33. Landgraf M, Baylies M, Bate M. Muscle founder cells regulate defasciculation and targeting of motor axons in the Drosophila embryo. Curr Biol 1999; 9(11):589-592.
34. Vactor DV, Sink H, Fambrough D et al. Genes that control neuromuscular specificity in Drosophila. Cell 1993; 73(6):1137-1153.
35. Grenningloh G, Rehm EJ, Goodman CS. Genetic analysis of growth cone guidance in Drosophila: Fasciclin II functions as a neuronal recognition molecule. Cell 1991; 67(1):45-57.

36. Nose A, Mahajan VB, Goodman CS. Connectin: A homophilic cell adhesion molecule expressed on a subset of muscles and the motoneurons that innervate them in Drosophila. Cell 1992; 70(4):553-567.
37. Lin DM, Fetter RD, Kopczynski C et al. Genetic analysis of Fasciclin II in Drosophila: Defasciculation, refasciculation, and altered fasciculation. Neuron 1994; 13(5):1055-1069.
38. Lin DM, Goodman CS. Ectopic and increased expression of Fasciclin II alters motoneuron growth cone guidance. Neuron 1994; 13(3):507-523.
39. Yu HH, Araj HH, Ralls SA et al. The transmembrane semaphorin sema I is required in Drosophila for embryonic motor and CNS axon guidance. Neuron 1998; 20(2):207-220.
40. Winberg ML, Mitchell KJ, Goodman CS. Genetic analysis of the mechanisms controlling target selection: Complementary and combinatorial functions of netrins, semaphorins, and IgCAMs. Cell 1998; 93(4):581-591.
41. Yu HH, Huang AS, Kolodkin AL. Semaphorin-1a acts in concert with the cell adhesion molecules fasciclin II and connectin to regulate axon fasciculation in Drosophila. Genetics 2000; 156(2):723-731.
42. Winberg ML, Noordermeer JN, Tamagnone L et al. Plexin A is a neuronal semaphorin receptor that controls axon guidance. Cell 1998; 95(7):903-916.
43. Fambrough D, Goodman CS. The Drosophila beaten path gene encodes a novel secreted protein that regulates defasciculation at motor axon choice points. Cell 1996; 87(6):1049-1058.
44. Sink H, Rehm EJ, Richstone L et al. Sidestep encodes a target-derived attractant essential for motor axon guidance in Drosophila. Cell 2001; 105(1):57-67.
45. Krueger NX, Van Vactor D, Wan HI et al. The transmembrane tyrosine phosphatase DLAR controls motor axon guidance in Drosophila. Cell 1996; 84(4):611-622.
46. Desai CJ, Gindhart Jr JG, Goldstein LS et al. Receptor tyrosine phosphatases are required for motor axon guidance in the Drosophila embryo. Cell 1996; 84(4):599-609.
47. Schindelholz B, Knirr M, Warrior R et al. Regulation of CNS and motor axon guidance in Drosophila by the receptor tyrosine phosphatase DPTP52F. Development 2001; 128(21):4371-4382.
48. Sun Q, Schindelholz B, Knirr M et al. Complex genetic interactions among four receptor tyrosine phosphatases regulate axon guidance in Drosophila. Mol Cell Neurosci 2001; 17(2):274-291.
49. Desai CJ, Krueger NX, Saito H et al. Competition and cooperation among receptor tyrosine phosphatases control motoneuron growth cone guidance in Drosophila. Development 1997; 124(10):1941-1952.
50. Johnson KG, Van Vactor D. Receptor protein tyrosine phosphatases in nervous system development. Physiol Rev 2003; 83(1):1-24.
51. Wills Z, Bateman J, Korey CA et al. The tyrosine kinase Abl and its substrate enabled collaborate with the receptor phosphatase Dlar to control motor axon guidance. Neuron 1999; 22(2):301-312.
52. Lanier LM, Gertler FB. From Abl to actin: Abl tyrosine kinase and associated proteins in growth cone motility. Curr Opin Neurobiol 2000; 10(1):80-87.
53. Chiba A, Hing H, Cash S et al. Growth cone choices of Drosophila motoneurons in response to muscle fiber mismatch. J Neurosci 1993; 13(2):714-732.
54. Chiba A, Snow P, Keshishian H et al. Fasciclin III as a synaptic target recognition molecule in Drosophila. Nature 1995; 374(6518):166-168.
55. Shishido E, Takeichi M, Nose A. Drosophila synapse formation: Regulation by transmembrane protein with Leu-rich repeats, CAPRICIOUS. Science 1998; 280(5372):2118-2121.
56. Nose A, Umeda T, Takeichi M. Neuromuscular target recognition by a homophilic interaction of connectin cell adhesion molecules in Drosophila. Development 1997; 124(8):1433-1441.
57. Nose A, Takeichi M, Goodman CS. Ectopic expression of connectin reveals a repulsive function during growth cone guidance and synapse formation. Neuron 1994; 13(3):525-539.
58. Kose H, Rose D, Zhu X et al. Homophilic synaptic target recognition mediated by immunoglobulin-like cell adhesion molecule Fasciclin III. Development 1997; 124(20):4143-4152.
59. Mitchell KJ, Doyle JL, Serafini T et al. Genetic analysis of Netrin genes in Drosophila: Netrins guide CNS commissural axons and peripheral motor axons. Neuron 1996; 17(2):203-215.
60. Kolodziej PA, Timpe LC, Mitchell KJ et al. Frazzled encodes a Drosophila member of the DCC immunoglobulin subfamily and is required for CNS and motor axon guidance. Cell 1996; 87(2):197-204.
61. Halfon MS, Hashimoto C, Keshishian H. The Drosophila toll gene functions zygotically and is necessary for proper motoneuron and muscle development. Dev Biol 1995; 169(1):151-167.
62. Rose D, Zhu X, Kose H et al. Toll, a muscle cell surface molecule, locally inhibits synaptic initiation of the RP3 motoneuron growth cone in Drosophila. Development 1997; 124(8):1561-1571.
63. Keleman K, Rajagopalan S, Cleppien D et al. Comm sorts robo to control axon guidance at the Drosophila midline. Cell 2002; 110(4):415-427.

64. Wolf B, Seeger MA, Chiba A. Commissureless endocytosis is correlated with initiation of neuromuscular synaptogenesis. Development 1998; 125(19):3853-3863.
65. Kolodkin AL, Matthes DJ, Goodman CS. The semaphorin genes encode a family of transmembrane and secreted growth cone guidance molecules. Cell 1993; 75(7):1389-1399.
66. Murray MJ, Merritt DJ, Brand AH et al. In vivo dynamics of axon pathfinding in the Drosophilia CNS: A time-lapse study of an identified motorneuron. J Neurobiol 1998; 37(4):607-621.
67. Davis GW, Schuster CM, Goodman CS. Genetic analysis of the mechanisms controlling target selection: Target-derived Fasciclin II regulates the pattern of synapse formation. Neuron 1997; 19(3):561-573.
68. Rose D, Chiba A. A single growth cone is capable of integrating simultaneously presented and functionally distinct molecular cues during target recognition. J Neurosci 1999; 19(12):4899-4906.
69. Keshishian H, Chiba A. Neuromuscular development in Drosophila: Insights from single neurons and single genes. Trends Neurosci 1993; 16(7):278-283.
70. Broadie K, Bate M. Innervation directs receptor synthesis and localization in Drosophila embryo synaptogenesis. Nature 1993; 361(6410):350-353.
71. Yoshihara M, Rheuben MB, Kidokoro Y. Transition from growth cone to functional motor nerve terminal in Drosophila embryos. J Neurosci 1997; 17(21):8408-8426.
72. Keshishian H, Chiba A, Chang TN et al. Cellular mechanisms governing synaptic development in Drosophila melanogaster. J Neurobiol 1993; 24(6):757-787.
73. Murray MJ, Whitington PM. Effects of roundabout on growth cone dynamics, filopodial length, and growth cone morphology at the midline and throughout the neuropile. J Neurosci 1999; 19(18):7901-7912.
74. Suzuki E, Rose D, Chiba A. The ultrastructural interactions of identified pre and postsynaptic cells during synaptic target recognition in Drosophila embryos. J Neurobiol 2000; 42(4):448-459.
75. Ritzenthaler S, Suzuki E, Chiba A. Postsynaptic filopodia in muscle cells interact with innervating motoneuron axons. Nat Neurosci 2000; 3(10):1012-1017.
76. Broadie KS, Bate M. Development of the embryonic neuromuscular synapse of Drosophila melanogaster. J Neurosci 1993; 13(1):144-166.
77. Saitoe M, Tanaka S, Takata K et al. Neural activity affects distribution of glutamate receptors during neuromuscular junction formation in Drosophila embryos. Dev Biol 1997; 184(1):48-60.
78. Broadie K, Bate M. Activity-dependent development of the neuromuscular synapse during Drosophila embryogenesis. Neuron 1993; 11(4):607-619.
79. Anderson MJ, Cohen MW, Zorychta E. Effects of innervation on the distribution of acetylcholine receptors on cultured muscle cells. J Physiol 1977; 268(3):731-756.
80. Davey DF, Cohen MW. Localization of acetylcholine receptors and cholinesterase on nerve-contacted and noncontacted muscle cells grown in the presence of agents that block action potentials. J Neurosci 1986; 6(3):673-680.
81. Dahm LM, Landmesser LT. The regulation of synaptogenesis during normal development and following activity blockade. J Neurosci 1991; 11(1):238-255.
82. Kopczynski CC, Davis GW, Goodman CS. A neural tetraspanin, encoded by late bloomer, that facilitates synapse formation. Science 1996; 271(5257):1867-1870.
83. Fradkin LG, Kamphorst JT, DiAntonio A et al. Genomewide analysis of the Drosophila tetraspanins reveals a subset with similar function in the formation of the embryonic synapse. Proc Natl Acad Sci USA 2002; 99(21):13663-13668.
84. Sink H, Whitington PM. Pathfinding in the central nervous system and periphery by identified embryonic Drosophila motor axons. Development 1991; 112(1):307-316.
85. Broadie K, Sink H, Van Vactor D et al. From growth cone to synapse: The life history of the RP3 motor neuron. Dev Suppl 1993; 227-238.
86. Jarecki J, Keshishian H. Role of neural activity during synaptogenesis in Drosophila. J Neurosci 1995; 15(12):8177-8190.
87. Chang TN, Keshishian H. Laser ablation of Drosophila embryonic motoneurons causes ectopic innervation of target muscle fibers. J Neurosci 1996; 16(18):5715-5726.
88. White BH, Osterwalder TP, Yoon KS et al. Targeted attenuation of electrical activity in Drosophila using a genetically modified K(+) channel. Neuron 2001; 31(5):699-711.
89. Yoon KS, Wells DG, Keshishian H. Activity-dependent remodeling of Drosophila NMJ deopends on presynaptic as opposed to postsynaptic electrical excitability. Soc Neurosci Abstr 2003; 457-21.
90. Ming G, Henley J, Tessier-Lavigne M et al. Electrical activity modulates growth cone guidance by diffusible factors. Neuron 2001; 29(2):441-452.
91. Hu H, Marton TF, Goodman CS. Plexin B mediates axon guidance in Drosophila by simultaneously inhibiting active Rac and enhancing RhoA signaling. Neuron 2001; 32(1):39-51.
92. Matthes DJ, Sink H, Kolodkin AL et al. Semaphorin II can function as a selective inhibitor of specific synaptic arborizations. Cell 1995; 81(4):631-639.

Metamorphosis and the Formation of the Adult Musculature

Devkanya Dutta and K. VijayRaghavan*

Abstract

T he somatic musculature of the adult fly consists of muscles that are morphologically and functionally very distinct from each other. How are these diverse muscle types generated during pupal development? This chapter summarizes the insights that have been gained into the genetic and molecular mechanisms that control different steps in patterning, differentiation and diversification of adult muscles.

Introduction

The main derivatives of the mesoderm in the adult fly include the somatic muscles of the head, thorax and abdomen; the visceral muscles; fat body; heart; gonadal mesoderm and the haemolymph. Adult somatic muscles are formed by the fusion of myoblasts that are set aside in the embryo and proliferate during larval life. At the onset of metamorphosis most larval muscles begin to histolyse. At this time adult myoblasts withdraw from the cell-cycle and migrate to specific ectodermal locations. There they fuse to form syncytial muscle fibers. Integrated into these events of myogenesis are muscle-specific events that specify the identity of each muscle fiber, its innervation type, and attachment to tendon cells. The architecture of adult somatic muscles is different from that of the larval somatic muscles. An example of such difference is in fiber number. In the larva each somatic muscle is a single multinucleate fiber whereas in the adult fly (akin to the situation in many skeletal muscles of vertebrates) each muscle is a collection of fibers. Understanding how such diversification is attained during the adult phase of myogenesis poses an interesting problem. We describe here our current understanding of the mechanisms behind the patterning of the two well-studied groups of adult somatic muscles: the flight muscles in the dorsal thorax, and the abdominal muscles.

Architecture of Adult Flight Muscles and Abdominal Muscles

The flight muscles can be classified according to their functions into two groups. The direct flight muscles (DFMs) are located near the wing hinge (Fig. 1B,F), and contraction of the DFMs directly causes a change in the orientation of the wing. The indirect flight muscles (IFMs) are the much larger muscles and constitute the bulk of the thoracic musculature. IFMs are subdivided into two groups, the dorsal longitudinal muscles (DLMs), and the dorso-ventral muscles (DVMs). Each hemisegment has one bundle of DLMs containing six fibers (Fig. 1A,D), and three bundles of DVMs. The DVMs consist of DVM I (3 fibers), DVM II (2 fibers) and DVM III (2 fibers) (Fig. 1A,E).[1,2] The IFMs move the wings indirectly by deformation of the thoracic exoskeleton rather than by direct action on the wing.

*Corresponding Author: K. VijayRaghavan—National Centre for Biological Sciences, Tata Institute of Fundamental Research, Bangalore 560065, India. Email: vijay@ncbs.res.in

Muscle Development in Drosophila, edited by Helen Sink. ©2006 Eurekah.com and Springer Science+Business Media.

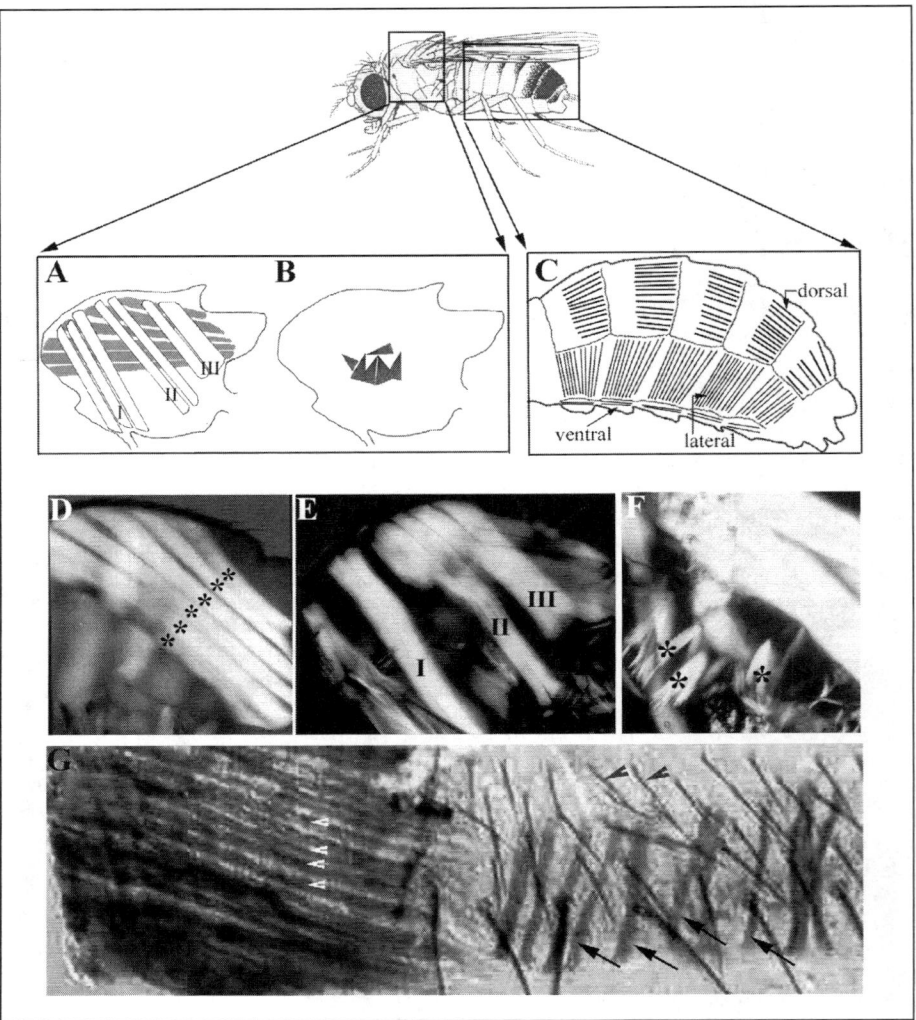

Figure 1. A-C) Diagram of the sagittal view of flight muscles and abdominal muscles in the hemisegment of an adult female fly. A) The indirect flight muscles (IFMs) are composed of a bundle of dorsal-longitudinal muscle (DLM) running antero-posteriorly and three bundles of dorso-ventral muscles (DVMs) oriented dorso-ventrally in the mesothorax. A DLM is comprised of six myotubes (grey) while the DVMs-I, II and III- consist of three, two and two myotubes (white) respectively. B) Direct flight muscles (DFMs) present near the wing hinge region. The prominent ones are shown in this schematic. C) The dorsal, lateral and ventral groups of abdominal muscles as present in the segments A2 to A6. The muscle patterns of A1 and the terminal segments are different from the rest and have been excluded. D-F) Hemithoracic preparations of adult flight muscles, observed using polarised optics. D) The six fibers of DLMs (marked by asterisks). E) The DVM bundles, labeled as I, II and III. F) Three DFM bundles (asterisks). G) Flat preparation of the abdominal A3 hemisegment of an adult fly showing the lateral muscles (a few of them are indicated by white arrowheads) and a subset of the dorsal muscle fibers (four of them indicated by black arrows). Muscles are visualized by the expression of *myosin heavy chain* (*mhc*)-*lacZ* transgene. Bristles present in the abdomen can be observed in the background (black arrowheads). In A-F, anterior is to the left and the dorsal, top. In G, the dorsal midline is to the right and anterior is to the top.

The DLMs contract the thorax and hence function as wing depressors while the DVMs expand the thorax, acting as wing elevators.

In the abdomen the muscles in each hemisegment are subdivided into three major clusters: dorsal, lateral and ventral muscles. Each cluster is comprised of a specific number of fibers (Fig. 1C,G).[3] In males an additional dorsal muscle, known as the male specific muscle or the "Muscle of Lawrence" (MOL), is present in the fifth abdominal segment. The muscles of the abdomen serve for functions that include movement, posture maintenance, copulation, and ovipositioning.

Developmental Profile of Adult Muscles

Preparation for adult myogenesis begins in the embryo where the adult myoblasts are specified in tandem with their embryonic counterparts (Fig. 2). After its inception the embryonic mesoderm is segmentally divided into domains expressing either high or low levels of the bHLH transcription factor Twist (Twi).[4,5] Inductive cues from the ectoderm in the form of Wingless, Decapentaplegic and the receptor tyrosine kinase-mediated signals demarcate the high Twi-expressing region into smaller clusters of cells that express Lethal of Scute.[6-8] From each smaller cluster of equivalent cells, by the influence of Notch-mediated lateral inhibition, a single cell adopts the fate of a "progenitor".[7,9,10] The progenitor cell divides asymmetrically to give rise to two daughter cells. One becomes the embryonic "founder cell" that serves as a "seed' for individual embryonic somatic muscles. The sibling can have either of two fates. It can either become another embryonic founder cell, or alternatively it can become a "precursor" for adult myoblasts[11]—a cell that divides to generate nests of myoblasts which will later form the adult muscles.

The Numb protein plays a crucial role in the acquisition of differential cell fate in the two daughter cells. Numb is asymmetrically localized in the progenitor cell and segregates asymmetrically as the progenitor divides. The daughter cell inheriting Numb down-regulates Notch signaling, loses expression of Twi, and functions to seed an embryonic somatic muscle. The cell that doesn't receive Numb, and consequently has active Notch signaling, maintains expression of *twi* and forms the adult muscle precursor.[11,12] Thus, by the end of embryogenesis, the adult muscle precursors are specified by a well-defined lineage (Fig. 2F).

In the embryonic abdomen each hemisegment has a definite pattern of adult precursor cells—a single cell present ventrally, two cells laterally and three cells dorsally.[13] Each cell remains closely linked with a specific branch of peripheral nerve. During larval life each cell proliferates to give rise to a clone of cells.[3] The abdomen of a third instar larva therefore, has six clusters of adult 8-15 myoblasts per hemisegment (a dorsal cluster is shown in Figure 2I. During early pupal stages proliferation continues and myoblasts spread out by migrating along the peripheral nerves. By 24hrs after puparium formation (APF) the larval abdominal muscles have histolysed completely, and adult myoblasts start fusing to form the muscle fibers (see schematic in Fig. 3).[3] Ablation studies have shown that cells of one cluster, that have arisen as a result of the division of one precursor, give rise to a specific set of muscles.[14] Hence ablation of a single precursor early during development results in the absence of a specific set of muscles. This suggests that the cells of a cluster are strictly channeled to form a particular muscle group and that these cannot be replaced by myoblasts from other clusters.

In the embryonic thorax adult precursors are present as a cluster of 6-7 cells in each hemisegment and remain tightly associated with the imaginal disc primordia (groups of cells, set aside in the embryonic and larval life, which give rise to adult epidermal structures like wings, legs etc during metamorphosis). From the beginning of the second larval instar these Twi-expressing adult precursor cells proliferate, and associate with the imaginal discs (Fig. 2H) and nerves.[2,3,13] With the onset of metamorphosis, the wing discs evaginate to release the myoblasts, which then migrate to specific epidermal positions and fuse to form muscle fibers (Fig. 3).[2] Clonal analysis studies show that the wing disc-associated adult myoblasts give rise to both DFMs and IFMs. In contrast myoblasts associated with the mesothoracic leg discs contribute to muscles of the second leg including the large tergal depressor of trochanter (or jump

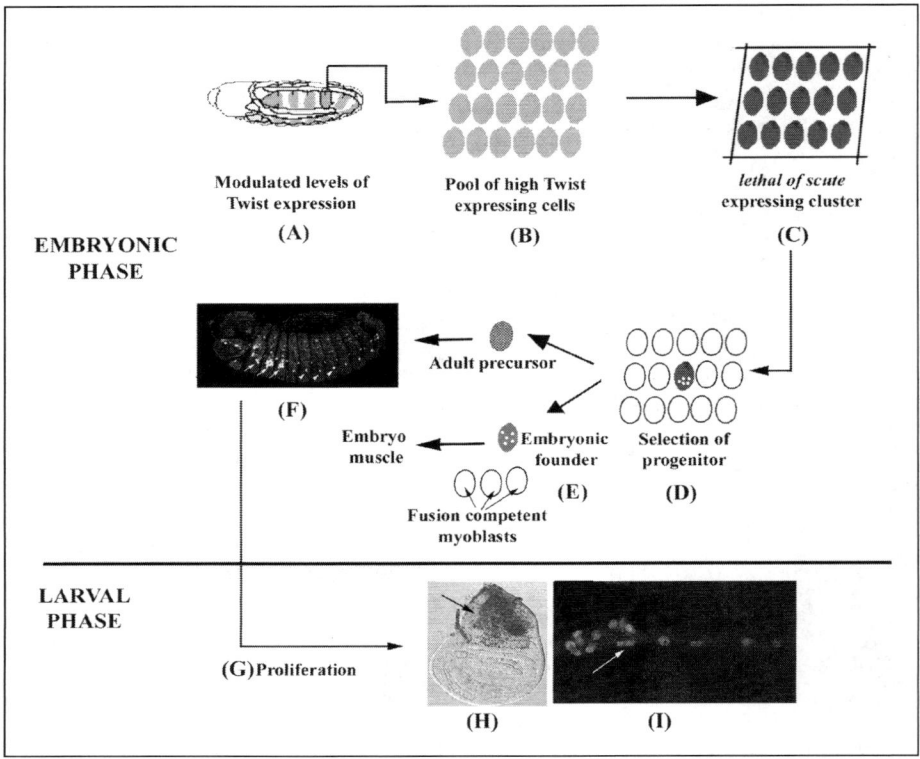

Figure 2. A simplified schematic illustrating the steps in the specification of adult muscle precursor in the embryo and generation of adult myoblast pool in the third instar larva. A) Cartoon of stage 10 embryo showing the alternate domains of high and low levels of Twist. Cells of high Twist expression (grey) will give rise to the somatic muscles and the adult muscle precursors. B) A group of high Twist expressing cells. C) Expression of the proneural gene, *lethal of scute* (*l'sc*) within the high Twist domain marks group of cells that have the potential to become muscle progenitor. D) Lateral inhibition, mediated by neurogenic genes, restricts *l'sc* expression to single cells that become the muscle progenitor (dark grey circle). The rest of the cells in the myogenic cluster behave as fusion-competent myoblasts (fcms) (white circles). Protein Numb (white dots) gets asymmetrically localized within the progenitor. E) Asymmetric division of the progenitor to form an embryonic muscle founder (light grey circle) and an adult precursor (hatched circle). The embryonic founders downregulate *twi*, fuse with the fcms and generate the embryonic somatic muscles. F) Twist expression pattern in a late embryo as revealed by staining with anti-Twist antibody. Patches of cells in the thoracic segments (indicated by white arrows) and single cells in the abdomen (three of them in adjacent segments indicated by white arrowheads) continue to express Twist and represent the adult precursor cells. G) During larval life adult precursors divide extensively to generate pools of myoblasts. H) *twi* expressing adult myoblasts (indicated by black arrow) on the presumptive notal region of the wing disc, assayed by X-Gal staining of *twi-lacZ* transgene. I) A cluster of *twi* expressing nerve-associated myoblasts (indicated by white arrow) in the abdomen of a third instar larva, visualized by anti-β-galactosidase staining in *twi-lacZ* larva. (K.G. Guruharsha provided Panel F).

muscle), another large muscle present in the mesothorax.[15] Indeed, myoblasts associated with each imaginal disc appear to contribute to muscles that are associated with the epidermal structure formed by that disc.[16] This suggests, at least in the head and thorax, a close association between developing muscles and the epidermis.

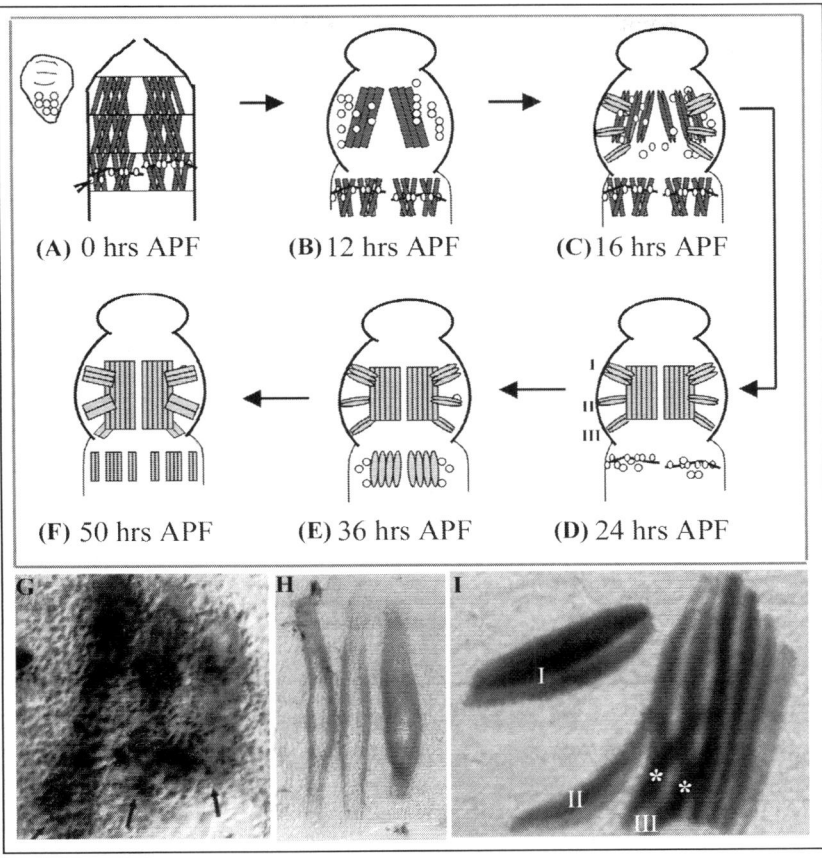

Figure 3. A-F) Schematic representation of the development of the IFMs and abdominal muscles during metamorphosis. A) Adult myoblasts (white circles) that will form the IFMs are attached to the wing discs during the larval life. The larval muscles are shown in dark grey. At the early stages of metamorphosis thoracic larval muscles begin to histolyse with the exception of three oblique muscles. B) By 12 hrs APF (after puparium formation), histolysis is complete and the three larval templates are clearly visible. By this stage myoblasts migrate into the muscle forming regions. The larval templates start splitting by 13 hrs APF. C) At 16 hrs APF splitting is in progress. At the same time DVMs (light grey fibers) form by the de novo fusion of myoblasts. The process of splitting is complete by 17-18 hrs APF. D) By 24 hrs APF, the DLMs and DVMs are complete. These muscles are shown in light grey. The three DVM bundles of one hemisegment are labeled as I, II and III. Muscle formation in the abdomen occurs later than in the thorax. The abdominal muscles develop from the adult myoblast pool (white circles, shown by black arrowhead in A) associated with the segmental and inter-segmental nerves (A). A-D) During early pupal stages (from 0-24 hrs APF), myoblasts proliferate and migrate out along the nerves. Histolysis of abdominal larval muscles begins at around 20 hrs APF. Fusion of the myoblasts begins by 28-30 hrs APF. E) 36 hrs APF. Fusion of the myoblasts and growth of the fibers are in progress. F) By 50 hrs APF, the pattern of adult muscles is largely complete. G-I) Development of IFMs, as revealed by the expression of *myosin heavy chain-lacZ* transgene (in G and H) and *actin(88)F-lacZ* transgene (in I). G) Un-histolysed larval templates (indicated by black arrows) at 10 hrs APF. H) A 16 hrs APF pupa undergoing the process of splitting of the larval templates. I) The DLM and the three DVMs (labeled I, II and III) in a 20 hrs APF pupa. Fibers of DVM III form below the DLM fibers and are indicated by white asterisks. In, G, H and I, dorsal midline is to the right and anterior is to the top. (Panel G is provided by J. Fernandes. I is reprinted, with permission, from: Roy S, VijayRaghavan K. BioEssays 1999; 21: 486-498).

Templates, Nerves, the Epidermis and Muscle Pattern

Muscle Templates

An interesting observation is that during metamorphosis three dorsal muscles of the larva in each mesothoracic hemi-segment are spared from histolysis (Fig. 3G). These three fibers function as "templates" for the future DLMs to which the incoming myoblasts fuse. Each template subsequently splits into two (Fig. 3H) to generate the six fibers that constitute the DLMs (Fig. 3I).[2] If templates are ablated, muscle fibers still form by the de novo fusion of myoblasts.[17] However, in this case an aberrant number of fibers form, showing that templates have an important role in controlling fiber number for the DLMs. The process of splitting of the larval templates itself seems to be initiated by interactions with the fusing myoblasts as depletion of the adult myoblast pool prevents splitting.[18] Perhaps signals from fusing myoblasts or the developing innervation provide cues which trigger components of the templates to initiate the process of splitting. These signals could involve the transcription factor Erect-wing as in an *erect-wing* allele the splitting of the larval templates does not take place.[18]

Nerves

Despite the histolysis of the larval muscles, most of the larval motoneurons persist through metamorphosis but are remodeled. They undergo extensive changes in morphology and target interactions to become motoneurons whose physiology and anatomy are specific for adult functions.[2,19-21] Nerves are utilized by myoblasts of both the abdomen and thorax as a substratum for attachment and migration. Nerves are also a possible source of trophic cues for myoblast proliferation, since denervation in the abdomen or thorax causes a distinct reduction in the size of myoblast pool.[22,23]

The formation of DVMs is strongly influenced by innervation.[23] Upon ablation of the DVM-innervating motoneuron the initial steps of myoblast segregation and fusion proceed normally, but the DVMs fail to form. This suggests that neural cues are required for some later aspects of DVM myogenesis. In contrast, for the abdominal muscles and DLMs the nerves do not appear to influence any fundamental aspect of patterning.[22,23] Denervation leads to a distinct reduction in the size of the muscles—which could be either because of the indirect effect of the depletion of the myoblast pool or a more direct effect on muscle growth—but the final plan of the muscle remains unchanged. Intriguingly, in the absence of larval templates the DLMs show a complete dependence on innervation for their development.[23] This compels one to propose that template-dependent formation is probably a superimposition on a common mode of de novo development.

The most dramatic effect of neuronal influence is observed for the special set of "Muscle of Lawrence" (MOL), present in the fifth abdominal segment (A5) of adult males. Innervation is absolutely required for the formation of MOL and denervating A5 leads to total failure of MOL formation.[22] More interestingly, the sex-specific development of MOL is governed uniquely by the "sexual" identity of the motoneuron and not by the epidermis or mesoderm. The MOL is formed in gynandromorphs which have a "male" neuronal genotype.[24] Also, mutations in the *fruitless (fru)* locus, that codes for a BTB domain containing transcription factor, affects MOL development.[25-27] A specific splice variant of the Fru protein is expressed in a subset of neurons in the adult male, and not in the female, and is thought to impart "maleness" to the neurons.[28] Ectopic expression of Fru in female neurons is sufficient to ectopically generate MOL in females. How Fru mediates MOL formation is still unclear. The MOL is relatively larger than the neighboring body wall muscles, and it has been proposed that Fru is involved in recruiting extra myoblasts to the developing MOL.[26]

Epidermis

The epidermis serves as another major source of inductive cues. In the thorax the distribution pattern and number of myoblasts associated with each imaginal disc is unique. For instance the myoblasts associated with the wing discs far exceed in number those associated with the haltere disc. This implies that the adult precursor myoblasts of T2 undergo greater number of divisions than the precursors of T3. Instruction for such segment specific prolif-eration and distribution comes from the epidermis[29] (described later in this review). Even within the wing disc, there probably are cues that restrict the localization of the myoblasts to the presumptive notum. The wing blade is transformed into a notum in *wingless* mutants and in this situation myoblasts are observed over the ectopic presumptive notum.[30] The epidermis also plays an important role in determining the migration pattern of the myo-blasts. When thoracic myoblasts adhered to their epidermal substratum (wing disc or leg disc) are transplanted into the abdomen, they do not migrate over to the abdominal epider-mis but instead remain associated with the thoracic epidermis. Yet if the transplantation site is thoracic, the myoblasts migrate over much longer distances to fuse with the host myoblasts.[31]

The molecular nature of the epidermal cues that govern myoblast proliferation or migration remains to be identified. Also not known is whether such cues emanate from specific regions of the epidermis. Certainly one defined subset of epidermal cells that is a strong candidate for governing muscle patterning is the set of muscle attachment cells or the tendon cells.[32] These cells develop closely juxtaposed to the developing muscles and, by analogy with their myogenic role in embryo,[33] might provide cues for the development of adult muscle fibers. Improper development of the tendon cells of the IFMs results in muscle defects that include improper splitting of the DLM templates and absence of DVM fibers.[34,35]

Vestigial, Cut and Apterous Impart Diversity within Thoracic Myoblasts and Canalize Them into Different Muscle Forming Pathways

Twist-expressing myoblasts associated with the wing disc give rise to both IFMs and DFMs in the mesothorax. For a long time these myoblasts were thought to be equivalent, and the mechanism by which they became allotted to different muscle forming programs was unclear. A study by Sudarsan et al. showed that these myoblasts are heterogeneous with respect to expression of two transcription factors Vestigial (Vg) and Cut (Ct) which divide this popula-tion into two distinct groups.[36] A larger proximal subset expresses Vg and low levels of Ct, and a smaller distal subset expresses high levels of Ct and no Vg (Fig. 4A-C). The high Ct-expressing myoblasts form the DFMs while Vg (and low Ct)-expressing myoblasts contribute to the for-mation of the IFMs. Higher level of Ct represses the expression of Vg in the myoblasts and prevents them from routing to the IFM pathway. On the other hand, in the IFM-forming myoblast subset Vg maintains low levels of Ct expression. Thus these two proteins seem to determine the differential fates of the myoblasts during pupal myogenesis.

In the pupal stage the DFM-forming myoblasts also express Apterous (Ap), a LIM homeodomain transcription factor (Fig. 4D-G).[37] Ap in turn is also involved in specifying the formation of the DFMs in the pupa. Expression of *ap* is repressed by Vg in the IFM-forming myoblasts.[38]

The Vg-Ct diversity originates in the embryo, where Vg starts expressing in a few cells of the adult myoblast precursor cluster of T2. It is further consolidated in the third instar larva when Ct expression is initiated.[36] The mechanisms that turn on the expressions of Vg in embryonic adult precursors and of Ct in adult myoblasts at the larval stage are not known. However, the maintenance of this diversity of myoblast groups is accomplished by Wingless (Wg) which is secreted from a thin strip of cells in the presumptive notum of the third instar larva.[39] The reach of Wg protein is restricted to the IFM forming subgroup of myoblasts. In these cells Wg maintains the expression of Vg, and perhaps via an indirect manner represses Ct.[36] Wg also functions in proliferation of the Vg-expressing cells. A reduction of Wg signal-ing results in a decrease in number of myoblasts in the region, in addition to a decrease in levels of Vg in the cells.[36]

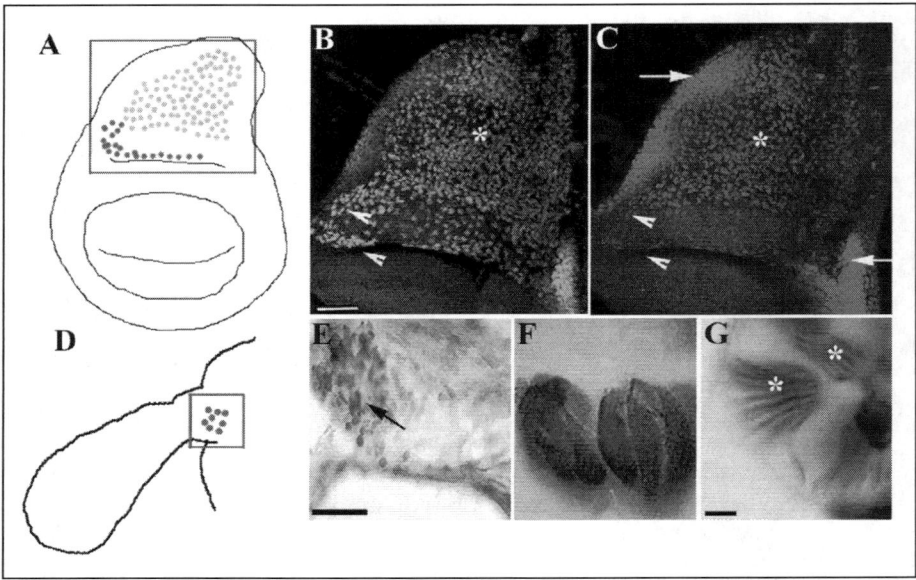

Figure 4. Expression of Vestigial (Vg), Cut (Ct) and Apterous (Ap) in adult myoblasts. A) Diagram of the wing disc to show the myoblasts overlying the epithelium of the notum. Yellow dots represent the myoblasts expressing Vg and low levels of Ct while green dots represent high Cut-expressing myoblasts. The blue box shows the approximate region shown in B and C. B,C) Confocal images of the notum of a wild-type third instar wing disc (area outlined in A) stained for Ct (green) (B) and Vg (red) (C). Proximal cells (asterisk) express both Vg and Ct (in lower levels). Distal cells (white arrowheads) lack Vg but express high levels of Ct. White arrows in C show Vg staining in the epidermal cells of the wing disc. D) Diagram of everting wing disc in the early metamorphic phase to show the approximate region where the DFM forming myoblasts (brown) migrate and subsequently fuse to form the DFMs. The region outlined by blue square is shown in E. E) Clusters of Ap expressing myoblasts (arrow) at 14 hrs APF, as revealed by anti-β-galactosidase staining in pupae containing the *ap*(muscle specific)–*lacZ* transgene. F) A completely formed DFM cluster at 36 hrs APF. G) Two bundles of DFMs (asterisks) in the adult fly, expressing *ap-lacZ*. In E-G, anterior is to the top and dorsal midline is to the right. Bars = 30 μm for B and C; 10 μm for E and F; 10 μm for G. (E-G are reprinted, with permission, from: Ghazi A, Anant S, VijayRaghavan K. Development 2000; 127:5309-5318.) A color version of this figure is available online at http://www.Eurekah.com.

The ventral population of myoblasts associated with the leg discs also exhibit similar molecular diversity. Most of the leg disc myoblasts are distributed as concentric rings on the disc, with a small population present as a cluster near the stalk. Vg (and low Ct) is expressed in the stalk myoblasts while the larger subset of myoblasts expresses Ct (and no Vg) (J. Nair and K.VijayRaghavan, unpublished data). One can hypothesize that such grouping reflects the division of the myoblasts into two groups: those that form the proximal leg muscles (the Vg-expressing myoblasts) and those that form the distal leg muscles (Ct-expressing cells). A role for ectodermal cues in specifying the molecular diversity of these myoblasts remains to be seen, but preliminary results suggest that Wg plays an important role here too (J. Nair and K.VijayRaghavan, unpublished data).

Together the results show that muscle identity in an adult fly is specified within a population of cells. This is in contrast to the situation in the embryo where muscle identity is specified by the individual founder cells. Yet the molecules employed remain conserved in both systems as Vg and Ap also specify subsets of muscles in the embryo.[10,40] The results also show that there are diverse myoblast types. Interestingly this feature is common in both *Drosophila* and vertebrate myogenesis.

Adult Founder Cells, Identified by *Dumbfounded*, Regulate Muscle Fiber-Number and Muscle Position

Somatic muscles in the adult fly are multi-fibered units. Each muscle consists of a bundle containing a specific number of myofibers. In this system, therefore, in addition to specifying the molecular property of each muscle, the correct number of myotubes needs to be set as well. This is achieved by a process that involves recruitment of special myoblasts for each muscle, whose number matches exactly the number of fibers formed for that muscle (Dutta et al., in press). Each of these special myoblasts "seeds" an individual myofiber, thereby generating the correct number of myotubes. These cells express *dumbfounded* (*duf*), a gene first identified in embryonic founder cells,[41] and are traceable using the *duf* enhancer trap line.

Prior to the formation of the DVMs, three, two and two high *duf-lacZ* expressing cells, corresponding to DVMs I, II and III respectively, are selected exactly at locations where the muscles will form (Fig. 5B). As each syncytium develops, the original nucleus continues to express high levels of *duf-lacZ* while low *duf-lacZ* levels are observed in the incoming nuclei. Similarly in the abdomen, a single *duf-lacZ* cell precedes every dorsal and lateral muscle fiber (Fig. 5C-F). These cells resemble the founder cells of the embryonic somatic muscles in terms of their spatial pattern of emergence (one per fiber) and expression of the gene *duf*. Hence they are termed "adult founder cells". Such myoblasts have been identified by anatomical criteria on the imaginal discs,[74] but *duf-lacZ* expression now shows a molecular expression pattern consistent with the cell being an adult founder cell.

That *duf-lacZ* positive cells in the adult (also seen in DFM development[67]) indeed have founder cell-like potential to seed fibers is confirmed in situations where fusion has been compromised (see Fig. 5G-J). Overexpression of a dominant-negative form of Rac1, a GTPase required for normal myoblast fusion in the embryo,[42] severely disrupts the process of fusion in the adult myoblasts. In such pupae the singular *duf-lacZ* positive cells differentiate to form mononucleate fibers that attach to correct epidermal locations and express normal differentiation markers. These results demonstrate that adult myoblasts are divided into two groups at the onset of differentiation: (i) the *duf-lacZ* positive adult founder cells that set the number and position of fibers to be formed for each muscle and (ii) the *duf-lacZ* negative fusion-competent myoblasts that fuse with the founders to develop the syncytial fibers.

The generation of a precise pattern of founder cells is a crucial determinant of muscle organization. Studies reveal that, unlike in the embryo, the lateral inhibition mechanism mediated by Notch is not involved in adult founder choice.[74] How are these founders selected? Do they have any additional roles? Future studies aimed at addressing these issues would further our understanding on molecular mechanisms of patterning.

Antagonistic Activities of Notch and D-Mef2 Regulate Myoblast Proliferation versus Differentiation

Adult myoblast precursors singled out during embryonic myogenesis maintain high levels of Twist.[13] Sustained Twi expression in the adult precursor is due to higher levels of Notch signaling.[11] Twi expression persists in the myoblasts as they divide in the larval stages and subsequently migrate to the muscle-forming locations in the early pupa.[2,3] Conditional removal of Notch in the larval stages results in premature decline in Twi levels, showing that Notch sustains Twi expression during myoblast proliferation.[45] Just prior to their fusion, the myoblasts down-regulate *twi* and no Twi immunoreactivity is observed in the developing myofibers.[2,45] Forced maintenance of Twi expression in the myoblasts by overexpressing Notch[intra] (an activated form of Notch) or Twi inhibits proper differentiation of the IFMs. This manifests as an absence of the expression of contractile proteins such as myosin and actin, and ultimately muscle degeneration.[45] The results confirm that Notch, via Twi, maintains the myoblasts in an undifferentiated state, and that a prerequisite for myogenesis to proceed is down-regulation of Notch signaling, and consequently Twi levels. In this context Twi behaves like its vertebrate counterpart where it is known to repress differentiation.[46,47]

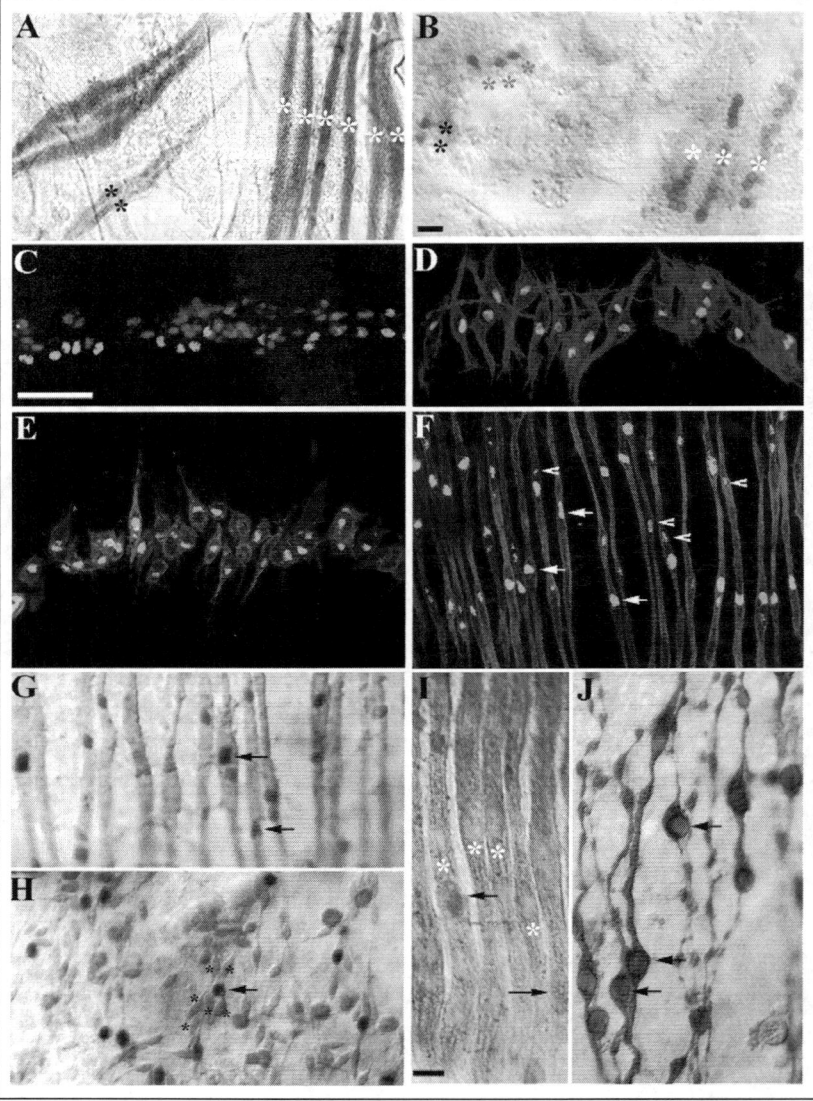

Figure 5. Presence of founder cells during adult myogenesis. A) Relative positions of DVM I (red asterisks), DVM II (black asterisks) and the DLM fibers (white asterisks), as visualized in a 24hrs APF *actin-lacZ* pupa. B) Founder cells preceding the formation of the DVMs. Three DVM I founders (red asterisks) and two DVM II founders (black asterisks), as revealed by *duf-lacZ* expression in a 12 hrs APF pupa. The nuclei of the three templates of DLMs also express *duf-lacZ* (white asterisks). The *duf-lacZ* line used for tracking founder cells has a P element—nuclear localizing *lacZ* inserted within the promoter region of *duf*,[41] hence the nuclear expression in the larval templates. C) Presence of founder cells for the dorsal muscles of the abdomen. A 28 hrs APF pupa double labeled for *duf-lacZ* (green) and Twist (red), showing the subset of myoblasts that have been specified as dorsal founders. D) Founder cells in the abdomen can also be visualized by staining with the monoclonal antibody 22C10. A 28 hrs APF *duf-lacZ* pupa (similarly aged as in C), double-labeled with anti-β-galactosidase (green) and 22C10 (red), reveals the dorsal founders. 22C10 beautifully accentuates the shape of the founder cells. (This figure legend is continued on the next page.)

Figure 5, continued. E) Founder cells for the lateral muscles in a 28 hrs APF *duf-lacZ* pupa. These cells, like in D, are double-labeled with anti-β-galactosidase (green) and 22C10 (red). F) The developing multi-nucleate lateral myotubes at 36 hrs APF, labeled for *duf-lacZ* (green) and 22C10 (red). Each fiber has one prominent *duf-lacZ* expressing nucleus (white arrows), presumably belonging to the single cell that preceded the multinucleate fiber. The remaining nuclei are recruited to express *duf-lacZ* at a lower level (a few of them shown by white arrowheads). G-J) Effects of dominant negative Rac1 (dnRac1) expression in adult myoblasts. G) Wild-type lateral muscle fibers expressing myosin (brown). Each fiber has a brightly stained *duf-lacZ* nucleus (black), indicated by arrows (for some fibers, these nuclei are not present within the field of view). H) *1151GAL4/UAS-Rac1N17* misexpression pupa showing the absence of syncytial fibers. The putative founder cells (black, one such cell indicated by black arrow) are correctly specified and express *duf-lacZ* and myosin. Unfused myoblasts (asterisks) also express myosin and in some cases can be seen clustered around *duf-lacZ* expressing founders. (*1151* is a GAL4-driver expressed in all adult myoblasts.) I,J) Magnified view of X-Gal (blue) and anti-MHC (brown) stained *duf-lacZ* pupa at 42 hrs APF. I) Wild-type lateral muscles showing one *duf-lacZ* expressing nucleus (black arrows) in each fiber. This nucleus corresponds to the high *duf-lacZ* expressing founder nucleus observed in F or G. White asterisks mark other nuclei within the syncytium. These nuclei cannot be detected by X-Gal staining because of lower level of β-galactosidase activity in them. J) *1151GAL4/UAS-Rac1N17* misexpression pupa. In the absence of normal fusion, thin mononucleate lateral fibers span the region. In A-D, dorsal midline is to the right and anterior is to the top. In E-J, dorsal midline is to the top and anterior is to the left. Bars = 18 μm in B; 50 μm in C-H; 20 μm in I and J. A color version of this figure is available online at http://www.Eurekah.com.

Surprisingly, adult myoblasts also express D-Mef2 (*Drosophila* Myocyte Enhancer Factor 2), a MADS box transcription factor that promotes muscle differentiation in the embryo.[48-51] Expression of D-Mef2 is initiated in the late third instar larval stage, and its activation is directly dependent on Twi.[52-53] Overexpression of D-Mef2 in wing disc associated myoblasts causes premature differentiation of these cells into myosin-expressing muscles on the wing disc, corroborating the myogenic role of D-Mef2 in adult (S. Roy and K.VijayRaghavan, unpublished data). Under normal conditions the function of D-Mef2 needs to be blocked, and this occurs through Notch signaling (S.Roy and K.VijayRaghavan, unpublished data). Simultaneous provision of high levels of Notch can suppress the premature myogenic activity of D-Mef2 overexpression. Notch, therefore, maintains the undifferentiated state of the myoblasts by keeping *twi* expression on, and also by blocking the function of D-Mef2. Higher levels of D-Mef2, provided by overexpressing the protein, probably override the function of Notch resulting in premature muscle differentiation.

Taken together, the results demonstrate a delicate control of myogenesis that involves a fine balance of stimulatory versus antagonistic activities of Notch, Twi and D-Mef2 in adult myoblasts. This ensures the accumulation of appropriate kinds of information in these cells such that they can execute their differentiation program at the right time (positive regulation), concomitantly ensuring that they are maintained in a proliferative and undifferentiated state, and that differentiation is not initiated inappropriately (negative regulation). But it is imperative to point out that these roles of Twi and Notch do not seem to hold true for the DFMs. DFMs remain normal in loss-of-function mutants of *twi* and *Notch* and in gain-of-functions situation where these genes are overexpressed in all imaginal myoblasts.[45] This demonstrates that DFMs, though closely related, might employ different regulatory proteins for their differentiation.

Homeotic Identity Control Muscle Organization in Adult Thoracic Segments

Each segment of the fly possesses a unique pattern of muscles. This segment specific diversification of muscles is controlled by the homeotic selector genes (or Hox genes) that are expressed along the anterior-posterior axis of the body.[54,55] Formation of muscles is not entirely governed by mesoderm autonomous properties and is influenced greatly by instructive cues from the neighboring tissues,[24,29,56,57,72] especially the ectoderm. Hence the Hox identity of the ectoderm is also of importance in determining muscle pattern. In *Drosophila* Hox expression domains in the mesoderm are expressed more posterior compared to the overlying ectoderm, so that in a given segment the ectoderm and mesoderm have different homeotic addresses.[1] Hence mutations in a Hox gene that alter the identity of the ectoderm need not necessarily change the identity of the mesoderm in the same segment and vice versa.

The role of homeotic genes in different stages of myogenesis has been studied in some detail for the adult flight muscles. Homeotic genes in the mesoderm determine early segregation of the adult precursor myoblasts in the embryo autonomously.[58] The pattern of the *twi*-expressing adult muscle precursor cells characteristic of the thorax can be converted to that seen in the abdomen by misexpressing the abdominal homeotic gene *abdominal-A* (*abd-A*) in the thoracic mesoderm. The wing disc associated adult myoblasts that will give rise to the flight muscles do not express any homeotic genes.[59] On the other hand, the haltere disc associated myoblasts of the adjacent segment express the gene *Antennapedia* (*Antp*).[59] The fusion property of these myoblasts is not restricted by their homeotic identity. When transplanted into the abdomen they can fuse with the forming myotubes of the abdomen and express abdominal specific differentiation genes.[60,61]

Proliferation and migration of the wing disc associated myoblasts are controlled entirely by the homeotic identity of the ectoderm. This is concluded from studies using the "four-winged" fly where an ectopic T2 epidermis is generated by homeotic transformation. A combination of three regulatory mutations in the homeotic gene *Ultrabithorax* (*Ubx*) results in the transformation of the T3 ectoderm to a T2 fate generating the "four-winged" phenotype. The adult myoblasts of homeotically transformed T3 (or HT3) however continue expressing *Antp*, thus maintaining their "T3-ness".[59] Despite *Antp* expression, the proliferation and migration of these HT3 myoblasts is identical to that of wild-type T2 myoblasts (Fig. 6E-G), suggesting that these events are governed entirely by the ectoderm.[29] However in the HT3 segment genesis of IFMs do not proceed to completion and no IFM fibers are observed.[29] Interestingly, and importantly, Rivlin at al.[57] conclude that IFMs are observed in the HT3. Such apparently contradictory results need to be reconciled, and perhaps the use of founder cell and IFM markers can resolve the reasons for the differences observed.

Our observations suggest that some aspect of IFM myogenesis require the correct homeotic identity of the mesoderm, and that is aborted due to the presence of *Antp* in HT3 myoblasts. Founders for DVMs or larval templates (founder like anlagens of DLMs) are not observed in HT3 of "triple mutant" pupae (D. Dutta, E.B. Lewis and K. VijayRaghavan, unpublished data). Yet when *Antp* hypomorphic mutations are introduced into the triple mutant background Antp levels in HT3 myoblasts are reduced and, more significantly, ectopic larval templates and DVM-like founders are observed in some of these pupae. In addition, adults of this genotype have ectopic IFM fibers in HT3 (D.Dutta, E. B. Lewis and K. VijayRaghavan, unpublished data). These results suggest that the selection of founder cells is dependent on the Hox identity of the myoblasts.

In contrast to the IFMs, formation of the DFMs seems to be shaped entirely by ectoderm since DFMs are observed in HT3 of triple mutant flies.[29] Consistent with this hypothesis is the observation that D-Wnt2, a glycoprotein related to Wingless, secreted from the wing disc is required for the patterning of DFMs in the pupal stage.[62] *D-Wnt2* is expressed strongly in the wing hinge region of the wing disc, in close vicinity to the DFM forming myoblasts. In mutants of *D-Wnt2* the pattern of a subset of DFMs is affected.

In addition to specifying muscle pattern in a segment specific manner along the anterior-posterior axis, Hox genes also seem to specify dorsal versus ventral muscle diversification in the thoracic segments of the adult. The leg disc associated myoblasts, which will give rise to the ventral set of muscle fibers in the three thoracic segments, express the homeotic gene *Sex combs reduced* (*Scr*) (S. Roy and K.VijayRaghavan, unpublished data; also see ref. 63). This dorsal-ventral division of homeotic identity is not observed in the embryo and appears to be a specialty of the adult myogenic program.

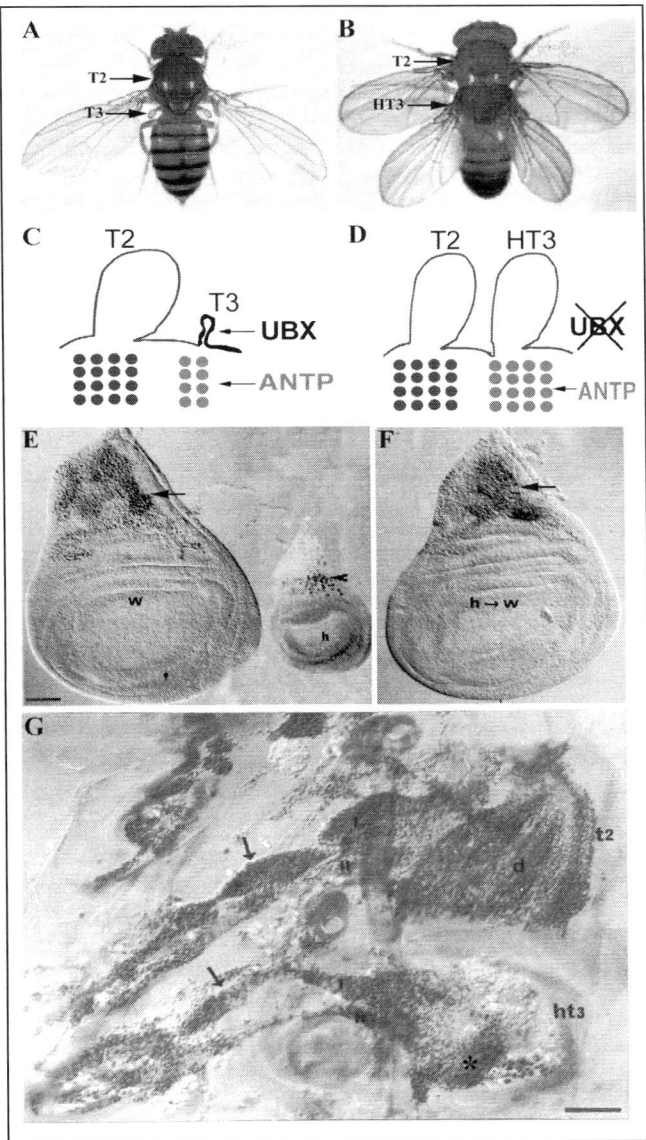

Figure 6. Homeotic control during adult thoracic myogenesis. A) Picture of a wild-type fly. T2 and T3 segments are indicated by arrows. B) Picture of a "four-winged" fly showing the normal T2 and the homeotically transformed T3 (HT3). C,D) Diagrams showing the expression of the homeotic genes in the thorax of the wild-type (C) and the triple mutant "four-winged" fly (D). C) Wild-type T2 mesoderm (black circles) does not express any homeotic gene. Wild-type T3 ectoderm expresses Ubx (thicker curve; indicated by black arrow) while T3 mesoderm expresses Antennapedia (Antp) (grey circles). D) In the "four-winged" fly, generated by three regulatory mutations in *Ubx*, the T3 ectoderm is transformed to a T2 fate. But the mesoderm of this transformed segment continues to express Antp (grey circles). (This figure legend is continued on the next page.)

Figure 6, continued. E-G) Proliferation and migration of HT3 myoblasts are transformed to a T2 pattern. E) Wing disc (w) and haltere disc (h) of wild-type third instar larva, stained with anti-Twist antibody to reveal the myoblasts associated with them. The number and spatial organization of the myoblasts on the wing disc (arrow) is different from that in the haltere disc (arrowhead). F) The haltere disc of the "four-winged" fly larva has transformed (h→w) and resembles the wing disc. The number and distribution pattern of the myoblasts (arrow) associated with this transformed disc is identical to that observed for T2 wing disc. G) Myoblast migration pattern in "four-winged" mutant pupa at 16 hrs, observed using anti-Twist antibody. Segregation of myoblasts forming the DVMs (I, II) and the jump muscles (arrow) are identical to that observed in T2. Considerable population of myoblasts have also migrated to the dorsal region of HT3 (black asterisk). In this figure, anterior is to the top and dorsal midline is to the right. Bars = 70 μm for E, F; 50 μm for G. (Panels E, F are reprinted with permission from: Fernandes J, Celniker SE, Lewis EB et al. Curr Biol 1994; 4: 957-964. Panel G is provided by J. Fernandes. Picture of the "four-winged" fly is kindly provided by E.B. Lewis). A color version of this figure is available online at http://www.Eurekah.com.

Stripe-Expressing Muscle Attachment Sites—Their Specification and Differentiation

Drosophila muscles are anchored to the epidermis via specialized cells called "tendon cells". These are analogous in function to the tendon cells of the vertebrates. *Drosophila* embryonic and adult tendon cells are characterized by the expression of *stripe* (*sr*), a gene encoding a zinc finger protein that is a member of the vertebrate Early Growth Response family of transcription factors.[32,64] Tendon cells for flight muscles first appear as five discrete *sr*-expressing domains in the presumptive notal region of the third instar larval wing disc (Fig. 7).[32] Expression of *sr* and positioning of each *sr*-expressing domain seems to be under the combinatorial action of different patterning genes and signaling molecules. Mutations in *ap* and *Notch* completely abolish *sr* expression, showing that initial expression of *sr* in all the domains is regulated by Notch and Ap.[37,65] Further demarcation into domains is achieved by the concerted activity of Wg and the transcription factor Pannier and its antagonist U-shaped (Ush) expressed in the notal region.[65] Though the molecular mechanism by which these genes function remains to be ascertained, the levels of each of these proteins play crucial roles in generating the myoblast clusters at their characteristic locations on the disc.[65]

In the metamorphosing pupa, *sr*-expressing domains are positioned in specific regions of the dorsal thorax and prefigure the future attachment sites. Cellular projections given out from the tendon cells come in contact with the growing end of the myofibers to form the myotendinous junction[32] Following muscle-epidermal contact components of the extracellular matrix of the myotendinous junction, such as the PS (position specific) integrins (*Drosophila* homologues of vertebrate integrins), are expressed.[32] A mature myotendinous junction of a DVM fiber typically comprises cellular projections from several tendon cells interdigitated with the myotube membrane.[66]

Proper differentiation of the myotendinous junctions requires the function of the *Broad complex* (*BR-C*), a primary response gene of the ecdysone cascade, in the tendon cells.[66,67] DVM attachments are disrupted in *BR-C* mutants belonging to the *reduced bristles on palps* (*rbp*) complementation group and the phenotype can be rescued by the BRC-Z1 isoform.[67,68] An ultrastructural study of *rbp* mutants revealed defective and delayed differentiation of tendon cells in the dorsal attachment sites.[66] These results lead to the conclusion that *BRC-Z1*, induced by the ecdysone cascade, regulates the expression of proteins that are involved in proper maturation of tendon cells.

In the embryo tendon cells are not merely insertion points for muscles. They play an active role in muscle patterning by providing essential attractive cues that guide muscle extension.[69,70] It is possible that adult tendon cells have similar muscle patterning properties. The early onset of *sr* on the larval wing disc is suggestive of its involvement in instructing the myoblasts, to which *sr*-expressing cells are tightly apposed on the disc, or to the developing myofibers to which they are closely juxtaposed in early pupa. Another conceivable function

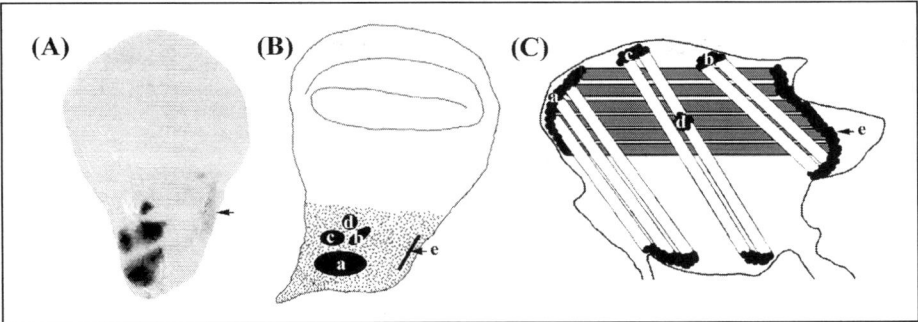

Figure 7. Muscle attachment sites. A) The five *stripe* (*sr*)- expressing domains on the presumptive notal region of the third instar wing disc, as revealed by β-Galactosidase staining in an enhancer trap *sr-lacZ* line. The arrow indicates a posterior- located band of cells having a fainter expression of *sr*. B,C) Schematic representation of the *sr* domains (black) in the wing disc and their fate map in the adult thorax. B) Four domains are present in the anterior compartment of the wing disc. One of these, 'a', is present in the medial region while the remaining, 'b-d', are present in the lateral region. The fifth domain, 'e' (indicated by arrow) is present as a narrow band in the posterior side. The dots indicate the myoblasts that are associated with the presumptive notum. C) Indirect flight muscles (DLMs: grey, DVMs; white) attached to the thoracic exoskeleton by *sr*-expressing tendon cells (black). The anterior attachment sites for DLMs are formed from 'a' while the posterior attachment cells are formed from 'e'. The dorsal attachment sites for DVMs I, II and III are formed by 'a', 'c' and 'b' respectively. 'd' forms the tendon cells for most of the DFMs (not labeled in the schematic). The ventral attachment sites for the DVMs comes from the *sr*-expressing cells of the ventral leg discs (not shown). In all panels, anterior is to the left; in C dorsal is to top. (Panel A provided by A. Ghazi. Panels B, C are reprinted, with permission, from: Ghazi A, Paul L, VijayRaghavan K. Mech Dev 2003; 120: 519-528).

of the different *sr*-expressing domains may be to attract a specific set of muscles, thereby ensuring the correct matching of the fibers to the correct cluster. Whether such interactions indeed exist remains to be verified.

Conclusions

The process of myogenesis is complex, and determining the factors that are responsible for establishing the adult muscle pattern is ongoing. From what we know it is becoming clear that, barring a few exceptions, adult myogenic program employs the same molecules that are utilized during embryonic myogenesis. However it adopts a number of innovative developmental designs and integrates the functions of these molecules within these designs to generate the final, uniquely adult muscle pattern. Major tasks in the future lie in unearthing molecular players to explain the various cellular events observed. But the more challenging task will be to identify the manner in which these molecules are molded into the adult program. Myogenesis in adult flies is similar to that of vertebrates in many respects.[71] Therefore the myogenic program of the adult fly has a wider biological significance, and will surely provide greater insights into understanding vertebrate myogenesis.

References

1. Bate M. The mesoderm and its derivatives. In: Bate M, Martinez-Arias A, eds. The Development of Drosophila melanogaster, Vol. 2. New York: Cold Spring Harbor Laboratory Press, 1993:1013-1090.
2. Fernandes J, Bate M, VijayRaghavan K. Development of the indirect flight muscles of Drosophila. Development 1991; 113:67-77.
3. Currie DA, Bate M. The development of adult abdominal muscles in Drosophila: Myoblasts express twist and are associated with nerves. Development 1991; 113:91-102.

4. Baylies MK, Bate M. Twist: A myogenic switch in Drosophila. Science 1996; 272:1481-1484.
5. Baylies MK, Bate M, Ruiz-Gomez M. The specification of muscle in Drosophila. Cold Spring Harb Symp Quant Biol 1997; 62:385-93.
6. Carmena A, Gisselbrecht S, Harrison J et al. Combinatorial signaling codes for the progressive determination of cell fates in the Drosophila embryonic mesoderm. Genes Dev 1998; 12:3910-3922.
7. Carmena A, Bate M, Jimenez F. Lethal of scute, a proneural gene, participates in the specification of muscle progenitors during Drosophila embryogenesis. Genes Dev 1995; 9:2373-2383.
8. Baylies MK, Bate M, Ruiz-Gomez M. Myogenesis: A view from Drosophila. Cell 1998; 93:921-927.
9. Baker R, Schubiger G. Autonomous and nonautonomous notch functions for embryonic muscle and epidermis development in Drosophila. Development 1996; 122:617-626.
10. Bate M, Rushton E, Frasch M. A dual requirement for neurogenic genes in Drosophila myogenesis. Dev Suppl 1993; 149-161.
11. Ruiz Gomez M, Bate M. Segregation of myogenic lineages in Drosophila requires numb. Development 1997; 124:4857-4866.
12. Carmena A, Murugasu-Oei B, Menon D et al. Inscuteable and numb mediate asymmetric muscle progenitor cell divisions during Drosophila myogenesis. Genes Dev 1998; 12:304-315.
13. Bate M, Rushton E, Currie DA. Cells with persistent twist expression are the embryonic precursors of adult muscles in Drosophila. Development 1991; 113:79-89.
14. Broadie KS, Bate M. The development of adult muscles in Drosophila: Ablation of identified muscle precursor cells. Development 1991; 113:103-118.
15. Lawrence PA. Cell lineage of the thoracic muscles of Drosophila. Cell 1982; 29:493-503.
16. VijayRaghavan K, Pinto L. The cell lineage of the muscles of the Drosophila head. Embryol Exp Morphol 1985; 85:285-294.
17. Farrell ER, Fernandes J, Keshishian H. Muscle organizers in Drosophila: The role of persistent larval fibers in adult flight muscle development. Dev Biol 1996; 176:220-229.
18. Roy S, VijayRaghavan K. Patterning muscles using organizers: Larval muscle templates and adult myoblasts actively interact to pattern the dorsal longitudinal flight muscles of Drosophila. J Cell Biol 1998; 141:1135-1145.
19. Fernandes J, VijayRaghavan K. The development of indirect flight muscle innervation in Drosophila melanogaster. Development 1993; 118:215-227.
20. Consoulas C, Restifo LL, Levine RB. Dendritic remodeling and growth of motoneurons during metamorphosis of Drosophila melanogaster. J Neurosci 2002; 22:4906-4917.
21. Tissot M, Stocker RF. Metamorphosis in Drosophila and other insects: The fate of neurons throughout the stages. Prog Neurobiol 2000; 62:89-111.
22. Currie DA, Bate M. Innervation is essential for the development and differentiation of a sex-specific adult muscle in Drosophila melanogaster. Development 1995; 121:2549-2557.
23. Fernandes JJ, Keshishian H. Nerve-muscle interactions during flight muscle development in Drosophila. Development 1998; 125:1769-1779.
24. Lawrence PA, Johnston P. The muscle pattern of a segment of Drosophila may be determined by neurons and not by contributing myoblasts. Cell 1986; 45:505-513.
25. Ito H, Fujitani K, Usui K et al. Sexual orientation in Drosophila is altered by the satori mutation in the sex-determination gene fruitless that encodes a zinc finger protein with a BTB domain. Proc Natl Acad Sci USA 1996; 93:9687-9692.
26. Taylor BJ, Knittel LM. Sex-specific differentiation of a male-specific abdominal muscle, the muscle of lawrence, is abnormal in hydroxyurea-treated and in fruitless male flies. Development 1995; 121:3079-3088.
27. Gailey DA, Taylor BJ, Hall JC. Elements of the fruitless locus regulate development of the muscle of Lawrence, a male-specific structure in the abdomen of Drosophila melanogaster adults. Development 1991; 113:879-890.
28. Usui-Aoki K, Ito H, Ui-Tei K et al. Formation of the male-specific muscle in female Drosophila by ectopic fruitless expression. Nat Cell Biol 2000; 2:500-506.
29. Fernandes J, Celniker SE, Lewis EB et al. Muscle development in the four-winged Drosophila and the role of the Ultrabithorax gene. Curr Biol 1994; 4:957-964.
30. Ng M, Diaz-Benjumea FJ, Vincent JP et al. Specification of the wing by localized expression of wingless protein. Nature 1996; 381:316-318.
31. VijayRaghavan K, Gendre N, Stocker R. Transplanted wing and leg imaginal discs in Drosophila melanogaster demonstrates interactions between epidermis and myoblasts in muscle formation. Dev Genes Evol 1996; 206:46-53.
32. Fernandes JJ, Celniker SE, VijayRaghavan K. Development of the indirect flight muscle attachment sites in Drosophila: Role of the PS integrins and the stripe gene. Dev Biol 1996; 176:166-184.

33. Volk T. Singling out Drosophila tendon cells: A dialogue between two distinct cell types. Trends Genet 1999; 15:448-453.
34. de la Pompa JL, Garcia JR, Ferrus A. Genetic analysis of muscle development in Drosophila melanogaster. Dev Bio 1989; 131:439-454.
35. Costello WJ, Wyman RJ. Development of an indirect flight muscle in a muscle-specific mutant of Drosophila melanogaster. Dev Biol 1986; 118:247-258.
36. Sudarsan V, Anant S, Guptan P et al. Myoblast diversification and ectodermal signaling in Drosophila. Dev Cell 2001; 1:829-839.
37. Ghazi A, Anant S, VijayRaghavan K. Apterous mediates development of direct flight muscles autonomously and indirect flight muscles through epidermal cues. Development 2000; 127:5309-5318.
38. Bernard F, Lalouette A, Gullaud M et al. Control of apterous by vestigial drives indirect flight muscle development in Drosophila. Dev Biol 2003; 260:391-403.
39. Bourgouin C, Lundgren SE, Thomas JB. Apterous is a Drosophila LIM domain gene required for the development of a subset of embryonic muscles. Neuron 1992; 9:549-561.
40. Phillips RG, Whittle JR. Wingless expression mediates determination of peripheral nervous system elements in late stages of Drosophila wing disc development. Development 1993; 118:427-438.
41. Ruiz-Gomez M, Coutts N, Price A et al. Drosophila dumbfounded: A myoblast attractant essential for fusion. Cell 2000; 102:189-198.
42. Luo L, Liao YJ, Jan LY et al. Distinct morphogenetic functions of similar small GTPases: Drosophila Drac1 is involved in axonal outgrowth and myoblast fusion. Genes Dev 1994; 8:1787-1802.
43. Dworak HA, Sink H. Myoblast fusion in Drosophila. BioEssays 2002; 24:591-601.
44. Taylor MV. Muscle differentiation: How two cells become one. Curr Biol 2002; 12:R224-8.
45. Anant S, Roy S, VijayRaghavan K. Twist and notch negatively regulate adult muscle differentiation in Drosophila. Development 1998; 125:1361-1369.
46. Spicer DB, Rhee J, Cheung WL et al. Inhibition of myogenic bHLH and MEF2 transcription factors by the bHLH protein Twist. Science 1996; 272:1476-1480.
47. Hebrok M, Wertz K, Fuchtbauer EM. M-twist is an inhibitor of muscle differentiation. Dev Biol 1994; 165:537-544.
48. Lin MH, Bour BA, Abmayr SM et al. Ectopic expression of MEF2 in the epidermis induces epidermal expression of muscle genes and abnormal muscle development in Drosophila. Dev Biol 1997; 182:240-255.
49. Bour BA, O'Brien MA, Lockwood WL et al. Drosophila MEF2, a transcription factor that is essential for myogenesis. Genes Dev 1995; 9:730-741.
50. Lilly B, Zhao B, Ranganayakulu G et al. Requirement of MADS domain transcription factor D-MEF2 for muscle formation in Drosophila. Science 1995; 267:688-693.
51. Olson EN, Perry M, Schulz RA. Regulation of muscle differentiation by the MEF2 family of MADS box transcription factors. Dev Biol 1995; 172:2-14.
52. Cripps RM, Black BL, Zhao B et al. The myogenic regulatory gene Mef2 is a direct target for transcriptional activation by twist during Drosophila myogenesis. Genes Dev 1998; 12:422-434.
53. Ranganayakulu G, Zhao B, Dokidis A et al. A series of mutations in the D-MEF2 transcription factor reveal multiple functions in larval and adult myogenesis in Drosophila. Dev Biol 1995; 171:169-181.
54. Michelson AM. Muscle pattern diversification in Drosophila is determined by the autonomous function of homeotic genes in the embryonic mesoderm. Development 1994; 120:755-768.
55. Hooper J. Homeotic gene expression in muscles of Drosophila larvae. EMBO J 1986; 5:2321-2329.
56. Volk T, VijayRaghavan K. A central role for epidermal segment border cells in the induction of muscle patterning in the Drosophila embryo. Development 1994; 120:59-70.
57. Rivlin PK, Gong A, Schneiderman AM et al. The role of Ultrabithorax in the patterning of adult thoracic muscles in Drosophila melanogaster. Dev Genes Evol 2001; 211(2):55-66.
58. Greig S, Akam M. Homeotic genes autonomously specify one aspect of pattern in the Drosophila mesoderm. Nature 1993; 362:630-632.
59. Roy S, Shashidhara LS, VijayRaghavan K. Muscles in the Drosophila second thoracic segment are patterned independently of autonomous homeotic gene function. Curr Biol 1997; 7:222-227.
60. Roy S, VijayRaghavan K. Homeotic genes and the regulation of myoblast migration, fusion, and fibre-specific gene expression during adult myogenesis in Drosophila. Development 1997; 124:3333-3341.
61. Lawrence PA, Brower DL. Myoblasts from Drosophila wing discs can contribute to developing muscles throughout the fly. Nature 1982; 295:55-57.
62. Kozopas KM, Nusse R. Direct flight muscles in Drosophila develop from cells with characteristics of founders and depend on DWnt-2 for their correct patterning. Dev Biol 2002; 243:312-325.

63. Glicksman MA, Brower DL. Expression of the sex combs reduced protein in Drosophila larvae. Dev Biol 1988; 127:113-118.
64. Lee JC, VijayRaghavan K, Celniker SE et al. Identification of a Drosophila muscle development gene with structural homology to mammalian early growth response transcription factors. Proc Natl Acad Sci USA 1995; 92:10344-10348.
65. Ghazi A, Paul L, VijayRaghavan K. Prepattern genes and signaling molecules regulate stripe expression to specify Drosophila flight muscle attachment sites. Mech Dev 2003; 120:519-528.
66. Sandstrom DJ, Restifo LL. Epidermal tendon cells require broad complex function for correct attachment of the indirect flight muscles in Drosophila melanogaster. J Cell Sci 1999; 112:4051-4065.
67. Sandstrom DJ, Bayer CA, Fristrom JW et al. Broad-complex transcription factors regulate thoracic muscle attachment in Drosophila. Dev Biol 1997; 181:168-185.
68. Restifo LL, White K. Mutations in a steroid hormone-regulated gene disrupt the metamorphosis of internal tissues in Drosophila: Salivary glands, muscle and gut. Roux's Arch Dev Biol 1992; 201:221-234.
69. Becker S, Pasca G, Strumpf D et al. Reciprocal signaling between Drosophila epidermal muscle attachment cells and their corresponding muscles. Development 1997; 124:2615-2622.
70. Frommer G, Vorbruggen G, Pasca G et al. Epidermal egr-like zinc finger protein of Drosophila participates in myotube guidance. EMBO J 1996; 15:1642-1649.
71. Roy S, VijayRaghavan K. Muscle pattern diversification in Drosophila: The story of imaginal myogenesis. BioEssays 1999; 21:486-498.
72. Lawrence PA, Johnston P. The genetic specification of pattern in a Drosophila muscle. Cell 1984; 36:775-782.
73. Rivlin PK, Schneiderman AM, Booker R. Imaginal pioneers prefigure the formation of adult thoracic muscles in Drosophila melanogaster. Dev Biol 2000; 222:450–459.
74. Dutta D, Anant S, Ruiz-Gomez M et al. Founder myoblasts and fibre number during adult myogenesis in Drosophila. Development 2004; 131:3761-3772.

CHAPTER 12

Molecular Basis of Muscle Structure

Jim O. Vigoreaux*

Abstract

The flight muscle myofibril is a precisely assembled cytoskeletal network of contractile proteins that produces high power to sustain flight. This chapter will focus on myofibrillar assembly during development of the indirect flight muscles. Studies in *Drosophila melanogaster* have combined genetics with microscopy to elucidate the series of steps that lead to the formation of the sarcomere and the role that individual myofibrillar proteins may play in this process. Despite much progress, the broad characterization of many of the mutants' effects has yet to uncover the mechanisms that regulate and dictate the assembly of the myofibril.

Introduction

Drosophila, as well as other Dipterans, has long been a favorite subject for research in muscle development, particularly the adult flight muscle. Up until the advent of recombinant DNA technology in the 1980's, the vast majority of the studies were of a descriptive nature, relying on light and electron microscopy to monitor the morphological and ultrastructural changes that take place in the developing flight muscle cell (see ref. 1 for review). In the early 1970s several groups of investigators pioneered the use of genetic screens to identify flightless *Drosophila* mutants (for examples, see refs. 2, 3; for review, see refs 4-6). These studies provided the foundation that established the indirect flight muscles (IFM) as a model system for muscle development and function. The characterization of these and subsequent mutant strains in the decades since, employing a wide variety of experimental approaches, has provide much insight into the functional properties of muscle proteins. While most of the interest has focused on their role in contractile activity, insight has been gained on their role in myofibril assembly. Studies have brought to light the fact that most, perhaps even all, proteins play fundamental roles in myofibril assembly that are separate from their roles in contractile mechanics.[5] This chapter will examine what is known about the roles of contractile proteins in myofibril assembly. Readers interested in more comprehensive coverage of flight muscle development are encouraged to see other recent reviews.[4-6]

Structure and Composition of the Myofibril

IFM fibers, like their vertebrate skeletal muscle counterparts, are syncytial cells with abundant mitochondria and myofibrils. The myofibril is a structural complex of approximately twenty proteins that compose distinct but interacting filament systems and cross-linking structures organized in repeating units known as sarcomeres (Fig. 1). The majority of these proteins (e.g., actin, myosin, tropomyosin) are highly conserved in striated muscles of invertebrates and vertebrates. The IFM also contain additional proteins, such as flightin, and variants of conserved proteins (arthrin, troponin H) that are unique to this tissue (Table 1).[5,6]

*Jim O. Vigoreaux—Department of Biology, University of Vermont, Burlington, Vermont 05405, U.S.A. Email: jvigorea@uvm.edu

Muscle Development in Drosophila, edited by Helen Sink. ©2006 Eurekah.com and Springer Science+Business Media.

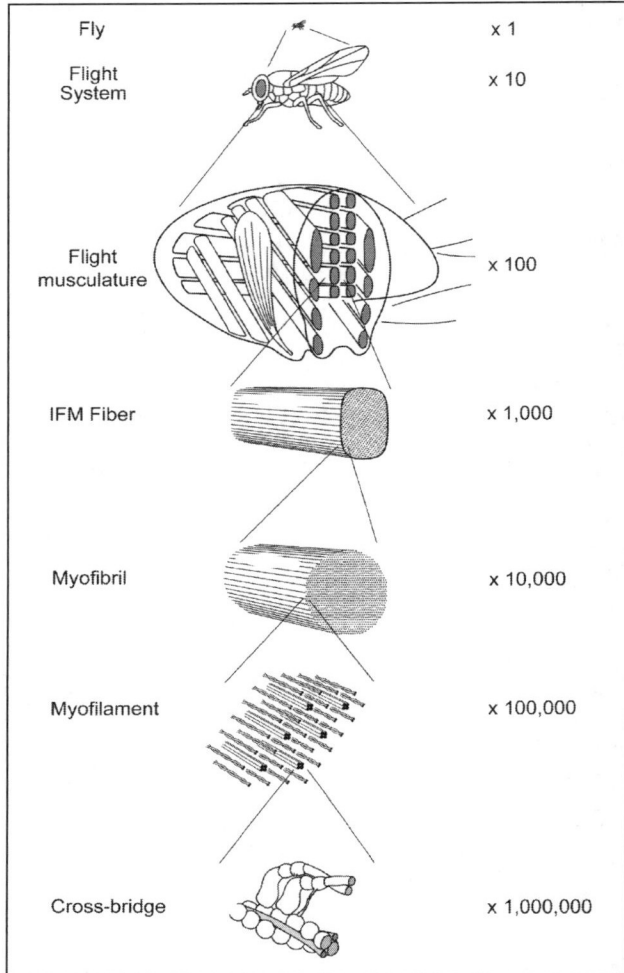

Figure 1. Organizational levels of the flight system of *Drosophila melanogaster*, in powers of ten. The flight musculature consists of the IFM (shown are the dorsolongitudinal muscles along the anterior-posterior axis, and the dorsoventral muscles oriented almost perpendicular to the dorsolongitudinal muscles) and the direct flight muscles that control wing movement (not shown). The leaf-shaped muscle is the tergal depressor of the trochanter that powers jump and the initiation of flight. Figure modified from Maughan DW, Vigoreaux JO. News Physiol Sci 1999; 14: 87-92.

On cross sections, thick filaments and thin filaments are arranged in a precise double hexagonal array, in a 1:3 ratio, with actin filaments at the dyad positions between thick filaments. Sarcomeres are ~3.2 μm and 1.5 μm wide (32-35 thick filaments across), and demarcated by solid, orderly Z bands that differ from their vertebrate muscle counterparts by having a hexagonal, instead of a tetragonal, lattice. M lines are present as in vertebrate muscle. IFM sarcomeres are also distinguished by their narrow I band (i.e., the 3 μm thick filaments span almost the entire length of the sarcomere). The assembly of this remarkably well-ordered and regular mechanical structure is still largely a puzzle. The regulatory mechanisms are undoubtedly complex, and involve control at the transcriptional, post-transcriptional, translational, and post-translational levels.[6]

Table 1. Role of myofibrillar proteins in IFM development

Protein	Size (kDa)	Component of	Function	Refs.
Projectin	1000	Connecting filaments	Not known	
Kettin	540, 700, 1000	Connecting filaments	Mutations lead to defective terminal Z discs	66
Stretchin-mlck Myostrandin P165	231, 225 165	Thick filaments	Not known	
Myosin heavy chain	200-220	Thick filaments	Essential for thick filament assembly	11, 13
Zetalin	210	Z band	Not known	
α-actinin	97-107	Z band	Required for normal morphogenesis of Z bands and terminal insertion sites	68
Paramyosin Mini-paramyosin	90-105 55	Thick filaments Thick filaments/M·line	Not known	
Troponin H	70, 78	Thin filaments	Not known	
Troponin T	47	Thin filaments	Required for myofibril integrity	54
Arthrin	49-56	Thin filaments	Not known	
Tropomodulin	45	Thin filaments	Required for thin filament length determination	65
Actin	43	Thin filaments	Essential for thin filament and Z band assembly	10, 62, 63
Tropomyosin	35-37	Thin filaments	Required for thin filament stability	64
ADP/ATP translocase	33	?	Not known	
Troponin I	25-29-35	Thin filaments	Required for sarcomere formation; regulate contractile force	55, 56
Glutathione-S-transferase 2	32-35	Thin filaments	Not known	
Myosin regulatory light chain	24-30	Thick filaments	Required for normal myofibril assembly	24
Flightin	20	Thick filaments	Required for thick filament length determination and stability	33
Myosin essential light chain	18	Thick filaments	Not known	
Troponin C	18	Thin filaments	Not known	

Ultrastructure of Developing Myofibrils

Reedy and Beall[7] conducted a detailed electron microscopy study of IFM during pupal stages of development that provided a broad scheme for myofibril assembly. Among their findings was the observation that the number of sarcomeres in each myofibril (~310) is determined early in development and remains constant throughout.[7] As in the adult myofibril, sarcomere dimensions exhibit a consistent regularity throughout all of pupal development. Sarcomeres, thick filaments, and thin filaments increase gradually and synchronously until achieving their final dimensions in the early hours of adult life.

Drosophila myofibrils form within temporary scaffolds of microtubules that are disassembled shortly before eclosion, but before the muscle reaches its final dimensions.[7] Aside from these microtubules, no other accessory or "chaperone" proteins have been invoked in IFM myofibrillar assembly. A *Drosophila* homologue of the nematode UNC-45, a chaperone involved in myosin folding, has been identified but there is no evidence yet that it participates in IFM development.[8] The mechanism of assembly is therefore largely dependent on the properties of the constituent proteins.

Genetics of Myofibrillar Proteins

Genetic analysis of IFM protein function has been a remarkably fruitful endeavor (for reviews see refs. 4-6, 9). Most of the studies, however, have focused on the adult ("developed") muscle and few studies have examined the effect of mutations on development per se. As a result, most of what is known about the role of myofibrillar proteins in muscle development has been inferred from studies of the adult muscle (Table 1). Figure 2 shows the possible arrangement and interactions of myofibrillar proteins in adult IFM.

The prevailing paradigm[10] is that thick filaments assemble independently of thin filaments, the organization of the former being dictated by M line proteins and that of the latter by Z band components (the role of connecting filaments has not been addressed). Interactions between thick filaments and thin filaments are likely responsible for their proper integration into a regularly spaced lattice and formation of the orderly sarcomere. This paradigm, based largely on studies of myofibrillar protein mutants, will serve as the focal point of this chapter. The discussion will be limited to those proteins for which mutants affecting IFM have been described.

Thick Filament Proteins

The major component of thick filaments is myosin. Paramyosin (PM) is found in low abundance and is believed to form an inner shell upon which myosin assembles (Fig. 2). Other components of the thick filament are mini-paramyosin (mPm), flightin, and myostrandin (Fig. 2 and Table 1).

Myosin Heavy Chain (MHC)

IFM from Mhc^7 null homozygotes lack thick filaments but show continuous arrays of presumably normal thin filaments anchored aperiodically within Z discs. The absence of MHC affects the expression or stability of at least 10 additional proteins shown to be missing or reduced in Mhc^7 IFM.[3,11] One of these proteins, myosin regulatory light chain (RLC), has been shown to be synthesized and degraded suggesting that the absence of MHC triggers proteolysis of thick filament associated proteins or, alternatively, a mechanism of regulated proteolysis may operate normally in the IFM to eliminate unassembled proteins.

MHC protein levels are dictated by gene copy number, with reduced levels of MHC in Mhc^7 heterozygotes resulting in myofibrillar abnormalities.[12,13] Small aggregates of well-organized arrays of thick and thin filaments are surrounded by out-of-register peripheral thin filaments.[13] The IFM of heterozygotes for null alleles of other myofibrillar protein genes (e.g., actin, RLC) also display out-of-register peripheral myofilaments surrounding a well-ordered core suggesting that myofibril assembly starts at the center and radiates out.[7]

Other studies have shown that increasing the gene copy number (by transgenic techniques) result in MHC overexpression and myofibrillar defects, namely excess thick filaments surrounding a well-formed myofibril core.[14] Homyk and Emerson[15] found that increasing the MHC gene copy to 3 using a chromosomal duplication had no effect on flight performance or wing posture (and presumably muscle structure, but this was not examined). Nevertheless, the duplication does affect muscle function when placed in a genetic background deficient for troponin T (TnT) but it is unclear how the two mutations interact.

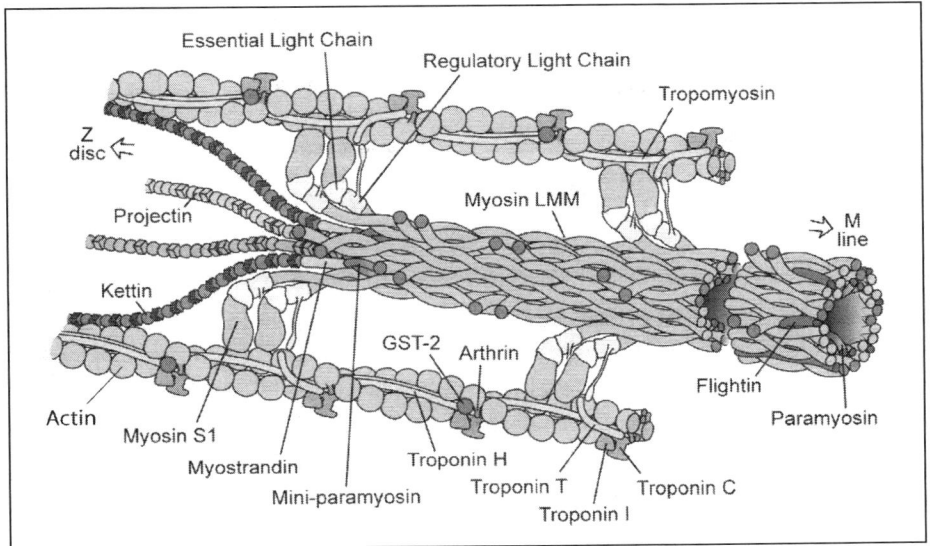

Figure 2. Schematic representation of an adult IFM myofibril. Shown is the region at the A-I junction. The proteins are drawn approximately to scale at their relative positions. For additional detail see reference 79. (Reprinted from ref. 79.)

Several missense *Mhc* alleles have been characterized and shown to have little or no noticeable effect on muscle development, but compromise the structural integrity of the muscle once it is recruited to power flight. Among these are two point mutations, *Mhc*[6] and *Mhc,*[13] that reside five amino acids apart in the myosin rod.[16] The structure of the developing myofibril is relatively unaffected, but the mutations lead to fiber hypercontraction and rapid breakdown of the sarcomere subsequent to eclosion. In addition, both mutations interfere with the normal accumulation of the IFM-specific protein flightin.[16] As discussed below, the myosin rod-flightin interaction is essential for the stability of thick filaments.

Another missense mutation, *Mhc*[9] results in glutamic acid 482 changed to lysine in a region of the myosin motor (globular head domain) close to the actin binding site.[17] In IFM, the mutation behaves like the null allele *Mhc,*[7] i.e., it prevents the accumulation of myosin (despite normal accumulation of *Mhc* RNA) and affects expression of the same subset of proteins.[18] Thick filaments and myofibrils are absent. A similar mutation in *C. elegans* myosin abolishes ATPase activity, raising the possibility that abnormal interaction with actin may lead to protein instability in the developing muscle.[19] However, the fact that the mutant myosin fails to accumulate in IFM that lack thin filaments argues against this possibility and leaves open the question of how the myosin motor domain participates in myofibril assembly.[19-20]

The role of the myosin motor domain was examined by Cripps et al who created transgenic lines expressing headless myosin.[20] They found that the myosin head is essential for normal sarcomere assembly though some aspects of assembly, such as integration of myofilaments into a double hexagonal lattice, appeared not to be affected by the absence of the motor domain. The observed defects implicate the myosin head in myofibril assembly. The possible mechanisms by which the head exerts its effect are discussed below.

Tissue-specific isoforms of MHC are generated via alternative splicing of five of the 18 exons in the *MHC* gene. Several studies have examined how isoform substitutions affect IFM properties. MHC isoforms are functionally interchangeably for development but not for physiology or, in some cases, muscle stability. Substitution of individual adult-specific alternative exons with their embryonic counterparts has no effect on IFM development.[21-22] Wells et al created transgenic *Drosophila* that expressed a single embryonic MHC isoform in all muscles.[23] This isoform permitted normal development of IFM indicating that the type of MHC isoform expressed does not dictate the structural qualities of the sarcomere. However, transformed flies are flightless. In addition, IFM myofibrillar structure deteriorates in aging adults. The aging effect on IFM ultrastructure is partly ameliorated by substituting the C-terminus of the embryonic MHC transgene with the adult C-terminus, suggesting that this region of the molecule is important for stability of IFM myofibrils.[23]

Myosin Regulatory Light Chains (RLC)

Flies that are heterozygous for an RLC null allele have reduced RLC expression and exhibit myofibrillar defects similar to those described for MHC null heterozygotes.[24] RLC has been shown to undergo multiple phosphorylations during the final stages of IFM development, at or near eclosion.[25] Replacement of two RLC serines that are phosphorylated by myosin light chain kinase (MLCK) with alanines had no effect on myofibrillogenesis.[26-27] Furthermore, heterozygotes for a deficiency that deletes the *mlck* gene fly normally, suggesting that phosphorylation of the two RLC serine residues has no role in development.[28] Similarly, removing the N-terminal extension of RLC ($Mlc2^{\Delta 2-46}$) has no effect on myofibril assembly.[29] There are no known mutants of the essential light chain (ELC).

Paramyosin and Mini-Paramyosin

Overexpression of mPm in the IFM had little to no effect on myofibril assembly, mPm sarcomeric distribution, thick filament diameter, sarcomere length, or on the expression of other myofibrillar proteins.[30] IFM ultrastructure of young and aged adults resemble wild-type. In contrast, transgenic flies expressing a chimeric Actin 88F (*Act88F*) promoter-mPm transgene resulted in higher levels of *mPm* expression, more developmental defects and age-dependant degeneration of the IFM than transgenic flies in which the *mPm* gene was regulated by its own promoter.[30] Subtle developmental defects in young flies, evident as wavy and disordered myofibrils, gave rise to irregular and disrupted sarcomeres in the aged (10 day old) adult. Thus, reprogramming the transcriptional activation of the *mPm* gene is more detrimental to muscle development than mPm overexpression which results in small developmental defects that are later magnified in the working muscle.[30] Studies of a hypomorphic mutant of PM showed that this protein is essential for thick filament formation and myofibril assembly of embryonic muscle.[31] Because the mutant is embryonic lethal, its affect on IFM development was not studied. PM and mPM undergo dephosphorylation during the transition from late stage pupa to flying adult[32] but the role, if any, of these phosphorylations on myofibril assembly has not been investigated. Overexpression of mPM does not affect its phosphorylation profile.[30]

Flightin

This myosin rod binding protein has been shown to be necessary for proper thick filament length determination as evidenced by the longer-than-normal thick filaments in late stage pupa of *fln⁰*, a flightin null mutant.[33] Sarcomeres are also, on average, longer than normal suggesting that thick filament length is an important determinant for sarcomere length in *Drosophila* IFM. The absence of flightin does not affect other properties of the sarcomere such as the double hexagonal array of thick and thin filaments, or the organization of the Z band. One difference between developing wild-type and *fln⁰* sarcomeres is the presence of electron dense stripes flanking the M line; they are more prominent and longer lasting in the latter strain.

Upon eclosion, the IFM of *fln⁰* flies undergoes rapid and irreversible degeneration, reminiscent of the process seen in *Mhc*,[13] the myosin rod mutant.[16,33] This process of 'hypercontraction' can be suppressed by mutations in the myosin motor domain and by coexpression of a headless myosin.[34] Thus, the thick filaments and sarcomeres that assemble in the absence of flightin are not competent to withstand contractile activity. All of the developmental and structural defects engendered by *fln⁰* can be rescued by expression of an *Act88F* promoter-*fln⁺* transgene. However, myofibrils have narrower diameters (i.e., less thick filaments) than wild-type which may result from either reduced levels of flightin expression or alterations in the timing of expression given that the *Act88F* promoter is activated early in myofibrillogenesis.[35] It seems possible that premature activation of flightin transcription may have an effect on thick filament number.[36]

Unlike MHC and RLC, flightin is not sensitive to gene dosage. Studies of transgenic flies with 4 copies of the flightin gene showed these flies do not overexpress flightin suggesting expression of this protein is regulated post-transcriptionally.[80] Heterozygotes for a deficiency that uncovers the *flightin* gene (*Df(3L)fln¹*) showed a ~20% reduction in flightin accumulation compared to wild-type IFM.[37] This reduction had no apparent effect on myofibril ultrastructure of developing pupa. However, adult myofibrils exhibit loosely organized myofilaments along the myofibril periphery that are easily dissociated with a detergent solution. These results suggest that flightin is necessary for maintaining the structural integrity of the lattice in the working muscle.

Flightin is first detected in the developing IFM at ~60 hours after puparium formation (APF), well after thick filaments assembly has been initiated.[38] Beginning shortly before eclosion, flightin (as well as RLC) undergoes extensive phosphorylation resulting in up to nine phosphovariants in adult IFM.[39] The role of these phosphorylation events is not yet known but they coincide with completion of sarcomerogenesis, raising the possibility that phosphorylation may be involved in establishing the final dimensions of the sarcomere. Preliminary studies of flightin phosphorylation site mutants are consistent with this interpretation.[36]

Thin Filament and Z Band Proteins

IFM thin filaments share many structural and biochemical characteristics with vertebrate striated muscle thin filaments.[40] However, IFM thin filaments are distinguished by the presence of a stably ubiquitinated actin (arthrin), a novel tropomyosin isoform with an extended C-terminus (troponin H), and glutathione-S-transferase (Table 1 and Fig. 2).

Actin

IFM from actin null (*Act88F^{KM88}*) homozygotes lack thin filaments and fail to assemble Z bands but still contain well aligned thick filaments and M lines.[13] Like the null mutant, some missense mutations in actin (e.g., *act88F^{V339I}*) prevent the assembly of thin filaments and no muscle structure is formed.[41] The absence of actin interferes with the expression of other myofibrillar proteins, primarily components of the thin filament. Similar to MHC, the absence of actin may trigger proteolysis of unassembled filament proteins. Unlike MHC mutants, some actin mutants trigger expression of heat shock proteins.[42] *Act88F^{KM88}*/+ heterozygotes are characterized by myofibrils that are only one-half to two-thirds the diameter of normal myofibrils. Central regions of precisely ordered hexagonal arrays of thick and thin filaments are surrounded by excess thick filaments that are not incorporated into the lattice.[13] Sarcomere length is not statistically different from wild-type. Other experiments have shown that increasing the *Act88F* gene copy number to four has no apparent effect on flight performance,[43] and presumably on muscle development, although this has not been tested.

One group of missense mutations in actin not only interferes with the assembly of the thin filament but also with assembly of the Z band. Two single amino acid mutations, *Act88F^{E334K}* and *Act88F^{E93K}*, form thin filaments but lack Z bands.[41,44] The latter mutant also has been shown to prevent the accumulation of α-actinin and other high molecular weight proteins. Mutations that convert glycine at position 6 to alanine and alanine at position 7 to threonine (*Act88F^{G6A A7T}*) result in multiple level Z bands that fail to cross the entire width of the myofibril. As a result, unanchored peripheral thin filaments of opposite polarities are present within the same half sarcomere.[45]

A second group of actin missense mutations (*Act88F^{R28C}*, *Act88F^{E334Q}*, and *Act88F^{G268D}*) affects fiber morphogenesis.[46] In some cases fibers detach from the cuticle followed by partial or complete shortening during development or after eclosion. The hypercontraction phenotype suggests a role for contractile forces in muscle development (see below). In contrast, the mutant $Act88F^{R95C}$, which maps to the myosin S1 binding site, has no effect on myofibril development but subsequently leads to highly disrupted and hypercontracted fibers after eclosion.[46,47] A third group of actin mutations, exemplified by $Act88F^{G245D}$, is characterized by causing myofibrillar degeneration preferentially in the center of the myofibril.[48]

Transgenic approaches have been used to assess the possible role of actin isoforms in muscle development. When the normally expressed Act88F isoform was substituted by human (beta)-cytoplasmic actin, which differs from IFM actin at 15 positions, several myofibrillar defects were evident, demonstrating that actin isoforms are not functionally interchangeable.[49] The cytoplasmic actin permitted assembly of myofibril-like bundles that lacked sarcomeric repeats. M lines were absent and amorphous, dense 'Z bodies' took the place of Z bands.[49] Many thin filaments failed to anchor in the Z bodies and out-of-register thick and thin filaments were often seen. Thick and thin filaments appeared shorter than normal, perhaps due to their inability to be in-register or due to the absence of Z bands. Nevertheless, the interaction between myosin crossbridges (the myosin motor domain) and thin filaments appeared to be normal and the deficient sarcomeres were likely to arise from failure of cytoplasmic actin to interact with one or more actin-binding proteins in the IFM.[49] In a similar study, Fyrberg et al[50] found that replacing IFM actin with *Drosophila* cytoplasmic (nonmuscle) actin, an isoform that differs at 18 positions, also resulted in abnormal sarcomeres. Transformants expressing actins with single isoform-specific amino acid replacements formed normal myofibrils.[50] These studies showed that while most individual amino acid changes are tolerated, replacements of multiple amino acids at once significantly alters the assembly and/or stability properties of actin. It remains to be established how changes in actin isoform interfere with IFM development. Also unclear is the role, if any, of post-translational modifications on filament formation and sarcomere assembly. Mutants that interfere with normal processing of the actin N-terminus are flight impaired but do not exhibit ultrastructural defects.[51] The role of ubiquitination of actin to generate arthrin has not been investigated.[52] The recent observation that arthrin fails to accumulate in the IFM of the TnI null mutant *heldup³* suggest that ubiquitination of actin may be linked to the formation of the troponin complex.[53]

Troponins T and I

These thin filament proteins are essential for normal myofibrillogenesis as evidenced by the absence of sarcomeres in mutants that prevent or reduced their expression. Mutants in TnT (*upheld²* and *upheld³*) display loosely organized bundles of thick filaments and randomly scattered I-Z-I "brushes" (electron dense structures interconnected by thin filaments).[54] Mutants that fail to express one or more IFM TnI isoforms (*heldup³*, *heldup⁴*, *heldup⁵*) assemble thin filaments and thick filaments but fail to form sarcomeres or myofibrils.[2,55,56] A recent study showed that the null allele *heldup³* fails to complete fiber elongation that normally takes place after fiber shortening at ~30 hrs APF.[53] Instead fibers progressively degrade and by eclosion, no fibers are evident. Reducing acto-myosin force production, through the introduction of a myosin headless construct or an MHC null allele, suppresses hypercontraction and improves fiber morphology but does not restore sarcomere assembly. These results suggest that unregulated acto-myosin forces are not the sole basis of the *heldup³* assembly defect. Indeed, it was found that the expression of other thin filament proteins and their mRNAs are reduced in *heldup³* IFM.[53]

Studies of the TnI mutant *heldup,²* in which alanine 116 is exchanged for valine, support the notion that contractile forces play a role in myofibril assembly. The developing IFM appears normal at 75 hrs after pupa formation (APF) but show signs of hypercontraction by 78 hrs APF.[34] The muscle continues to deteriorate resulting in complete loss of sarcomere integrity in the adult fly. That acto-myosin generated force was the culprit of muscle degeneration had been established by studies in which a headless myosin transgene, as well as several myosin motor domain missense mutations, suppress the hypercontraction phenotype.[34,57] The hypercontraction phenotype is also suppressed by an intragenic mutation[58] and a tropomyosin mutation.[59] The headless myosin and the motor missense mutations also suppress the hypercontraction phenotype of *upheld,*[101] a TnT point mutation that shows similar characteristics to *heldup.*[2]

Like flightin, RLC, and PM/mPM, changes in TnT phosphorylation have been documented to occur during the final stages of *Drosophila* flight muscle development but the significance of these events has not been investigated.[60,61]

Tropomyosin and Troponin H

IFM tropomyosin, encoded by the *tropomyosin2* (*TmI*) gene, is essential for normal muscle structure. Various mutants have been examined and shown to have defective sarcomere structure; however, studies have examined adult flies only and it is not known if the defective ultrastructure arises from errors in development or use-induced atrophy. One example is the *TmI* hypomorph *Ifm(3)3* which shows various sarcomeric defects including Z bands that are irregular in size and shape, and disordered arrays of peripheral myofilaments as seen in mutant heterozygotes of actin, MHC, and other sarcomere proteins.[62,63] Sarcomeres are, on average, shorter than in wild-type (2.7 μm vs 3.4 μm). Other *TmI* single base change mutant alleles show similar phenotypes.[64] Like *Act88F* and *Mhc*, the *TmI* gene is haploinsufficient for flight but it is not yet clear if the haploinsufficiency arises from decreased tropomyosin protein levels or other, secondary factors (e.g, alterations in the expression or assembly of interacting proteins).[64] Unlike *TmI* the *TmII* gene, which encodes two IFM-specific isoforms of troponin H, is not haploinsufficient.[64] Myofibrils of TnH heterozygotes have normal structure but slightly smaller diameter (~31 thick filaments across the midline vs 36 for wild-type).

The defects engendered by *Ifm(3)3* are rescued by the expression of a non-IFM tropomyosin isoform (Scm-TmI, normally expressed in other muscles of the larva and adult).[62] These results demonstrate that the function of tropomyosin in IFM development in not isoform specific.

Tropomodulin

The role of this protein (TMOD, encoded by the *sanpodo* gene) in myofibril assembly was ascertained by transient overexpression during development using a heat shock promoter.[65] An IFM-specific, 45 kD isoform is present in adult muscles where it is believed to function as a permanent cap at the thin filament pointed end. In contrast, a 43 kD isoform is expressed throughout development acting as a dynamic cap that allows continuous elongation from the pointed end. Heat-shock induced overexpression caused the 43 kD isoform to behave as a permanent cap, instead of a dynamic one and prevented core thin filaments from growing even after TMOD expression had returned to normal levels. In addition, the switch over to the 45 kD isoform that normally takes place after eclosion did not occur. Transient overexpression of TMOD affected myofibril structure but did not affect sarcomere elongation, length of thick filaments, addition of peripheral thin filaments, or the length of thin filaments assembled after the levels of TMOD had declined.[65] These results demonstrate that TMOD is involved in the control of thin filament elongation but is not a determinant of final sarcomere length.

Z Band Proteins

Mutations in Z band proteins kettin and α-actinin result in myofibrillar defects. The *kettin* hypomorphic allele *ket⁵* showed incomplete Z bands that did not intersect the entire width of the sarcomere[66] while the hypomorphic α-actinin allele *fliA³* showed Z bands that were thin and punctate.[67] The defective Z bands in *fliA³* disintegrated as the fly aged. However, the etiology of this phenotype is not completely understood since another α-actinin mutant (*fliA⁴*) in which α-actinin expression is similarly reduced, does not show age-dependent degeneration. More importantly, however, is the observation that myofibril structure in newly eclosed *fliA³* is surprisingly normal with thick filaments and thin filaments precisely packed in double hexagonal arrangements.[67] These results suggest that α-actinin is not essential for initiating or organizing filament assembly during myofibril formation; instead, its principal role is stabilizing or anchoring thin filaments once localized within the sarcomere.[67] In contrast, *ket⁵* homozygotes showed myofibrillar defects before eclosion suggesting that kettin is essential for sarcomere assembly. Kettin is not haploinsufficient as heterozgotes for the loss-of-function *ket[14]* allele show normal IFM structure at late state pupal stages.[66] The terminal Z band is also susceptible to mutations in α*-actinin*[68] and *kettin.*[66]

Myofibril Assembly

Several lessons have been learned from the studies reviewed here. The first is the that muscle defects seen in sarcomere protein gene aneuploids (i.e., nondiploid) can be ascribed to quantitative imbalances in cases where protein expression levels vary in direct proportion with gene copy number. This is true for actin, MHC, and RLC but is not universally true for all sarcomeric proteins, particularly those like flightin whose accumulation is not directly dictated by gene copy number. Nevertheless, there is strong evidence that correct stoichiometry of the major contractile proteins is important for normal IFM development.[13,14] Alterations in the actin:myosin protein ratio result in abnormal assembly, as evidenced by unintegrated peripheral myofilaments and occasional fraying of the hexagonal lattice, and nonfunctional muscle that leads to flight impairment. The *Act88F* null heterozygote and the *Mhc* null heterozygote display more myofibrillar defects than the *Act88F:Mhc* double null heterozygote. One interpretation of these results is that filament imbalances, rather than absolute deficit of one myofilament type, is the underlying cause of the myofibrillar defects.[13]

The second lesson is that accessory proteins such as flightin and TMOD are intimately involved in determining the precise and final length of thick filaments and thin filaments, respectively. The coordinated lengthening of thin and thick filament length during IFM development is not consistent with models that invoke a "molecular ruler".[69,70] In vertebrate striated muscle, the giant protein titin has been proposed to serve as a ruler given its span of the sarcomere and its association with numerous myofibrillar proteins.[70] In *Drosophila*, the *sallimus* gene encodes variably sized titin/kettin isoforms; the predominant isoform in IFM, the 540 kD kettin, is too short to serve as a ruler.[71,72] Mutations that affect *Drosophila* "titin" result in defective myoblast fusion and abnormal myofilament organization in embryonic muscles, but the role of this protein in adult muscle development has not been explored.[73,74]

The presence of Z bands also appears to be essential for dictating the final length of thick filaments and thin filaments, as evidenced by the shorter filaments present in the IFM of transgenic flies expressing a human cytoplasmic actin isoform.[49] Hence, the third lesson is that anchoring structures (or specific Z band proteins like kettin) play a role alongside accessory proteins (flightin and TMOD) and interfilament connections (myosin motor) in dictating sarcomere length. Z bands and M lines, as anchoring structures for thin filaments and thick filaments respectively, would be expected to participate in the "seeding" or organization of myofilaments during the initial stages of myofibril assembly. However, no studies have addressed these issues yet.

The fourth lesson is that the myosin motor domain plays a central role in myofibrillogenesis. Though its precise function has not been defined, it is evident that contractile force is generated during myofibril assembly and is perhaps necessary for it. One possibility is that contractile force is needed to align growing myofilaments. Another possibility is that the motor domain plays a passive, structural role to dictate the proper spacing and/or alignment of thick and thin filaments. The parallel increase in length of thick and thin filaments suggests that interactions between myofilaments play a prominent role in establishing filament and sarcomere length. A mutant "dead-head", i.e., a myosin with normal motor structure that is biochemically incapacitated, may help to distinguish between these two possibilities.

Thick filament length appears to be the primary determinant for sarcomere length in *Drosophila* IFM. Since thin filaments grow longer in *fln⁰*, it is possible that the length of thin filaments is established through coordinated interactions with the thick filament, perhaps by the myosin head or other thick-to-thin filament cross-connecting structure. The presence of normal length thick filaments in TMOD overexpression lines supports this conclusion.

The fifth lesson, and perhaps the one that deserves the closest scrutiny in future investigations, relates to the recurring observation that many IFM mutants affect the expression of multiple proteins.[5] The significance of this is two-fold. First, the analysis of mutant proteomes provides information about functional protein interactions since the proteins whose expression is affected by a particular mutation are likely to interact with the product of the mutant gene.[75] Second, the developmental (and functional) defects that are engendered by a particular mutation may arise from the effect of the mutation on interacting proteins more so than on the intrinsic properties of the mutant protein.

Concluding Remarks

Progress has been made in understanding the molecular basis of muscle structure, but many of the fundamental questions that were raised decades ago still remain unanswered. We still have little understanding of the regulatory mechanisms that coordinate contractile protein gene expression, synthesis and assembly, how the polymerization of proteins is orchestrated, how length of myofilaments is determined, and how lattice spacing is defined. Not surprisingly, most of the progress made addresses the role of particular proteins but the sum of these particulars falls short of providing a comprehensive explanation of myofibril assembly. The available mutants need to be further exploited to establish the stage at which myofibril assembly is impaired and the molecular processes that are perturbed by each mutation.

As prodigious as genetics has been in the study of muscle development, other approaches are needed to complement the study of mutants. Cell culture and in vitro approaches have been particularly enlightening in the study of vertebrate skeletal and cardiac myofibrillogenesis but have found limited use in flight muscle development. Genomic approaches have been used in *Drosophila* to identify a cluster enriched for genes expressed in terminally differentiated muscle and clearly these approaches hold great promise for future studies, especially when combined with large-scale functional genomic approaches such as insertional mutagenesis and enhancer detection systems.[76-78]

Acknowledgements

I want to thank David Maughan and Gary Nelson for the artwork and NSF (MCB-0090768 and MCB-0315865) for their support.

References

1. Crossley AC. The morphology and development of the Drosophila muscular system. In: Ashburner M, Wright TRF, eds. The Genetics and Biology of Drosophila, Vol. 2b. London: Academic Press, 1978:499-560.
2. Deak II, Bellamy PR, Bienz M et al. Mutations affecting the indirect flight muscles of Drosophila melanogaster. J Embryol Exp Morphol 1982; 69:61-81.
3. Mogami K, Hotta Y. Isolation of Drosophila flightless mutants which affect myofibrillar proteins of indirect flight muscle. Mol Gen Genet 1981; 183:409-417.
4. Bernstein SI, O'Donnell PT, Cripps RM. Molecular genetic analysis of muscle development, structure and function in Drosophila. Int Rev Cytol 1993; 143:63-152.
5. Vigoreaux JO. Genetics of the Drosophila flight muscle myofibril: A window into the biology of complex systems. Bioessays 2001; 23:1047-1063.
6. Vigoreaux JO, Swank DM. The development of the flight and leg muscle. In: Gilbert LI, Iatrou K, Gill S, eds. Comprehensive Molecular Insect Science. Oxford: Elsevier, 2004:in press.
7. Reedy MC, Beall C. Ultrastructure of developing flight muscle in Drosophila. I. Assembly of myofibrils. Develop Biol 1993; 160:443-465.
8. Yu Q, Hipolito LC, Kronert WA et al. Characterization and functional analysis of the Drosophila melanogaster unc-45 (dunc-45) gene. Mol Biol Cell 2003; 14(S):45a.
9. Swank DM, Wells L, Kronert WA et al. Determining structure/function relationships for sarcomeric myosin heavy chain by genetic and transgenic manipulation of Drosophila. Microsc Res Tech 2000; 50(6):430-442.
10. Fyrberg E, Beall C. Genetic approaches to myofibril form and function in Drosophila. TIG 1990; 6(4):126-131.

11. Chun M, Falkenthal S. Ifm(2)2 is a myosin heavy chain allele that disrupts myofibrillar assembly only in the indirect flight muscle of Drosophila melanogaster. J Cell Biol 1988; 107:2613-2621.
12. Mogami K, O'Donnell PT, Bernstein SI et al. Mutations of the Drosophila myosin heavy-chain gene: Effects on transcription, myosin accumulation, and muscle function. Proc Natl Acad Sci USA 1986; 83(5):1393-1397.
13. Beall CJ, Sepanski MA, Fyrberg EA. Genetic dissection of Drosophila myofibril formation: Effects of actin and myosin heavy chain null alleles. Genes Dev 1989; 3:131-140.
14. Cripps RM, Becker KD, Mardahl M et al. Transformation of Drosophila melanogaster with the wild-type myosin heavy-chain gene: Rescue of mutant phenotypes and analysis of defects caused by overexpression. J Cell Biol 1994; 126(3):689-699.
15. Homyk T, Emerson CP. Functional interactions between unlinked muscle genes within haploinsufficient regions of the Drosophila genome. Genetics 1988; 119:105-121.
16. Kronert WA, O'Donnell PT, Fieck A et al. Defects in the Drosophila myosin rod permit sarcomere assembly but cause flight muscle degeneration. J Mol Biol 1995; 249:111-125.
17. Bernstein SI, Milligan RA. Fine tuning a molecular motor: The location of alternative domains in the Drosophila myosin head. J Mol Biol 1997; 271(1):1-6.
18. O'Donnell PT, Collier VL, Mogami K et al. Ultrastructural and molecular analyses of homozygous-viable Drosophila melanogaster muscle mutants indicate there is a complex pattern of myosin heavy-chain isoform distribution. Genes Dev 1989; 3(8):1233-1246.
19. Kronert WA, O'Donnell PT, Bernstein SI. A charge change in an evolutionarily-conserved region of the myosin globular head prevents myosin and thick filament accumulation in Drosophila. J Mol Biol 1994; 236(3):697-702.
20. Cripps RM, Suggs JA, Bernstein SI. Assembly of thick filaments and myofibrils occurs in the absence of the myosin head. EMBO J 1999; 18(7):1793-1804.
21. Swank DM, Bartoo ML, Knowles AF et al. Alternative exon-encoded regions of Drosophila myosin heavy chain modulate ATPase rates and actin sliding velocity. J Biol Chem 2001; 276(18):15117-15124.
22. Swank DM, Knowles AF, Kronert WA et al. Variable N-terminal regions of muscle myosin heavy chain modulate ATPase rate and actin sliding velocity. J Biol Chem 2003; 278(19):17475-17482.
23. Wells L, Edwards KA, Bernstein SI. Myosin heavy chain isoforms regulate muscle function but not myofibril assembly. EMBO J 1996; 15(17):4454-4459.
24. Warmke J, Yamakawa M, Molloy J et al. Myosin light chain-2 mutation affects flight, wing beat frequency and indirect flight muscle contraction kinetics in Drosophila. J Cell Biol 1992; 119:1523-1539.
25. Takano-Ohmuro H, Takahashi S, Hirose G et al. Phosphorylated and dephosphorylated myosin light chains of Drosophila fly and larva. Comp Biochem Physiol 1990; 95B:171-177.
26. Tohtong R, Yamashita H, Graham M et al. Impairment of muscle function caused by mutations of phosphorylation sites in myosin regulatory light chain. Nature 1995; 374:650-655.
27. Dickinson MH, Hyatt CJ, Lehmann F-O et al. Phosphorylation-dependent power output of transgenic flies: An integrated study. Biophys J 1997; 73:3122-3134.
28. Tohtong R, Rodriguez D, Maughan D et al. Analysis of cDNAs encoding Drosophila melanogaster myosin light chain kinase. J Muscle Res Cell Motil 1997; 18(1):43-56.
29. Moore JR, Dickinson MH, Vigoreaux JO et al. The effect of removing the N-terminal extension of the Drosophila myosin regulatory light chain upon flight ability and the contractile dynamics of indirect flight muscles. Biophys J 2000; 78:1431-1440.
30. Arredondo JJ, Mardahl-Mesnil M, Cripps RM et al. Overexpression of miniparamyosin causes muscle dysfunction and age- dependant myofibril degeneration in the indirect flight muscles of Drosophila melanogaster. J Muscle Res Cell Motil 2001; 22(3):287-299.
31. Liu H, Mardahl-Dumesnil M, Sweeney ST et al. Drosophila paramyosin is important for myoblast fusion and essential for myofibril formation. J Cell Biol 2003; 160(6):899-908.
32. Maroto M, Arredondo J, Goulding D et al. Drosophila paramyosin/miniparamyosin gene products show a large diversity in quantity, localization, and isoform pattern: A possible role in muscle maturation and function. J Cell Biol 1996; 134(1):81-92.
33. Reedy MC, Bullard B, Vigoreaux JO. Flightin is essential for thick filament assembly and sarcomere stability in Drosophila flight muscles. J Cell Biol 2000; 151:1483-1499.
34. Nongthomba U, Cummins M, Clark S et al. Suppression of muscle hypercontraction by mutations in the myosin heavy chain gene of Drosophila melanogaster. Genetics 2003; 164(1):209-222.
35. Fernandes J, Bate M, Vijayraghavan K. Development of the indirect flight muscles of Drosophila. Development 1991; 113(1):67-77.
36. Barton BE, Ayer G, Cajigas IJ et al. Defects in flight muscle ultrastructure and function in transgenic Drosophila with mutations of phosphorylation sites in flightin. Mol Biol Cell 2002; 13S:319a.

37. Vigoreaux JO, Hernandez C, Moore J et al. A genetic deficiency that spans the flightin gene of Drosophila melanogaster affects the ultrastructure and function of the flight muscles. J Exp Biol 1998; 201:2033-2044.
38. Vigoreaux JO, Saide JD, Valgeirsdottir K et al. Flightin, a novel myofibrillar protein of Drosophila stretch-activated muscles. J Cell Biol 1993; 121(3):587-598.
39. Vigoreaux JO, Perry LM. Multiple isoelectric variants of flightin in Drosophila stretch-activated muscles are generated by temporally regulated phosphorylations. J Muscle Res Cell Motil 1994; 15:607-616.
40. Cammarato A, Hatch V, Saide J et al. Drosophila muscle regulation characterized by electron microscopy and three-dimensional reconstruction of thin filament mutants. Biophys J 2004; 86(3):1618-1624.
41. Drummond DR, Hennessey ES, Sparrow JC. Characterisation of missense mutations in the Act88F gene of Drosophila melanogaster. Mol Gen Genet 1991; 226:70-80.
42. Okamoto H, Hiromi Y, Ishikawa E et al. Molecular characterization of mutant actin genes which induce heat-shock proteins in Drosophila flight muscles. EMBO J 1986; 5(3):589-596.
43. Hiromi Y, Okamoto H, Gehring WJ et al. Germline transformation with Drosophila mutant actin genes induces constitutive expression of heat shock genes. Cell 1986; 44(2):293-301.
44. Sparrow J, Reedy M, Ball E et al. Functional and ultrastructural effects of a missense mutation in the indirect flight muscle-specific actin gene of Drosophila melanogaster. J Mol Biol 1991; 222:963-982.
45. Reedy MC, Beall C, Fyrberg E. Formation of reverse rigor chevrons by myosin heads. Nature 1989; 339:481-483.
46. Nongthomba U, Cummins M, Clark S et al. Suppression of the muscle hypercontraction phenotype by mutations in the myosin heavy chain gene of Drosophila melanogaster. Genetics 2003; 164:209-222.
47. An H, Mogami K. Isolation of 88F actin mutants of Drosophila melanogaster and possible alterations in the mutant actin structures. J Mol Biol 1996; 260:492-505.
48. Sakai Y, Okamoto H, Mogami K et al. Actin with tumor-related mutation is antimorphic in Drosophila muscle: Two distinct modes of myofibrillar disruption by antimorphic actins. J Biochem (Tokyo) 1990; 107(3):499-505.
49. Brault V, Reedy MC, Sauder U et al. Substitution of flight muscle-specific actin by human (beta)-cytoplasmic actin in the indirect flight muscle of Drosophila. J Cell Sci 1999; 112(Pt 21):3627-3639.
50. Fyrberg EA, Fyrberg CC, Biggs JR et al. Functional nonequivalence of Drosophila actin isoforms. Biochem Genet 1998; 36(7-8):271-287.
51. Schmitz S, Clayton J, Nongthomba U et al. Drosophila ACT88F indirect flight muscle-specific actin is not N- terminally acetylated: A mutation in N-terminal processing affects actin function. J Mol Biol 2000; 295(5):1201-1210.
52. Ball E, Karlik CC, Beall CJ et al. Arthrin, a myofibrillar protein of insect flight muscle, is an actin-ubiquitin conjugate. Cell 1987; 51:221-228.
53. Nongthomba U, Clark S, Cummins M et al. Troponin I is required for myofibrillogenesis and sarcomere formation in Drosophila flight muscle. J Cell Sci 2004; 117(9):1795-1805.
54. Fyrberg E, Fyrberg CC, Beall C et al. Drosophila melanogaster troponin-T mutations engender three distinct syndromes of myofibrillar abnormalities. J Mol Biol 1990; 216:657-675.
55. Barbas JA, Galceran J, Torroja L et al. Abnormal muscle development in the heldup3 mutant of Drosophila melanogaster is caused by a splicing defect affecting selected troponin I isoforms. Mol Cell Biol 1993; 13(3):1433-1439.
56. Beall CJ, Fyrberg E. Muscle abnormalities in Drosophila melanogaster heldup mutants are caused by missing or aberrant troponin-I isoforms. J Cell Biol 1991; 114:941-951.
57. Kronert WA, Acebes A, Ferrus A et al. Specific myosin heavy chains mutations suppress troponin I defects in Drosophila muscles. J Cell Biol 1999; 144(5):989-1000.
58. Prado A, Canal I, Barbas JA et al. Functional recovery of troponin I in a Drosophila heldup mutant after a second site mutation. Mol Biol Cell 1995; 6(11):1433-1441.
59. Naimi B, Harrison A, Cummins M et al. A tropomyosin-2 mutation suppresses a troponin I myopathy in Drosophila. Mol Biol Cell 2001; 12(5):1529-1539.
60. Benoist P, Mas JA, Marco R et al. Differential muscle-type expression of the Drosophila troponin T gene. A 3-base pair microexon is involved in visceral and adult hypodermic muscle specification. J Biol Chem 1998; 273(13):7538-7546.
61. Domingo A, Gonzalez-Jurado J, Maroto M et al. Troponin-T is a calcium-binding protein in insect muscle: In vivo phosphorylation, muscle-specific isoforms and developmental profile in Drosophila melanogaster. J Muscle Res Cell Motil 1998; 19(4):393-403.

62. Miller RC, Schaaf R, Maughan DW et al. A nonflight muscle isoform of Drosophila tropomyosin rescues an indirect flight muscle tropomyosin mutant. J Muscle Res Cell Motil 1993; 14(1):85-98.
63. Karlik CC, Fyrberg EA. An insertion within a variably spliced Drosophila tropomyosin gene blocks accumulation of only one encoded isoform. Cell 1985; 41:57-66.
64. Kreuz AJ, Simcox A, Maughan D. Alterations in flight muscle ultrastructure and function in Drosophila tropomyosin mutants. J Cell Biol 1996; 135(3):673-687.
65. Mardahl-Dumesnil M, Fowler VM. Thin filaments elongate from their pointed ends during myofibril assembly in Drosophila indirect flight muscle. J Cell Biol 2001; 155(6):1043-1053.
66. Hakeda S, Endo S, Saigo K. Requirements of Kettin, a giant muscle protein highly conserved in overall structure in evolution, for normal muscle function, viability, and flight activity of Drosophila. J Cell Biol 2000; 148:101-114.
67. Roulier EM, Fyrberg C, Fyrberg E. Perturbations of Drosophila a-actinin cause muscle paralysis, weakness, and atrophy but do not confer obvious nonmuscle phenotypes. J Cell Biol 1992; 116:911-922.
68. Fyrberg E, Kelly M, Ball E et al. Molecular genetics of Drosophila alpha-actinin: Mutant alleles disrupt Z disc integrity and muscle insertions. J Cell Biol 1990; 110:1999-2011.
69. Whiting A, Wardale J, Trinick J. Does titin regulate the length of muscle thick filaments? J Mol Biol 1989; 205(1):263-268.
70. Gregorio CC, Granzier H, Sorimachi H et al. Muscle assembly: A titanic achievement? Curr Opin Cell Biol 1999; 11(1):18-25.
71. Machado C, Sunkel CE, Andrew DJ. Human autoantibodies reveal titin as a chromosomal protein. J Cell Biol 1998; 141(2):321-334.
72. Kulke M, Neagoe C, Kolmerer B et al. Kettin, a major source of myofibrillar stiffness in Drosophila indirect flight muscle. J Cell Biol 2001; 154(5):1045-1057.
73. Zhang Y, Featherstone D, Davis W et al. Drosophila D-titin is required for myoblast fusion and skeletal muscle striation. J Cell Sci 2000; 113(Pt 17):3103-3115.
74. Machado C, Andrew DJ. D-Titin: A giant protein with Dual roles in chromosomes and muscles. J Cell Biol 2000; 151(3):639-652.
75. Henkin J, Vigoreaux JO. Mapping myofibrillar protein interactions by mutational proteomics. In: Vigoreaux JO, ed. Nature's Versatile Engine: Insect Flight Muscle Inside and Out. Georgetown: Landes Bioscience, 2006:270-284.
76. White KP, Rifkin SA, Hurban P et al. Microarray analysis of Drosophila development during metamorphosis. Science 1999; 286(5447):2179-2184.
77. Arbeitman MN, Furlong EE, Imam F et al. Gene expression during the life cycle of Drosophila melanogaster. Science 2002; 297(5590):2270-2275.
78. Horn C, Offen N, Nystedt S et al. Piggybac-based insertional mutagenesis and enhancer detection as a tool for functional insect genomics. Genetics 2003; 163(2):647-661.
79. Maughan D, Vigoreaux J. Nature's strategy for optimizing power generation in insect flight muscle. In: Sugi H, ed. Mysteries About the Sliding Filament Mechanism: Fifty Years After its Proposal. New York: Kluwer/Plenum Press, 2004:in press.
80. Barton B, Ayer G, Heymann N et al. Flight muscle properties and aerodynamic performance of Drosophila expressing a flightin transgene. J Exp Biol 2005; 208:549-560.

CHAPTER 13

Whole Genome Approaches to Studying *Drosophila* Muscle Development

Eileen E.M. Furlong*

Abstract

With the development of microarray technology, and other whole genome approaches, it is now possible to systematically screen an entire genome for genes that are differentially expressed during two different stages of development and/or two different types of cells. This can be readily applied to muscle development by using genetic manipulations that either overexpress a key gene during development, or by using loss-of-function mutants. This chapter describes the genomic approaches that can be used to identify the vast majority (if not all) of the genes that are normally expressed in muscle cells, and can give an indication of their function during muscle development. These approaches include microarray analysis (transcriptome analysis), RNAi screens (phenome mapping), protein localization studies (localizome mapping), and protein-protein interaction studies (interactome mapping). As microarray analysis is the genomic approach that has been applied the most extensively to *Drosophila* muscle development, the main emphasis of this chapter will be on gene expression profiling during muscle development.

Introduction

Over a decade of research on muscle development in *Drosophila* has uncovered genes that are essential for all aspects of muscle development, including myoblast specification, migration, fusion, mytobube attachment to the epidermis, and the attraction of specific motoneurons. These studies utilized genetic and nongenetic approaches to form the basis of our current knowledge on how different aspects of muscle development are genetically regulated. The recent completion of the *Drosophila* genome project, and that of many other species including humans, has opened the possibility of using whole genome approaches to understand the genetic regulation of muscle development on a global level. Consequently, the technologies and reagents are now available to systematically identify the complete set of genes that are associated with the progression of a particular developmental event. These large-scale approaches should not be viewed as a replacement of genetic approaches, but rather genetic and genomic approaches should be used to complement each other. In order to gain a comprehensive understanding of how muscle development occurs in the coming years, the integration of genetic, genomic, biochemical and bioinformatic approaches will be required.

A number of different types of functional genomic approaches have been used in *Drosophila*, some of which have been specifically applied to muscle development. *Drosophila* is an excellent model organism for functional genomics due to the relative small size of its genome, the reduced incidence of gene duplication compared to vertebrates, and the development of

*Eileen E.M. Furlong—Developmental Biology and Gene Expression Programmes, EMBL, Meyerhofstr 1, 69117 Heidelberg, Germany. Email: Eileen.Furlong@embl.de

Muscle Development in Drosophila, edited by Helen Sink. ©2006 Eurekah.com and Springer Science+Business Media.

essential community reagents by the Berkeley *Drosophila* Genome Project (BDGP). This chapter will discuss the use of microarray analysis (transcriptome analysis), RNAi screens (phenome mapping), protein localization studies (localizome mapping) and protein-protein interaction studies (interactome mapping); the first two of which have been specially applied to *Drosophila* muscle development.

Transcriptome Analysis

The Use of Microarrays for Gene Expression Profiling

Microarrays allow the user to measure the relative levels of mRNA for many or all of the genes in a genome, in parallel. Therefore, if the experiment is properly controlled, microarrays provide a very powerful and efficient method to obtain a genome-wide view of which mRNAs are differentially expressed between two samples. There are many different types of microarrays and more are being developed all the time. The most commonly used microarrays can be categorized into three main types:

cDNA Arrays

cDNA arrays are produced from PCR products of ESTs (Expressed Sequence Tags) or one exon of a gene. The purified PCR products are printed onto a coated microarray slide using a microarray spotting robot. RNA samples, either Poly A+ or total RNA, from the two different conditions to be compared are converted to cDNAs, which are labeled with two different dyes—Cy3 (green) and Cy5 (red) being the dyes most commonly used. For example, RNA from wild-type embryos can be used to synthesize Cy3-labelled cDNA and the RNA from mutant embryos can be used to produce Cy5-labelled cDNA. The labeled samples are then pooled and hybridized to a microarray. As this is a competitive hybridization, if a particular RNA is more abundant in the mutant embryo, the hybridized spot will appear red in color. If the RNA is more abundant in the wild-type embryo the spot will appear green in color, and if the RNA is at equal abundance in both samples the spot will appear yellow (Fig. 1). The fluorescence intensities for each channel are subsequently quantified using a microarray slide scanner.

Long Oligonucleotides

Long oligonucleotides, commonly 70mers, are being widely used for expression profiling in different species. The signal intensity from long oligonucleotides is generally lower than that of cDNA arrays. However, as they are designed to contain regions of little or no homology to other genes, there is very little cross-hybridization. Therefore, despite the lower fluorescence intensities, these arrays are often more sensitive than their cDNA counterparts. Because of the potential increase in sensitivity, as well as making large-scale PCR reactions redundant, long oligonucleotides are slowly replacing the use of cDNAs for studies in many model organisms.

As cDNA and long oligonucleotide arrays are hybridized in a competitive manner, these platforms are usually referred to as two-channel arrays. Two-channel arrays measure the relative abundance of an mRNA species between the two samples and therefore the measurements are always taken as a ratio of sample A over sample B.

Affymetrix Arrays

Affymetrix arrays for *Drosophila* consist 14 pairs of 25bp olgionucleotides for each gene, one perfect match oligonucleotide and a 1bp mismatch oligonucleotide. These oligonucleotides are synthesized in situ on the microarray slide by photolithography and solid-phase chemistry. In contrast to two-channel arrays, Affymetrix arrays are hybridized by initially amplifying the RNA sample and then hybridizing labeled RNA to the array. Each sample to be analyzed is hybridized to a separate array, and therefore this platform is often referred to as single-channel arrays. The results obtained are the absolute intensity values for each gene under a given condition, rather than a relative ratio value.

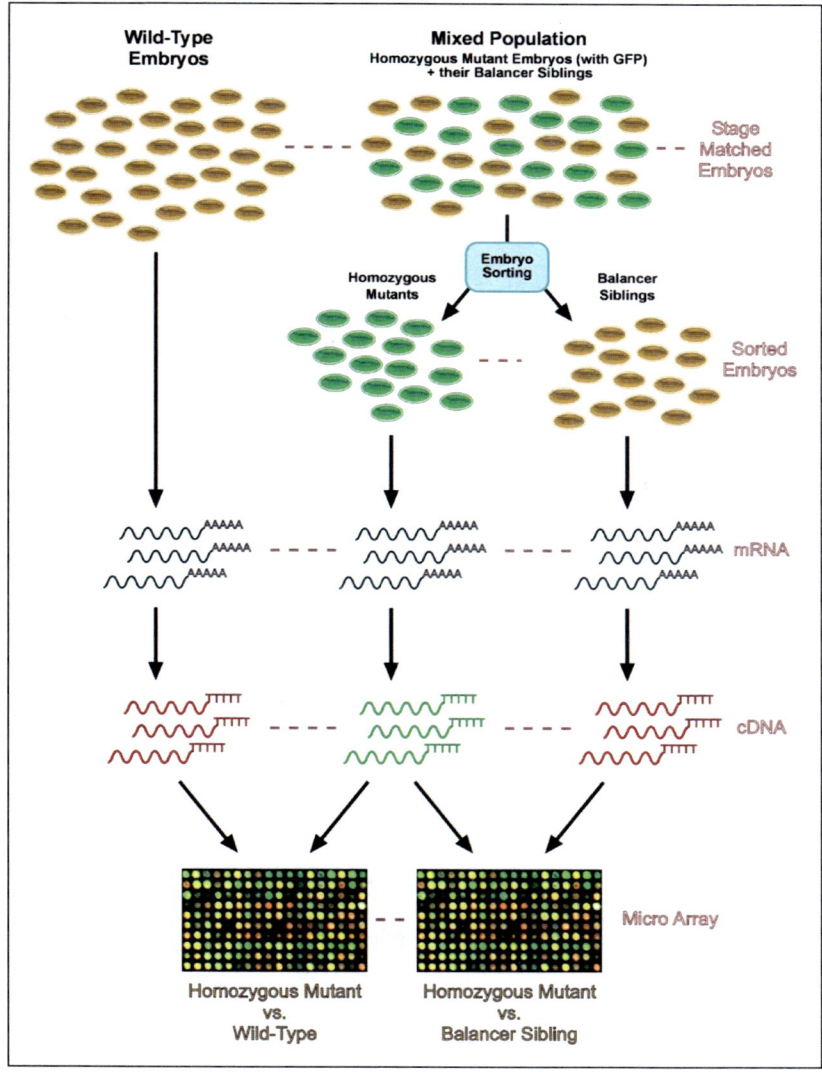

Figure 1. Gene Expression Profiling of *Drosophila* Homozygous Mutant Embryos. Homozygous mutant embryos are isolated from their balancer siblings based on the presence or absence of GFP. The sorting can either be done by hand or by using an automated embryo sorter. The left hand side of the figure shows a comparison of the gene expression profiles of homozygous mutant embryos to stage matched wild-type embryos. The wild-type embryos are collected and processed in an identical manner to the mutant embryo collection. The right hand side of the figure shows a comparison of homozygous mutant embryos to their balancer sibling embryos. The balancer siblings are obtained from the same embryo collection. The RNA is isolated from all embryo collections using an identical protocol. The RNA is then converted to cDNA that is fluorescently labeled. In the example shown in the figure, the RNA obtained from the homozygous mutant embryos is converted to Cy3 labeled cDNA and hybridized to a microarray with Cy5 labeled cDNA obtained from either staged matched wild-type embryos or from the balancer siblings. The relative changes in gene expression between the two samples is quantitated by measuring the fluorescent intensities of each spot on the array.

After any microarray experiment, the raw data must be normalized to remove systematic biases due to differences in the specific activities in the labeled samples, differences in the ability of the scanner to detect the intensities from each channel, quenching, positional effects on the array etc. It is very important to perform repeat experiments, which can be of two types. Technical repeats are preformed by labeling and hybridizing the same sample to two different microarrays. The differences in the ratio between two technical repeats give an indication of the variation between different labeling reactions, different hybridizations, and different microarrays. Usually this is expressed as the correlation coefficient, and for technical repeats should be in the >0.95 range. Biological repeats are performed using independently collected and prepared samples. The variation in biological repeats is the sum of both the biological variation and the technical variation. For any microarray experiment, it is essential to do a number of biological repeats so that some statistical analysis can be performed on the data.

Within the *Drosophila* community, different research groups are using a number of different microarray platforms. Most of the large-scale *Drosophila* microarray-using laboratories are using two channel microarray platforms, which are made in-house. In order to improve the ability to cross compare datasets from different *Drosophila* research groups, there is currently an effort to design a set of 70 mer oligonucleotides for every predicted gene in the genome. The hope is that these long oligonucleotides will ultimately become the standardized platform for *Drosophila* microarray experiments. Such standardization should vastly improve the evaluation and integration of datasets from different sources, and would therefore allow the results obtained from microarray experiments to be used in a controlled manner to help annotate the predicted function of uncharacterized *Drosophila* genes.

Applying Transcriptome Mapping to Drosophila Mutant Embryos

In order to identify genes involved in a particular type of muscle cell's development, the gene expression profiles of wild-type embryos can be compared to mutant embryos which fail to develop that cell type. Rigorous genetic studies have generated loss-of-function and gain-of-function mutant strains for a number of key genes involved in muscle development. We are now at the exciting time where these genetic reagents can be combined with genomic approaches to identify genes that are normally expressed in a particular muscle type at a particular stage in development. However, there are several technical issues that must be addressed when analyzing the expression profiles of homozygous mutant embryos.

The first issue is the presence of the balancer chromosome. *Drosophila* recessive lethal mutations are maintained over balancer chromosomes. These are modified chromosomes that allow the stable maintenance of *Drosophila* stocks by repressing the recovery of recombinants. Homozygous mutant embryos are obtained from heterozygous (i.e., balancer chromosomes carrying) parents. Therefore in an embryo collection the homozygous mutant embryos are mixed with their balancer chromosome bearing, heterozygous siblings. Before analyzing the affect of a particular mutant on gene expression, a pure population of homozygous mutant embryos must be isolated.

By placing a green fluorescent protein (GFP) marker gene on either the balancer chromosome, or by recombining it onto the mutant chromosome, it is possible to recognize the homozygous mutants in living embryos, and thereby separate them from their balancer chromosome carrying, heterozygote siblings. The embryo sorting can be done by hand picking the homozygous embryos of interest from their heterozygous siblings. This works well when examining a single mutant at one developmental time point. However, when collecting homozygous mutant embryos from a number of developmental stages, with three or four biological repeats for a number of different mutants, it is easier and more consistent to use an automated approach. The *Drosophila* embryo sorter is an automated machine that can sort *Drosophila* embryos based on the presence or absence of a genetic marker (e.g., GFP).[1] The machine can sort at an accuracy of >99% while having little or no affect on viability. The presence of large numbers of unfertilized embryos in the homozyous mutant population will affect the gene expression profiles obtained. This problem can be avoided by placing the GFP onto the same chromosome as the mutant rather than on the balancer chromosome, and is therefore the preferred method for sorting using the embryo sorter (Fig. 1).

The second consideration when analyzing the expression profiles of homozyous mutant embryos is the genetic background in which the mutant is maintained. The balancer chromosome is not only very effective in maintaining the mutation of interest, but will also stably maintain any other spurious mutations that have accumulated on that chromosome. Therefore, when comparing homozygous mutant embryos to stage matched wild-type embryos, the changes in gene expression will reflect that of the known mutation and also any other unwanted mutations that have accumulated on the same chromosome. In order to avoid this two approaches can be taken: (i) Sorting trans-heterozgous mutant embryos from two different null alleles of the gene of interest. (ii) Alternatively, in cases where only one characterized null allele is available for the gene of interest, the background of the mutation can be genetically 'cleaned up'. For this, the mutation undergoes two outcrosses, where females are selected to allow for two successive recombination events. The resulting 'cleaned up' strain is then established from a single male.

Finally, the embryo staging between the wild-type and the mutant embryos must be verified to ensure that they are tightly matched. Microarray experiments on *Drosophila* embryos work very well because of the ability to obtain tightly staged collections during embryogenesis. When comparing the expression profile of wild-type embryos to stage matched homozygous mutant embryos, it is advisable to fix a small percentage of both the wild-type and mutant population for each embryo collection. The rest of the embryo is collection snap-frozen for RNA isolation. Immunostaining the fixed population of embryos with a muscle marker will allow for tighter stage verification, rather than just looking at morphological landmarks, and will also verify the presence of the muscle defect in the mutant population. Only tightly stage matched embryo collections should be used for RNA isolation and microarray analyses.

A detailed molecular analysis of the *Drosophila* lifecycle has been conducted, comparing the expression profiles of 66 consecutive time periods spanning the entire lifecycle.[2] This includes 30 consecutive one-hour time points during embryogenesis, the first six of which are overlapping one-hour intervals. All 30-embryo time points were carefully stage verified and recollected where necessary. The expression profiles of all stages of the lifecycle were compared using a common reference sample, which was composed of a mixture of RNA from all stages of the lifecycle. This study identified groups of genes whose transcript abundance is tightly coregulated during different stages of embryogenesis. These groups of genes or clusters can therefore serve as molecular fingerprints for each stage of embryogenesis, and could therefore help to molecularly stage mutant embryo collections.

When analyzing homozyous mutant embryos the expression patterns can either be compared to stage-matched wild-type embryos or to the balancer sibling population (Fig. 1). The advantage of comparing the mutant embryos to their balancer siblings is the fact that the embryos will be exactly stage-matched, as they come from the same embryo collection. However, this may not always be the case if the mutant is developmentally delayed. An additional disadvantage of this approach is that the balancer sibling population will also contain homozygous balancer embryos, which develop abnormally and therefore have aberrant gene expression profiles.

The Use of Microarray Analysis to Molecularly Characterize Muscle Development

To date, the combination of the embryo sorter and DNA microarrays has been used successfully to compare the expression profiles of wild-type embryos to stage matched embryos that are missing all muscle cells and to embryos that produce ectopic myoblasts.[3]

Total Muscle Analyses

Twist is a bHLH transcription factor that is essential for several aspects of the early stages of muscle development including mesoderm gastrulation, proliferation, migration, segmentation and finally specification.[4-7] In twist homozygous mutant embryos no muscle of any type develops. In our study, a molecularly characterized loss-of-function mutation for *twist* was genetically 'cleaned up' by out-crossing the mutation of interest and establishing a stock from a single male. The embryo sorter was used to isolate large populations of twist homozygous mutant embryos at three consecutive one-hour time points in development. The gene expression profiles of *twist* homozygous mutant embryos were compared to stage matched wild-type embryos.

In total, 360 mesodermal genes were identified, of which 294 had not been previously characterized.[3] The expression pattern of 67 of these genes was verified by in situ hybridization, 61 of which are expressed in *twist* regulated tissues. This represents a success rate of 91%. This shows that microarrays are sensitive enough to detect tissue specific changes in gene expression within the context of the entire embryo, and therefore provide a proof of principle for this type of analysis. Furthermore, there are dozens of mutants that affect different aspects of muscle development that could be analyzed in a similar manner.

Depending on which of the three stages of development the gene expression changes were detected, the new muscle genes could be divided into three groups of genes. Group 1 genes had reduced expression levels in all three of the developmental time-points and represents genes that are expressed in the mesoderm at all stages examined e.g., *dMef2*. The second group of genes only had reduced expressed in the early time and therefore represents genes that are expressed in the early mesoderm e.g., *stumps*. While the third group of genes had reduced expression only in the later time point and therefore genes that are only expressed during later stages of muscle development (> stage 11) e.g., *blown fuse*. By combining mutant embryo analysis with embryo collections at different developmental stages it is not only possible to obtain spatial information about what tissues genes are expressed in, but also temporal information. This can yield valuable insight into the likely roles of new genes in the development of that tissue. This temporal information is often distorted in over expression experiments.

Using the current annotation, the proteins encoded by the newly identified mesoderm genes include 11 putative transcription factors, 5 signal transduction molecules, 3 protein kinases, 8 cell adhesion proteins and 89 proteins of unknown function. At the time when the study was conducted, the *Drosophila* genome sequence had not been published. Consequently, the study was conducted on microarrays containing all of the available ESTs at that time, which represents about 35% of the *Drosophila* genome. The large number of uncharacterized genes identified in this study, which will likely more than double with whole genome analysis, indicate how much we still have to learn about mesoderm specification and muscle development.

The main advantage of studying the expression profiles of homozygous mutants embryos is that the embryos are intact at the time of RNA isolation. Although some non cell autonomous signaling may be missing due to the absence of the cells of interest, many of cells in the embryo continue to develop normally. However, the disadvantage of this approach is its limited sensitivity. Eventually we will hit a detection limit where it will not be possible to accurately detect changes in gene expression in very small numbers of cells within the context of the entire embryo. This detection limit will depend on how many cells are of interest and at what level the genes are being expressed within those cells.

Analyses of Rare Muscle Cell Populations

Genetic manipulations can be used to trick the embryo into producing an ectopic population of one type of muscle cell, usually at the expense of another. The gene expression profiles of embryos enriched in a particular muscle cell type, can then be compared to stage-matched wild-type embryos (Fig. 2A). This has been done in the case of the founder cells and the fusion-competent myoblasts.[8] There are thirty different founder cells in each hemisegment of

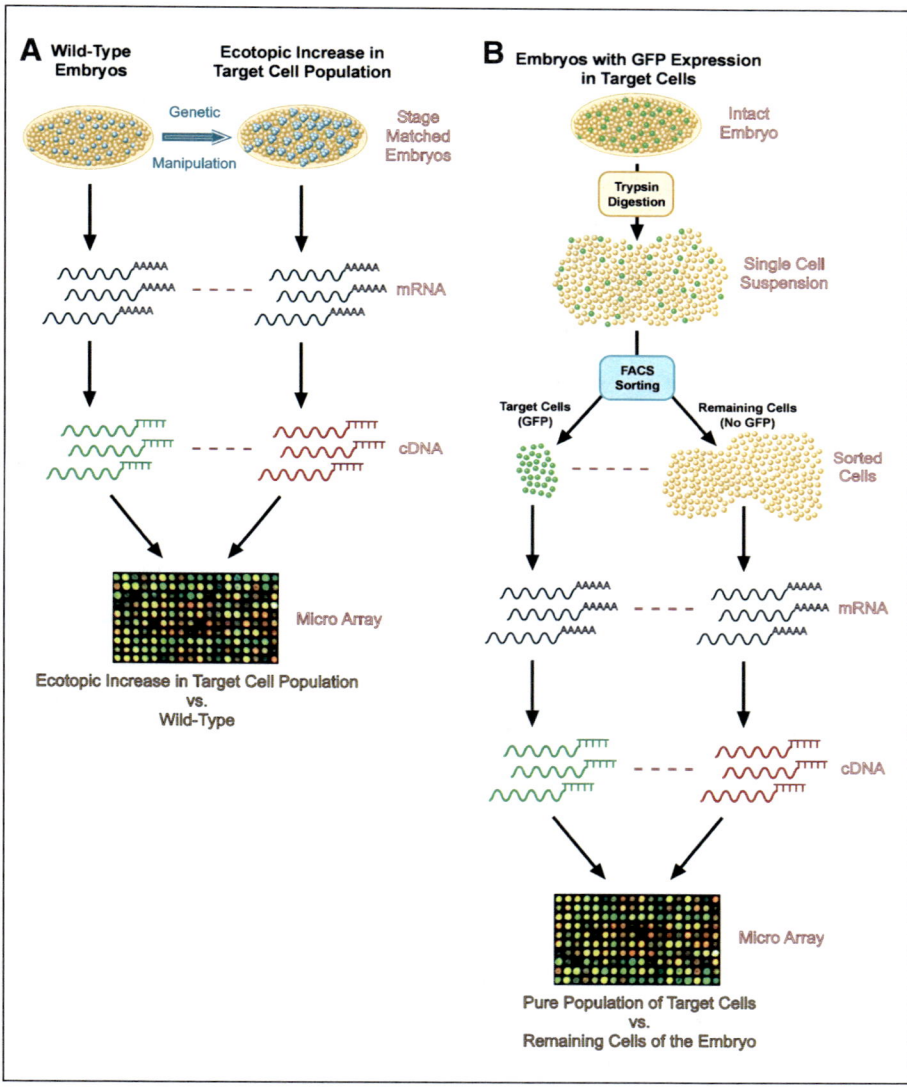

Figure 2. Gene expression profiling of rare populations of cells in *Drosophila* embryos. A) Where possible, genetic manipulations can be used to increase a population of rare cells within the embryo, usually at the expense of another cell population. RNA can then be isolated from the entire embryo with the ectopic increase in target cells and from stage matched wild-type embryos. The RNA is converted to fluorescently labeled cDNA and hybridized to an array. At least some of the genes that are normally expressed within the target cell population are likely to have an increased expression in the embryos that are enriched in this cell type, compared to wild-type embryos. B) Where specific enhancers are available, GFP can be selectively expressed in a rare cell population. After dissociation of the embryos a pure population of the target cell type can be obtained by FACS sorting. The gene expression profiles of the sorted rare cell population can then be compared to dissociated cells from the rest of the embryo. In other words, the flow-through from the FACS machine. Due to the very small amount of sample obtained from this approach, amplification of the RNA will usually be required prior to microarray analysis.

the *Drosophila* embryo, which represents roughly 1% of the total number of cells of the embryo. It is therefore not possible to accurately detect changes in gene expression in these cells within the context of the entire wild-type embryo. In order to address this we used a muscle specific enhancer to drive the expression of an activated form of Ras in the mesoderm. This led to the production of ectopic founder cells. Similarly, overexpression of an activated form of Notch was used to drive the production of ectopic fusion-competent myoblasts, while blocking the production of founder cells.[8] Both the overexpression of Ras and Notch were done in the genetic background of Toll[10B] mesoderm-enriched embryos. This served to increase the population of immature myoblasts that can respond to these signals, thereby increasing the number of myoblasts that could be converted to a founder cell fate and therefore boosting the signal. As Toll[10B] embryos do not develop any ectoderm and nervous tissue, this genetic background also reduces the tissue complexity of these embryos.

The gene expression profiles of founder cell-enriched embryos compared to fusion-competent myoblast-enriched embryos identified 83 genes with differential expression between the two types of myoblasts. Many of these are normally only expressed in a subset of founder cells in wild-type embryos, indicating that it is possible to detect changes in gene expression in whole animal experiments for transcripts that are normally only expressed in a fraction of 1% of the cells in the embryo. For example, *asense*, *tartan* and *nidogen* were all confirmed by confocal microscopy to be transiently expressed in a subset of founder cells (Fig. 4 J-R in ref. 8). One of the more surprising genes identified in this study was *phyllopod*. Phyllopod is a novel adaptor protein which is know to play an essential role in other tissues in targeting proteins for degradation via the E3 ubiquitin protein ligase complex.[9] *phyllopod* plays an essential role in photoreceptor and external sensory cell fate determination by targeting the degradation of the transcriptional repressor Tramtrack.[10-12] However, there was no known function for *phyllopod* in muscle development. The microarray experiment described above predicts that Phyllopod is enriched in founder cells. Subsequent confocal analysis showed that Phyllopod is transiently expressed in a subset of founder cells in somatic muscle. Analysis of loss-of-function *phyllopod* mutant embryos showed that *phyllopod* is an essential regulator of somatic muscle differentiation,[8] providing a strong link between the E3 ubiquitination pathway and muscle patterning.

Using genetic manipulations to increase the population of muscle cells of interest within the context of the intact embryo has the advantage that the cell's extracellular matrix (ECM) contacts are intact. Therefore spurious gene expression changes due to loss of the ECM can be avoided. However, the cellular context of the myoblasts within these embryos is abnormal, and so transcripts whose expression is controlled by neighboring cell-cell interactions or by nonautonomous effects may not be detected. An alternative approach is to overexpress GFP in the population of muscle cells of interest, dissociate the embryos and FAC sort out the GFP expressing muscle cells. The expression profiles of the FAC sorted muscle cells can be examined using RNA amplification followed by microarray analysis (Fig. 2B).

Recently a direct in vivo method has been developed to isolate expressed mRNAs from particular populations of cells within a tissue.[13,14] This technique exploits the prior knowledge of an endogenous Poly-A+ binding protein specific to the cells of interest, or the ability to overexpress an epitope-tagged Poly-A+ binding protein in the cells of interest. The cells or embryos are then cross-linked to complex the Poly-A+ binding protein with the expressed mRNAs. The expressed mRNAs are specifically enriched from the tissue of interest by immunoprecipitating the complex of mRNA/Poly-A+ binding protein. The mRNAs enriched by the immunoprecipitation should represent the genes expressed in that tissue and can be identified using microarray analysis. Therefore, the RNA is immunoprecipitated from intact cells within their normal physiological surrounding. This is a very promising method to accurately determine the expression profiles of specific populations of cells, and has already been used in intact *C. elegans* embryos to identify genes that are enriched in both muscle and neurons.[14] Given the relative ease at which transgenic *Drosophila* strains can be generated and the large number of existing tissue-specific enhancers this approach should be readily applicable to *Drosophila*. Although the sensitivity of this technique still remains to be determined, it may offer an alternative to FAC sorting for expression profiling rare populations of cells.

In order to obtain the most comprehensive list of muscle development genes as possible, a combination of both loss-of-function and gain-of-function experiments will be required for a large number of different mutants at all stages of muscle development. This is a formidable task, but one that can be achieved within the coming years.

Phenome Mapping

New functions were assigned to previously uncharacterized *C. elegans* genes using whole genome RNAi screens.[15] In *C. elegans* the double stranded RNA (dsRNA) is administered by feeding transformed bacteria to the worms or by simply bathing the worms in solution. Unfortunately for *Drosophila* embryos the most efficient method of dsRNA administration is by injecting each embryo by hand. Even then, the severity of the phenotype observed varies greatly within the injected population. To complicate matters, some genes that have a clear phenotype with a characterized loss-of-function mutation, don't seem to give a phenotype at all using RNAi. In an extreme example, *nautilus* was shown to give a strong muscle phenotype by RNAi[16] but was largely wild-type in the null mutation.[17]

Despite these difficulties in *Drosophila,* a large RNAi screen for genes that affect heart development has been performed. dsRNAi probes for 5,800 genes (~ 36% of the predicted genes in the genome) were hand injected into embryos expressing the *beta-galactosidase* gene under the control of the dMef2 heart-expressing enhancer[18] This was a huge effort, and led to the successful identification of 132 genes that showed heart defects in the dsRNA-injected embryos. The heart phenotypes observed for a proportion of these genes are likely indirect effects due to their essential roles in other processes. However, the phenotypes of some of the new genes identified are the result of a specific requirement of that gene in heart development. An interesting gene identified in this study, *simjang*, is essential for the specification of two of the four *tinman*-expressing cardioblasts. Simjang is a zinc finger protein. It is homologous to vertebrate proteins that are components of the chromatin-remodeling complex recruited by methyl-CpG-DNA, suggesting a requirement for epigenetic regulation in cardioblast specification in *Drosophila.*

Systematic RNAi screens have been carried out in *Drosophila* tissue culture systems to identify genes involved in the Hedgehog signaling pathway[19] as well as in the control of cell morphology.[20] RNAi works at a very high efficiency in established cells lines[21] and is relatively easy to do in tissue culture on a whole genome scale. RNAi screens of this type can clearly identify genes involved in basic cellular processes, however it is unclear how much established cell lines mimic muscle development during embryogenesis. Schneider SL2 cells are perhaps the best characterized and seem to at least partially activate the myogenic program in response to the ectopic expression of *daughterless.*[22] A more physiological, but very labor intensive, approach would be to use primary culture of immature myoblasts. This approach is most likely to succeed when studying a single process or step of muscle development.

Tissue culture based RNAi screens using either primary or established cell lines may also miss some genes due to the reduced complexity of these systems. However, for the genes identified by RNAi, the screen will give an indication of whether the gene is essential for muscle development and for what aspect of muscle development the gene function is required. Both of these pieces of information are not usually inherent by simply looking at a gene's expression pattern.

Localizome Mapping

There are a number of other approaches that can systematically identify new genes involved in muscle development. These can be used to validate the roles of genes identified from microarray based experiments, serve in a complementary fashion to identify new roles for genes in muscle patterning and give an indication for the predicted function of new muscle genes.

The BDGP have initiated a project to do in situ hybridizations for every predicted *Drosophila* gene. To date the expression patterns for 3012 genes has been carefully documented, giving both the developmental stage and tissue in which the gene is expressed. This project will likely take several more years to be completed, but will serve as a very important resource, whose value will extend beyond the *Drosophila* community.

Individual groups have begun efforts to create *Drosophila* strains that contain GFP fusions for every *Drosophila* protein. These fusion proteins will not only identify the tissue in which a gene is expressed, but also provide the subcellular localization of the encoding proteins, thereby giving an indication of the possible function of uncharacterized genes. While GFP fusions have recently been generated for every gene in the yeast genome by homologous recombination,[23,24] in *Drosophila* the task is much more daunting as each fusion strain is generated by screening for protein trapping events. An artificial exon encoding GFP flanked by splice acceptor and donor sequences is contained within a transposable element. Fusion proteins are randomly generated by mobilizing the P-element and selecting strains where the GFP has inserted into an intron, and is incorporated into the mature transcript in the correct reading frame.[25] The frequency of obtaining a viable GFP fusion event in the correct reading frame is approximately 1:1000. Due to the large number of embryos that have to been screened, this approach has been greatly facilitated by the automated embryo sorter. However, it will take a huge effort to generate strains that contain fusion proteins covering the majority of the *Drosophila* genome. Some proteins will never be targeted in this manner due a lack of introns or as a result of the inaccessibility of the chromatin. However, the GFP fusion strains that are generated will be a very valuable resource to characterize the role of these proteins in vivo. To date, GFP fusions have already been generated for Tropomyosin II and kettin.[25]

Interactome Mapping

Whole genome protein-protein interaction screens have been conducted in yeast and *C. elegans* using Tap-tag[26] and yeast-two-hybrid approaches.[27] This has led to the identification of large protein interaction networks, revealing new potential functions for unknown genes that physically interact with proteins known to be involved in a particular cell function. The first attempts at a protein interaction map for *Drosophila* has recently been published where 7048 proteins were analyzed using a yeast-two-hybrid approach.[28] A statistical model, based on a training set of known interacting and noninteracting proteins, was used to select 4780 unique interactions (4679 proteins) to create a higher confidence interaction map. This covers approximately 30% of the predicted genome. As with all large-scale approaches, protein interaction datasets are inherently noisy and therefore contain many false positives and negatives. However, they can still give useful indications of a protein's binding partner. In the case of newly identified genes that are expressed in muscle, this may give an indication of their functional role in muscle development.

Conclusion

All the whole genome high throughput approaches discussed in this chapter have their inherent caveats. They will each miss the roles of some genes in muscle development due to the intrinsic biases of the assay, and the occurrence of false negatives. They will also all contain some incorrect data due to the presence of false positives. It is, therefore, very important to combine the data from different types of experiments. In the case of microarray experiments there should be enough biological repeats for statistical analysis for a given genotype or condition. To increase the confidence of the role of a group of genes in a given aspect of muscle development, multiple genotypes or conditions should be used, e.g., loss-of-function and gain-of-function, or loss-of-function of two factors acting in the same pathway.

Data arising from any single high throughput approach will only provide a rough indication of the functional role of a protein. In order to get a comprehensive picture of how a cell develops into a myoblast and becomes part of a multicellular muscle, the integration of multiple approaches is needed, ranging from expression studies, protein localization studies, functional assays and protein-protein interaction studies.

It is the acquisition and combination of these whole genome datasets with relation to muscle development that is the major challenge in the coming years. Much progress has been made in recent years in the development and implementation of computational approaches to group together genes that behave similarly in large datasets from different types of high throughput experiments. These data integration methods will allow the parts list of muscle development to be ordered into functional groups of genes or modules, each of which will be required to drive the progression of a particular aspect of muscle development.

Mutations in vertebrate homologs of many genes that play essential roles in *Drosophila* muscle development cause human diseases. For example, mutations in *Mef2A* result in an inherited form of coronary artery disease[29] and *NKX2-5* mutations cause congenital heart disease.[30] Therefore the identification of functional modules for muscle development in *Drosophila* will also help our understanding of how defects in muscle development occur, and may provide possible insights into how muscle diseases can be treated.

References

1. Furlong EE, Profitt D, Scott MP. Automated sorting of live transgenic embryos. Nat Biotechnol 2001; 19(2):153-156.
2. Arbeitman MN, Furlong EE, Imam F et al. Gene expression during the life cycle of Drosophila melanogaster. Science 2002; 297(5590):2270-2275.
3. Furlong EE, Andersen EC, Null B etal. P atterns of gene expression during Drosophila mesoderm development. Science 2001; 293(5535):1629-1633.
4. Thisse B, el Messal M, Perrin-Schmitt F. The twist gene: Isolation of a Drosophila zygotic gene necessary for the establishment of dorsoventral pattern. Nucleic Acids Res 1987; 15(8):3439-3453.
5. Leptin M, Grunewald B. Cell shape changes during gastrulation in Drosophila. Development 1990; 110(1):73-84.
6. Borkowski OM, Brown NH, Bate M. Anterior-posterior subdivision and the diversification of the mesoderm in Drosophila. Development 1995; 121(12):4183-4193.
7. Baylies MK, Bate M. Twist: A myogenic switch in Drosophila. Science 1996; 272(5267):1481-1484.
8. Artero R, Furlong EE, Beckett K et al. Notch and ras signaling pathway effector genes expressed in fusion competent and founder cells during Drosophila myogenesis. Development 2003; 130(25):6257-6272.
9. Li S, Xu C, Carthew RW. Phyllopod acts as an adaptor protein to link the sina ubiquitin ligase to the substrate protein tramtrack. Mol Cell Biol 2002; 22(19):6854-6865.
10. Li S, Li Y, Carthew RW et al. Photoreceptor cell differentiation requires regulated proteolysis of the transcriptional repressor tramtrack. Cell 1997; 90(3):469-478.
11. Pi H, Wu HJ, Chien CT. A dual function of phyllopod in Drosophila external sensory organ development: Cell fate specification of sensory organ precursor and its progeny. Development 2001; 128(14):2699-2710.
12. Tang AH, Neufeld TP, Kwan E et al. PHYL acts to down-regulate TTK88, a transcriptional repressor of neuronal cell fates, by a SINA-dependent mechanism. Cell 1997; 90(3):459-467.
13. Tenenbaum SA, Carson CC, Lager PJ et al. Identifying mRNA subsets in messenger ribonucleoprotein complexes by using cDNA arrays. Proc Natl Acad Sci USA 2000; 97(26):14085-14090.
14. Roy PJ, Stuart JM, Lund J et al. Chromosomal clustering of muscle-expressed genes in Caenorhabditis elegans. Nature 2002; 418(6901):975-979.
15. Kamath RS, Fraser AG, Dong Y et al. Systematic functional analysis of the Caenorhabditis elegans genome using RNAi. Nature 2003; 421(6920):231-237.
16. Misquitta L, Paterson BM. Targeted disruption of gene function in Drosophila by RNA interference (RNA-i): A role for nautilus in embryonic somatic muscle formation. Proc Natl Acad Sci USA 1999; 96(4):1451-1456.
17. Balagopalan L, Keller CA, Abmayr SM. Loss-of-function mutations reveal that the Drosophila nautilus gene is not essential for embryonic myogenesis or viability. Dev Biol 2001; 231(2):374-382.
18. Kim YO, Park SJ, Balaban RS et al. A functional genomic screen for cardiogenic genes using RNA interference in developing Drosophila embryos. Proc Natl Acad Sci USA 2004; 101(1):159-164.

19. Lum L, Yao S, Mozer B et al. Identification of hedgehog pathway components by RNAi in Drosophila cultured cells. Science 2003; 299(5615):2039-2045.

20. Kiger A, Baum B, Jones S et al. A functional genomic analysis of cell morphology using RNA interference. J Biol 2003; 2(4):27.

21. Caplen NJ, Fleenor J, Fire A et al. dsRNA-mediated gene silencing in cultured Drosophila cells: A tissue culture model for the analysis of RNA interference. Gene 2000; 252(1-2):95-105.

22. Wei Q, Marchler G, Edington K et al. RNA interference demonstrates a role for nautilus in the myogenic conversion of schneider cells by daughterless. Dev Biol 2000; 228(2):239-255.

23. Huh WK, Falvo JV, Gerke LC et al. Global analysis of protein localization in budding yeast. Nature 2003; 425(6959):686-691.

24. Ghaemmaghami S, Huh WK, Bower K et al. Global analysis of protein expression in yeast. Nature 2003; 425(6959):737-741.

25. Morin X, Daneman R, Zavortink M et al. A protein trap strategy to detect GFP-tagged proteins expressed from their endogenous loci in Drosophila. Proc Natl Acad Sci USA 2001; 98(26):15050-15055.

26. Gavin AC, Bosche M, Krause R et al. Functional organization of the yeast proteome by systematic analysis of protein complexes. Nature 2002; 415(6868):141-147.

27. Reboul J, Vaglio P, Rual JF et al. C. elegans ORFeome version 1.1: Experimental verification of the genome annotation and resource for proteome-scale protein expression. Nat Genet 2003; 34(1):35-41.

28. Giot L, Bader JS, Brouwer C et al. A protein interaction map of Drosophila melanogaster. Science 2003; 302(5651):1727-1736.

29. Wang L, Fan C, Topol SE et al. Mutation of MEF2A in an inherited disorder with features of coronary artery disease. Science 2003; 302(5650):1578-1581.

30. Schott JJ, Benson DW, Basson CT et al. Congenital heart disease caused by mutations in the transcription factor NKX2-5. Science 1998; 281(5373):108-111.

CHAPTER 14

Comparison of Muscle Development in *Drosophila* and Vertebrates

Michael V. Taylor*

Abstract

There are many fundamental similarities in the biology of *Drosophila* and vertebrates, and *Drosophila* has become a prominent model organism for studies of animal development. Here the development of the different vertebrate muscle types (skeletal, cardiac and smooth) is compared with their anatomical counterparts in *Drosophila* (somatic, heart/cardiac and visceral). The similarities are highlighted and attention is drawn to any differences. The chapter emphasizes, in particular, the impact of *Drosophila* on the genetic analysis of muscle development. The body of research covered herein also allows an assessment of the extent of the similarity between different aspects of *Drosophila* and vertebrate muscle development. One outcome of this is an evaluation of the usefulness of *Drosophila* to inform a variety of studies of clinical significance.

Introduction

There are fundamental similarities between the biology of the fruit fly, *Drosophila melanogaster*, and that of vertebrate species, including ourselves. While this statement may be readily accepted today, not so many years ago the extent of the similarity was less clear. Similarity at the molecular genetic level has now been brought into sharp focus through the extensive genome sequencing and annotation of the Genome Projects. Such projects have revealed that 60% of *Drosophila* proteins share sequence similarity with human proteins.[1] This genetic conservation is one reason why *Drosophila* is regarded by many as a super model organism for the analysis of animal development. Other key reasons are the wealth of knowledge accrued in one hundred years of laboratory-based research, the range of available experimental techniques, and the speed and relatively low cost of experimentation. Importantly, genetic analysis of gene function is greatly facilitated in *Drosophila* because its genome is simpler in composition than its vertebrate counterparts.

A crucial question for muscle biologists, given the experimental advantages of *Drosophila*, is how similar are the many facets of muscle development between *Drosophila* and vertebrate species. Fundamental similarities highlight key events and underscore the utility of using a simpler model organism with a view to understanding other animals, and to developing new clinical approaches. Differences indicate how animals use similar gene sets and developmental mechanisms to produce different morphological outcomes. They also caution against simple extrapolation from one species to another.

*Michael V. Taylor—Cardiff School of Biosciences, Cardiff University Main Building, Park Place, Cardiff CF10 3TL, U.K. Email: TaylorMV@cf.ac.uk

Muscle Development in Drosophila, edited by Helen Sink. ©2006 Eurekah.com and Springer Science+Business Media.

Vertebrate muscle is categorized into three major muscle types defined by their structural and functional properties: skeletal, cardiac and smooth. The latter includes blood vessel muscle and the visceral muscle surrounding the gut. In this chapter I will restrict myself to these three major muscle types and for this comparison I will equate them to their anatomical counterparts in *Drosophila*—the somatic (or body wall) muscle, heart, and visceral muscle respectively. There is no *Drosophila* counterpart of vertebrate blood vessel smooth muscle. The extent to which the different muscle types are physiological equivalents in *Drosophila* and vertebrates is beyond the scope of this chapter. Nevertheless, and although there are differences in the anatomy and some of the cell biology, it remains apparent that when the development of each muscle type is analyzed there are many similarities between *Drosophila* and vertebrates.

The coverage of the broad topic of this chapter is necessarily selective. Firstly, it will focus mainly on skeletal/somatic muscle, reflecting the area of most research. Secondly, much of the chapter will focus on the development in the *Drosophila* embryo of the larval muscles, because this is where most is known. However, references will be made to the *Drosophila* adult because although some aspects of the development of the adult tissue may essentially be a recapitulation of the embryonic development, other aspects differ and serve as interesting comparisons with vertebrate species. Finally, in making comparisons with *Drosophila* only specific, illustrative aspects from the wide range of vertebrate species studied are included. The aim here is to highlight some areas of interest, rather than to be comprehensive.

Overview of *Drosophila* Muscle Development

Drosophila somatic muscle, cardiac muscle, and the visceral muscle all develop from the mesoderm,[2] which is the layer of cells between the endoderm and ectoderm in the gastrula. It is specified in the ventral region of the syncytial blastula embryo. At gastrulation the mid-ventral cells invaginate and spread in a layer under the ectoderm to form the mesoderm proper. At this stage they are not committed to a specific cell fate. Then, during the next few hours, the embryo becomes segmented and each mesodermal segment becomes subdivided along the anterior-posterior (A/P) and dorsal-ventral (D/V) axes (Fig. 1). The cells proliferate, diversify, and commit to different cell fates that include the three major muscle types together with other mesodermal derivatives (e.g., the fat body, which is equivalent to vertebrate liver). Concurrently, the mesoderm begins to separate into two cell layers, classically known as the somatic and splanchnic mesoderm.[3] The somatic muscles derive from the external, somatic mesoderm and the visceral muscles derive from the internal, splanchnic mesoderm.[3] The cardiac muscle derives from the most dorsal, external mesodermal cells. Progenitor populations of each of these derivatives develop at specific positions along the A/P and D/V axes in each segment. How the heart and visceral muscle subsequently develop from these cells is described later. The process for the somatic muscle can be summarized as follows.[2,4-6]

In each abdominal hemisegment of the *Drosophila* embryo a stereotypic pattern of thirty distinct larval somatic muscles develops. Each muscle has characteristic properties, including its size, shape and innervation. Within the somatic mesoderm individual muscle progenitors are singled out from their neighbors, while the other cells become the "fusion-competent myoblasts". Progenitors divide asymmetrically to make two "founder" cells, or in certain cases one founder cell and one adult muscle precursor. The adult muscle precursors will later proliferate and differentiate into the adult abdominal muscles. Other adult muscle precursors in the thoracic segments will form the intricate and diverse adult thoracic musculature. The founder cells in the developing embryo have a critical role. Each seeds the development of a specific muscle and endows it with specific characteristics through the expression of "muscle identity" genes. Each founder cell attracts and fuses with fusion-competent myoblasts to form mature individual multinucleate myotubes, the final syncytial muscles, which attach to specific sites on the epidermis.

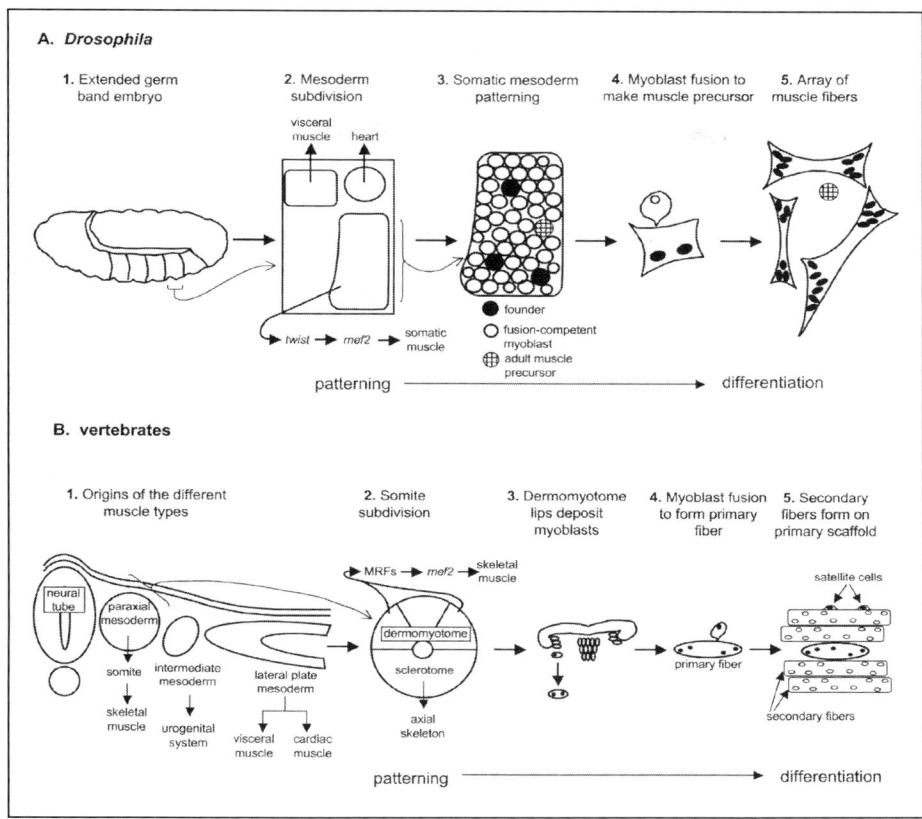

Figure 1. Comparative overview of somatic/skeletal muscle development in *Drosophila* and vertebrates. A) *Drosophila*. 1) Extended germ band *Drosophila* embryo (approximately 6 hours after egg laying at 25°C, anterior to left). 2) A mesodermal hemisegment in the *Drosophila* abdomen is subdivided into the progenitors of the different muscle types under the influence of signals from adjacent cells and other patterning information. 3) Founder cells, fusion-competent myoblasts, and adult muscle precursors are produced within the somatic mesoderm. 4) Founder cells fuse with fusion-competent myoblasts to form bi- or tri-nucleated muscle precursors. 5) At the end of embryogenesis there is a fully differentiated array of 30 distinct syncytial muscle fibers in each abdominal hemisegment. Part is shown together with a single Adult Muscle Precursor cell. Inspired by ref 290. B) Vertebrates. 1) Developmental origins of the different muscle types from the mesoderm are shown in a cross-section of one half of the dorsal region of a typical vertebrate embryo. 2) A somite is subdivided into the progenitors of different cell types under the influence of signals from adjacent cells. 3) Myoblasts are deposited from the medial and lateral lips of the dermomyotome. The sclerotome develops as a loose assembly of centrally located cells. 4) Myoblasts fuse to form small primary muscle fibers. 5) Secondary muscle fibers form on a scaffold of primary fibers. Satellite cells, which mediate postnatal muscle growth, are indicated.

Overview of Vertebrate Muscle Development

As in *Drosophila*, vertebrate skeletal, cardiac and visceral muscle develop from the mesoderm. During gastrulation the mesoderm forms by the recruitment and ingression of cells from the epiblast through the primitive streak, and later through the tail bud.[7,8] Then, again as in *Drosophila*, the mesoderm is progressively subdivided and different regions form the progenitors of the different muscle types.

Skeletal muscle arises from the paraxial mesoderm that is present either side of the neural tube in two wide strips of loose unconnected cells or mesenchyme.[7] The intermediate mesoderm lies lateral to the paraxial mesoderm and gives rise to the urogenital system (Fig. 1). Lying most laterally is the lateral plate mesoderm. The inner, splanchnic layer of the lateral plate mesoderm becomes the visceral muscle,[7] while the cardiac muscle develops from bilaterally symmetrical regions of the lateral plate mesoderm that eventually come together at the midline.[9]

All skeletal muscles of the vertebrate body, together with some of the head, are derived from the somites.[8,10] This chapter will focus on the trunk and limb musculature. The somites are transient structures that form from the paraxial mesoderm. They reveal an underlying segmentation, echoing the overt segmentation of *Drosophila*. However, unlike *Drosophila* in which segments appear simultaneously, in vertebrates the somites form sequentially from the anterior end of the paraxial mesoderm at regular time intervals.[11]

The next step in somitogenesis is the conversion from mesenchymal tissue to an epithelial ball. The cells in this epithelial somite are multipotent and progressively acquire specific fates under the influence of signals received from nearby cells. The somite subdivides into a dorsal dermomyotome and ventral sclerotome (Fig. 1). As the dermomyotome develops, its medial lip deposits myoblasts underneath into the myotome.[12] These are the progenitors of the epaxial muscle, the deep dorsal muscles around the backbone, and they differentiate immediately. At the lateral lip muscle precursors are also deposited. These will make the hypaxial muscle, ie. all the ventrally, laterally and superficially located muscles. At inter-limb levels they again immediately differentiate and eventually make the abdominal body wall muscles. However, at limb level, they do not differentiate immediately, but instead migrate to the limbs where they first proliferate and then differentiate.[7,13]

In the development of the skeletal musculature, the first step is the formation of a scaffold of relatively small primary fibers.[10,14-16] This requires that myoblasts fuse to form syncytia. Subsequently, secondary fibers are added alongside the primary fibers as the muscles grow (Fig. 1). A population of cells (satellite cells) is also put aside as a reservoir for subsequent growth and repair. Vertebrate skeletal muscles have many constituent muscle fibers. For example, small eye muscles have hundreds of fibers, whereas limb muscles of large animals may have a thousand times as many.[10] Similarly, adult *Drosophila* somatic muscles are also composed of multiple fibers,[17] although in contrast each larval *Drosophila* muscle fiber is composed of only a single syncytial fiber.

Somatic/Skeletal Muscle Development

The following sections describe and compare the different stages of somatic muscle development in *Drosophila* and skeletal muscle development in vertebrates.

Subdivision of the Vertebrate Somite and Drosophila Mesoderm

The subdivision of the vertebrate somite has been extensively reviewed elsewhere.[8,12,18,19] Here the aim is to compare this process with the subdivision of a mesodermal segment of the *Drosophila* embryo. For the vertebrate examples, I will focus on the chick, where embryological manipulations have been used to analyze the process, and the mouse where genetic analyses are more readily undertaken. There are some species differences that will not be discussed here, but which are highlighted in other reviews.[8,20]

Experimental manipulations show that when both the vertebrate somites and the *Drosophila* mesoderm first form, the constituent cells can contribute to a broad range of mesodermal derivatives.[7,8,21] However, cell fate subsequently becomes restricted, and both the somite and a *Drosophila* mesodermal segment become subdivided into groups of cells that will develop differently.

In the vertebrate somite there is a dorsal/ventral difference. The ventral region called the sclerotome will make cartilage and bone, the axial skeleton. The dorsal epithelial region called

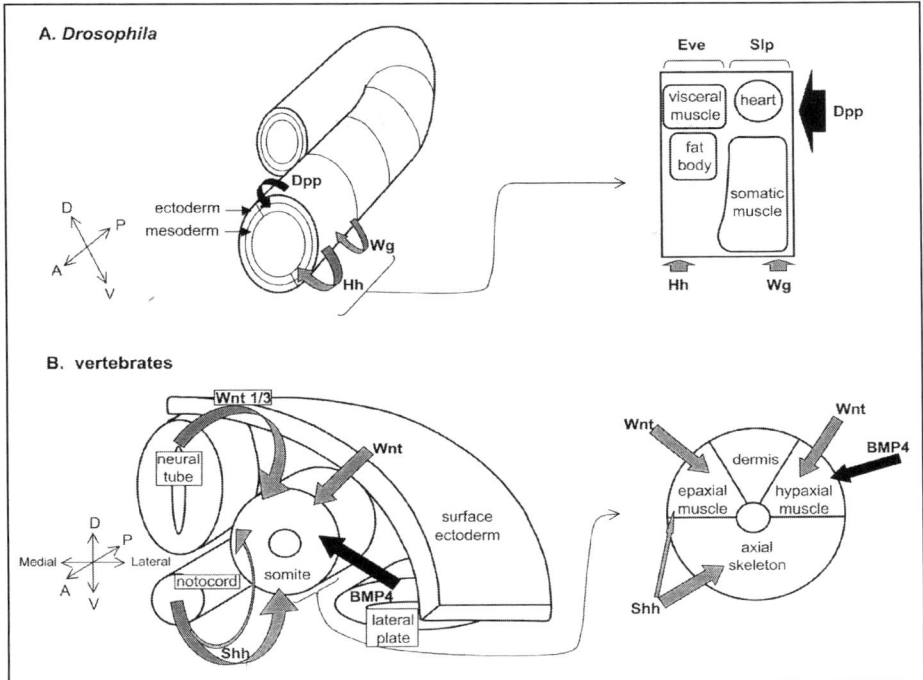

Figure 2. Similar signals pattern *Drosophila* mesodermal segments and vertebrate somites. Cross-sections of an extended germ band *Drosophila* embryo (A) and one half of the dorsal region of a vertebrate embryo (B) are compared. Similar molecules from the surrounding tissues signal to both the *Drosophila* mesoderm and to the developing vertebrate somite. *Drosophila* Dpp is related to BMP4, Wg to Wnt, and Hh to Shh. Also depicted are simplified cross-sections of a single *Drosophila* mesodermal abdominal hemisegment (A) and of a vertebrate somite (B) to illustrate their subdivision into progenitor populations for different cell types under the influences of these signals. Inspired by reference 290.

the dermomyotome underlies the surface ectoderm and gives rise to muscle and dermis (Fig. 2). Within the dermomyotome there is also a medio-lateral difference. The central region makes dermis, the mesenchymal connective tissue of the back skin. The medial region (closest to neural tube) makes epaxial muscle, and the lateral region (furthest from neural tube) makes hypaxial muscle.

A *Drosophila* mesoderm segment is subdivided too (Fig. 2). Along the A/P axis this is into two domains, one expressing the Even-skipped (Eve) transcription factor, the other expressing the Sloppy paired (Slp) transcription factor. It is also subdivided in the D/V axis. In the Slp domain, the most dorsal cells will develop into the heart, and the remainder will form the somatic musculature. In the Eve domain, the most dorsal cells form visceral muscle, and those more ventrally, the fat body.[22-24]

In both vertebrates and *Drosophila* there is a crucial role for the adjacent tissues in the development of these different cell fates. In the chick, ablation experiments and the surgical insertion of barriers have shown the importance of the surface ectoderm, neural tube, notocord and lateral plate mesoderm, and also that these adjacent tissues produce diffusable signals.[25-27] These signals may be antagonistic. For example, the specification of the lateral somitic lineage results from the antagonistic actions of a diffusable medializing signal from the neural tube and a diffusable lateralizing signal from the lateral plate mesoderm.[25,27]

In insects, surgical ablation experiments were also used to indicate the role of the adjacent ectoderm.[2,28] More recently, it was found that the adjoining ectoderm provides an important signal(s) for *Drosophila* mesoderm development in experiments where gastrulation was arrested,[29] or mesoderm migration after gastrulation was inhibited.[30-32]

Signals That Subdivide

Having shown that adjacent tissues provide crucial signals, the next step is to identify them. In vertebrates, the general approach has been to analyze the effect of specific signals selected on the basis of their expression patterns. A substantial body of evidence, mainly from chick and mouse, has produced a model in which Wnt, Bone Morphogenetic Protein (BMP) and Hedgehog (Hh) signals are needed for the development of epaxial muscle, hypaxial muscle and sclerotomal cells (Fig. 2).

The first step in these analyses was to ask what specific signals *could* do. The combination of BMP4 from lateral plate mesoderm and a Wnt from the surface ectoderm was found to signal hypaxial muscle development.[25,27,33] Whereas the combinatorial action of either Wnt 1 or Wnt 3 from the neural tube together with Sonic hedgehog (Shh) from the floor plate/ notocord, can signal epaxial muscle development.[12,26,34] Shh alone can signal sclerotome development.[35,36] However, there is much complexity to the subdivision of the somite and the subsequent development of different cell types. There are multiple steps and the details of the roles of the different signaling molecules are still unclear. For example, it is unclear if each of these signals is required for the initiation and/or the maintenance of the myogenic pathway of differentiation.[8,37]

Nevertheless, this body of work shows that BMP4, Wnts 1 and 3, and Shh *can* signal the development of specific somite compartments. The important question is whether they actually do so in vivo. Answering this requires loss-of-function studies, and is complicated by vertebrates having multiple, closely-related genes with complex, dynamic and often overlapping patterns of expression, e.g., the Wnt family. However, there have been some definitive genetic experiments. For example in the mouse, a double "knock-out" of Wnt 1 and Wnt 3a shows that these signals regulate the formation of the medial compartment of the dermomyotome that makes the epaxial muscle.[38] A mouse "knock-out" also shows a role for Shh in epaxial muscle determination.[39] In the chick, the introduction of a "dominant negative" construct to "knock down" function shows a role for Wnt 5b in the early steps of myogenesis.[37] This is an alternative approach to assessing the in vivo role, and could be an informative general strategy if each dominant negative construct can be shown to be specific for a single signaling molecule.

Drosophila has great advantages for this type of study because genetic analyses can be readily undertaken and there are often only single copies of key genes. Indeed, a series of elegant genetic experiments has identified essential signals in the subdivision of the *Drosophila* mesoderm. A simplified version of one mesodermal segment is shown in Figure 2. Strikingly, the signals that are important in the development of different fates in the *Drosophila* mesoderm are similar to those implicated in the vertebrate somite. They are Decapentaplegic (Dpp), which is related to the vertebrate BMP family, Wingless (Wg), which is a Wnt family member, and Hedgehog (Hh), which is related to Shh.

Signaling from ectodermal Dpp is crucial in subdividing the mesoderm in the D/V axis,[40,41] and is required for the development of some dorsal somatic muscles, visceral muscle and heart. The subdivision along the A/P axis is more complex. Each mesodermal segment is subdivided into two domains by the two aforementioned transcription factors, Slp and Eve. Somatic muscle and heart develop from the Slp domain where Slp cooperates with Wg signaling in the development of muscle and heart progenitors.[24,42] Visceral muscle and fat body develop from the Eve domain. This requires *eve* function and is partially mediated by Hh signaling.[23]

Detailed comparisons between the subdivision of a *Drosophila* mesodermal segment and a vertebrate somite are problematic because of their different developmental anatomy and because signals are used at multiple times and for different events. Nevertheless, there are general similarities in both the molecular players utilized and the developmental strategies adopted.

Comparison of *Drosophila* and Vertebrates

In *Drosophila*, as in vertebrates, the development of progenitors for specific cell types is dependent on the activities of more than one signal. For example, dorsal somatic muscle progenitors in *Drosophila* require Dpp plus Wg, while those for vertebrate hypaxial muscle require BMP4 plus Wnt. In *Drosophila,* the activity of these signals is also dependent on intersecting with cells expressing specific complements of transcription factors. It is not yet apparent to what extent this is the case in vertebrates too. Some signals have opposite actions. For example, in vertebrates, dorsalizing Wnt signaling can antagonize ventralizing Shh signaling,[12] while in the A/P subdivision in *Drosophila*, Wg signaling has effects opposite to those of Hh.[23]

Different skeletal/somatic muscles have different signaling requirements. In vertebrates, epaxial muscle develops in response to different signals than hypaxial muscle, while in *Drosophila*, *wg* is required for only a subset of muscle founder cells and hence a subset of muscle fibers.[43,44] In another parallel, *hh* has a major role in the development of derivatives other than somatic muscle. In the case of *Drosophila*, this is for the visceral muscle and fat body, while in vertebrates it is for the sclerotome.

In both *Drosophila* and vertebrates, Wnt signaling from outside the mesoderm is critical. However, in chick there is also an early role for Wnt 5a expressed in the presomitic mesoderm before somites have budded off.[37] This can be compared with *Drosophila* where although the major source of Wg is the ectoderm, a contribution from mesodermal *wg* expression early in mesoderm subdivision cannot be excluded.[23,43,45]

Wnts are the major dorsalizing signal in vertebrate somites.[12] This contrasts with *Drosophila* where the key signal regulating the D/V subdivision is Dpp. However, there is evidence in the chick that the related signal (BMP) upregulates Wnt 1 and Wnt 3a in the dorsal neural tube.[26]

In *Drosophila*, in addition to the A/P subdivision within a mesodermal segment, there are also variations along the A/P axis between different segments in both the pattern of muscles and the characteristics of some muscles.[46-48] This is under the control of the Hox genes, which encode the transcription factors required in all animals to establish segment specific differences. This is mirrored in vertebrates, where Hox genes confer the predisposition of somites at limb levels to produce migratory muscle precursors that will populate the limb buds and produce limb muscle.[49]

Conversion of Mesoderm Subdivision to Muscle Differentiation

The combination of molecular mechanisms described above serves to subdivide the mesoderm into groups of cells that develop into the skeletal musculature of vertebrates and the broadly equivalent somatic muscles of *Drosophila*. How is this information interpreted to make functional muscle? This issue can be divided into two aspects: (i) the earliest steps in which the patterning events initiate the first steps of muscle development, and (ii) later events that actually produce differentiated muscle. In this section I will address the first aspect. The second will be addressed subsequently.

The Vertebrate Myogenic Regulatory Factor Family

Progress in understanding muscle development was catalyzed by findings in a landmark paper that described the identification of MyoD in mice.[50] This single factor can convert many cell types, including those of nonmesodermal origin, into myoblasts.[51] The myoblasts can then differentiate into muscle. MyoD is a skeletal muscle-specific, basic helix-loop-helix (bHLH) transcription factor that directly activates gene expression by binding a conserved sequence found in many muscle gene promoters and enhancers. Because MyoD can trigger the muscle program in a different cell type, this gave rise to the idea that it was a master regulatory gene, one that coordinately regulates a cohort of genes to produce a specific phenotype. However, it quickly became apparent that there is more to muscle than MyoD. One aspect to this is that mice have a family of four closely related proteins with similar properties: Myf5, myogenin, MRF4, plus MyoD.[51] Together they are often known as myogenic regulatory factors (MRFs).

Myf5 and MyoD are generally the first two MRFs expressed in vertebrate muscle development, around the time of somite formation. Genetic analyses in mice make it clear that Myf5 and MyoD both function in muscle specification. In a *Myf5* "knock-out", muscle development is essentially normal, although MyoD activation is delayed.[52-54] Similarly, muscle development is essentially normal in a *MyoD* "knock-out", although *Myf5* expression is up-regulated.[55] However, and strikingly, a *Myf5/MyoD* double mutant makes no muscle at all.[56] Evidently, there is some sort of redundancy between *MyoD* and *Myf5* and/or compensatory mechanisms operate.

In contrast, myogenin functions later in muscle differentiation. *myogenin* "knock-outs" have a striking phenotype with many mononucleate myoblasts, but very little differentiated muscle.[57,58] Muscle progenitors develop, but differentiation does not proceed. MRF4 was also generally considered to function only in the later steps of differentiation, but recent findings in mice indicate an early role in muscle determination too,[59] which now awaits further analysis.

It is clear that Myf5 and MyoD occupy pivotal points in muscle development and it is therefore important to understand what is upstream and downstream of these regulators, i.e., how they themselves are regulated and what their targets are. Evidence is accumulating that points of integration of the patterning information that subdivides the mesoderm are the promoters/enhancers of the MRF gene family. For Myf5, many regulatory elements dispersed throughout a large genomic locus have been described.[60-62] For MyoD, enhancers and promoters have also been defined in a number of species,[63-65] and upstream regulators identified. For example, analysis of both mouse mutants and chick embryonic manipulations indicates that three of these regulators are Myf5, MRF4, and the paired class transcription factor, Pax3.[53,59,66] As well as cross-regulation by other MRFs, a role for autoregulation of MyoD has been suggested by cell culture experiments,[67] and has received support from work in *Xenopus*.[68] Patterning signals also regulate MyoD. Notable amongst these in the mouse are the up-regulation of MyoD expression by Wnt signaling,[37,69] and the down-regulation of MyoD expression by Notch signaling.[70] Another pathway in *Xenopus* is the direct activation of MyoD by embryonic Fibroblast Growth Factor (eFGF), which can be linked directly to the earliest signals that shape the body plan.[71] This differs from the mouse in that MyoD activation precedes somite formation in frogs, as it does in zebrafish. The challenge now is to establish how all these regulatory signals and transcription factors link up with the enhancers/promoters already identified.

Understanding the targets of MyoD and Myf5 is crucial to understanding their profound effects on muscle development. It is now possible to look for and analyze targets on a genome-wide scale. It has already been found that there are subprograms within muscle development, with MyoD regulating different targets at different times during myogenesis and with distinct mechanisms of control.[72]

What Is a *Drosophila* MRF?

The seminal MRF work in vertebrates raised the question of which gene(s) in *Drosophila* plays the role(s) of the MRFs. *Drosophila* appears to have only one gene whose sequence is closely related to vertebrate MRFs.[73,74] It is called *nautilus* and the encoded protein has a bHLH domain that shares approximately 90% sequence identity with that of vertebrate MRFs. Outside this domain, it is highly divergent. Biochemically, Nautilus shares properties with the MRFs. It can convert 10T1/2 fibroblasts into myoblasts in combination with the ubiquitously expressed bHLH protein Daughterless.[75] It is also required for the activation of the myogenic program in response to Daughterless in *Drosophila* SL2 cultured cells.[76] Moreover, conversion of cardiac muscle to somatic muscle is reported in *Drosophila* embryos in response to Nautilus.[77]

Despite these physical and functional similarities between Nau and MRFs, genetic analysis has established that *nautilus* does *not* play an equivalent role to MRFs in embryonic muscle development. *nautilus* has a restricted pattern of expression in developing somatic muscle and is not essential for myogenesis. In *nautilus* mutants, most muscle development is unaffected, with only a small subset of muscle fibers missing.[78] Nautilus is an example of how a similar molecule linked to a similar developmental pathway nevertheless has different functions in different species.

The *Drosophila* gene whose function is most closely related to the vertebrate MRFs is *twist*. It also encodes a bHLH transcription factor. At gastrulation *twist* is expressed uniformly throughout the mesoderm. Its expression then modulates. Within each segment there is a domain of low Twist expression that gives rise to visceral muscle, and a domain of high Twist expression that gives rise to somatic muscle. Slp may directly activate *twist* expression in the latter.[6] Genetic analysis of *twist* function in muscle development is complicated by its essential role in gastrulation and mesoderm formation. However, a subsequent requirement in somatic muscle development was revealed using a temperature-sensitive allelic combination to reduce Twist activity after gastrulation.[79] Twist also has a dominant effect on myogenesis, similar to that of the vertebrate MRFs in culture. Ectopic *twist* expression in the ectoderm or mesoderm induces cells to express muscle marker genes and form small syncytia.[79,80] It pushes cells down the somatic muscle differentiation pathway. Twist homodimers have this effect, which contrasts with MRFs, where the major active species is a heterodimer with the ubiquitously expressed bHLH proteins.[81-83] Note that possible in vivo roles for MRF homodimers detected in vitro have been suggested.[84-86] The effect of Twist in *Drosophila* embryos appears more striking than that of MRFs in vertebrate embryos, where, in both frog and mouse, ectopic expression of MyoD can activate muscle genes in ectodermal cells, but not the entire myogenic program.[87-89]

Another parallel between Twist and MyoD is the links to signals that pattern the mesoderm. Notch signaling represses *twist* expression during mesoderm subdivision,[90] as it does MyoD in vertebrates,[70] while Wg positively influences *twist* expression through the *slp* gene,[6,24,42] and Wnts up-regulate MyoD.[37,69] Lastly, like the MRFs, which are first expressed in the proliferating, undifferentiated cells of the somite medial wall, Twist is expressed in the subdividing mesoderm in cells that will become the larval somatic muscle and which still divide.[2]

Mef2—A Key Muscle Differentiation Factor

Although Twist in *Drosophila* and the MRF family in vertebrates are clearly pivotal to muscle development, other factors are required to make skeletal muscle. The Myocyte enhancer factor-2 (Mef2) transcription factors, which belong to the MADS family, are one of the other key players in skeletal/somatic muscle differentiation.[91,92] This is an area of muscle development where research in vertebrates and in *Drosophila* has progressed together. Like the MRFs, the initial identification and characterization of Mef2 was through mammalian cell culture studies. Mef2 regulates muscle gene transcription through a binding site found in the promoter of nearly every known muscle specific gene.[92] A striking finding is that Mef2 and MRFs synergize in the myogenic conversion of fibroblasts. They also synergistically activate muscle gene expression, either through binding to separate sites on the enhancer, or by physical interaction between Mef2 and an MRF at a single site.[93]

A general problem in developmental biology after finding that a specific molecule can activate a certain pathway, is to determine whether it actually does so in vivo. This question can often be most clearly and readily addressed in *Drosophila*. This has been the case for Mef2. Mice have four *Mef2* genes with overlapping patterns of expression that complicates the "knock-out" analysis. *Mef2c* is the earliest of these four genes expressed in skeletal muscle development.[94] However, *Mef2c* mutants are early embryonic lethal, and reported effects on skeletal muscle development are limited to an observation of a differentiation defect indicated by a reporter transgene.[95] In contrast, *Drosophila* has a single *mef2* gene. Genetic analysis has revealed a striking phenotype and advanced knowledge significantly. In *mef2* mutants, correctly patterned somatic muscle precursors cells are produced, but these cells fail to fuse and differentiate further.[91,96-98] The defect is therefore relatively late in the differentiation process, and in some ways this phenotype resembles the *myogenin* "knock-out" in mice. The role of *mef2* during skeletal muscle development is therefore established in *Drosophila*. At present this remains to be clearly demonstrated in vertebrates, although the use of a "dominant negative" Mef2 construct has shown that Mef2 family function is required for MRFs to convert cultured cells into skeletal muscle.[99] *Drosophila mef2*, like vertebrate *Mef2* genes, is expressed in developing skeletal, heart and smooth muscle cells. While the focus here is on skeletal/somatic muscle, *mef2* also plays a crucial role in the differentiation of both heart and visceral muscles. In fact, the analysis of *Drosophila mef2* mutants identified *mef2* as the first gene shown to control differentiation in multiple muscle cell types.[91]

There are direct links between Mef2 and the earliest steps of muscle specification described in the previous section. A variety of experiments in both *Drosophila* and vertebrates show that Twist and MRFs activate *mef2* gene expression. Thus, in *Drosophila*, *mef2* expression is activated by Twist,[100,101] and MRFs directly activate *Mef2c* expression in skeletal muscle.[102,103] There is also accumulating evidence from gene expression studies that Mef2 and MRFs activate and maintain the expression of each other. This picture is incomplete, but it may be very important for maintaining the muscle phenotype. This echoes the possible auto-activation and cross-activation of MyoD described earlier. In *Drosophila*, maintenance of *mef2* expression in developing somatic muscle results from a direct positive feedback mechanism, in which Mef2 activates the *mef2* gene.[104] In mice, a more complex autoregulation occurs via *myogenin*, whose expression is controlled by MRFs and Mef2,[105,106] together with a Mef3 site that binds Six/sine oculis homeoproteins.[107] Taken together, these results point to evolutionarily conserved gene transcription loops to maintain the muscle phenotype.

To understand the place of Mef2 in muscle differentiation, one will need to understand not only how *mef2* gene expression is regulated, but also the genes whose expression Mef2 controls. This will shed light on how Mef2 coordinates muscle differentiation. However, so far rather few genes are identified as direct targets of Mef2 in *Drosophila* somatic muscle development. They include: *TmI*, *β3 tubulin*, and *Actin57B*.[108-110] An aspect that remains largely unexplored is that there are effects of Mef2 on gene expression earlier than generally supposed, before overt differentiation is underway. These targets include *Dmeso18E* and *Actin57B*.[110,111]

Lastly, the *levels* of transcription factors have received little attention when trying to understand how a relatively small number of transcription factors can coordinate all the temporal and spatial aspects of gene expression necessary for proper development. However, the level of transcription factors expressed at a certain time may be critical for normal myogenesis. For example, for the MRFs in mice, skeletal muscle development is sensitive to Myogenin levels,[112] while for Mef2, experiments in *Drosophila* indicate that different levels of Mef2 are required for different aspects of muscle development.[113]

Muscle Patterning

Different animals have different and characteristic patterned arrays of muscles. In *Drosophila*, examples include the thirty distinct larval muscles in each abdominal hemisegment,[114] and the intricate arrangement of at least seventy-eight muscles in the adult thorax.[17] In vertebrates, an example is the forty muscles in the tetrapod limb.[115] There are developmental characteristics shared by all these somatic/skeletal muscles. They all have a network of transcription factors to activate the muscle genes that encode the proteins that produce functional syncytial fibers with ordered assemblies of contractile proteins and attachments. However, different muscles also have distinct characteristics. Examples include their size and shape, their points of attachment, and their physiological/biochemical properties, e.g., slow- or fast-contracting.

There are two components to consider in the analysis of the development of these muscle arrays. First, is the pattern itself. How is the reproducible and distinctive pattern of muscles produced? How is a muscle directed to develop at a particular place? Second, how are different elements of this pattern conferred with their specific properties? The best understood example is the development of the abdominal larval muscle pattern during *Drosophila* embryogenesis. Central to this is the role of so-called "founder cells".[2,114,116-118]

Drosophila Embryo Muscle Patterning

Analysis of grasshopper muscle development indicated how a muscle pattern could develop. Grasshoppers have single muscle pioneer cells that prefigure the muscle pattern. They are large cells that span the future muscle territories and act as a scaffold on which the muscle pattern is assembled.[119] The pioneers also seed the muscle, that is, if the pioneer is ablated the muscle does not form, even though the myoblasts that would normally contribute are still present.[120] In *Drosophila*, it was also found that the final larval abdominal somatic muscle pattern is prefigured by small muscle precursors, each comprising two or three fused myoblasts.[114] Together they form a scaffold for the development of the larval musculature. Each precursor derives from a single founder cell, which itself is produced by the asymmetric division of a muscle progenitor.

Subsequent genetic analysis revealed two crucial properties of these founder cells that propelled *Drosophila* to prominence in consideration of how muscle patterns develop during animal development. First, in mutant embryos where myoblast fusion does not occur, founder cells differentiate into "mini muscles", with the characteristics of the full size muscle that it would normally give rise to, including its innervation.[118,121] Founder cells are thus revealed as a distinct population of myoblasts with intrinsic muscle pattern information.

Second, it was found that a number of genes, many encoding transcription factors, were expressed in small numbers of founder cells. These include *Kruppel, S59/slouch, apterous* and *ladybird*. They are not just convenient markers, their specific expression also translates into a specific function. Thus, muscle identity is specified by the autonomous function of these transcription factors.[122-125] In normal development, founder cells fuse with the fusion-competent myoblasts to form the final syncytial muscles, and recruit the incoming nuclei to the characteristic pattern of identity gene expression of that muscle.[2,117,118] However, it is not a case of one gene, one founder cell, as identity genes are expressed in overlapping patterns in multiple founder cells. Therefore, some form of combinatorial model for muscle identity is attractive, but how this might operate remains to be established.

Much also remains to be uncovered about how the events that pattern the mesoderm direct the activation of specific identity genes in specific founder cells. However, in the case of a dorsal muscle progenitor, detailed analysis has identified how many of these patterning signals are integrated at an identity gene enhancer to produce precisely localized gene expression.[126,127] In summary, founder cells are central to *Drosophila* somatic muscle development. Each seeds the muscle, directs its fusion, and endows the developing muscle with characteristic properties. It is therefore an intriguing and pressing question whether this conceptually elegant paradigm has parallels in vertebrate muscle development.

Primary and Secondary Fibers in Vertebrate Myogenesis

In vertebrate development, there are two major waves of myogenesis.[14,15] First, primary myofibers form from the fusion of mononucleate myoblasts. These primary fibers are the anlage for all future muscles and control the site of the subsequent assembly of secondary myofibers, which adds mass to the muscles.[10,16] The primary fibers are small, but extend from tendon to tendon of the embryonic muscle and become innervated.[128,129] These characteristics are similar to those of *Drosophila* embryonic founder cells revealed in the absence of myoblast fusion.[118] Furthermore, like founders, the crucial role of primary fibers is to define the type, shape and location of a muscle.[16] They serve as a scaffold to organize subsequent secondary fiber formation in a way that recalls *Drosophila* embryo muscle precursors or grasshopper pioneers. It is not yet known how the number and location of these primary fibers is regulated. However, comparisons with the development of the adult *Drosophila* musculature are likely to be helpful, because in some ways this more closely resembles the development of vertebrate skeletal muscle than does the development of embryonic *Drosophila* muscle.

Adult *Drosophila* Myogenesis

During *Drosophila* metamorphosis almost all larval muscles degenerate and are replaced by a set of adult muscles.[2,130] These muscles differ greatly in size and strength according to the number and size of their constituent fibers.[17] The fact that some adult *Drosophila* muscles consist of many fibers makes them more like vertebrate skeletal muscles, and contrasts with the single fiber muscles of the *Drosophila* embryo. Other similarities that contrast with the *Drosophila* embryo include the migration of myoblasts to make the muscles and the physiological differences between muscles. In the *Drosophila* thorax, myoblasts migrate from the imaginal discs to specific positions near the epidermis,[131] while in vertebrates subpopulations of progenitor cells undergo long range migrations to form muscle masses in the limbs, diaphragm and tongue. There are also physiological differences between asynchronous and synchronous muscles in *Drosophila*,[132,133] and between fast- and slow-contracting muscles in vertebrates.

The *Drosophila* adult muscles develop in characteristic positions with a characteristic number of constituent fibers. To reiterate the questions posed at the beginning of this section. How is this patterning regulated and how are the different characteristics of the muscles conferred? The example of the adult *Drosophila* flight muscles illustrates progress towards answering these questions. These thoracic muscles are grouped into the Direct Flight Muscles (DFMs) and Indirect Flight Muscles (IFMs). The latter have two sub-groups: the Dorso-Ventral Muscles (DVMs), and the Dorsal Longitudinal Muscles (DLMs). The DLMs are the largest adult muscles and are prominent examples of multifiber muscles.

The first muscle organizing features identified were for the DLMs. The DLMs assemble on templates provided by a small set of persistent larval muscles, which, in contrast to the other larval muscles, do not degenerate during metamorphosis.[2,130,131,134] Incoming myoblasts fuse with these templates to produce the DLMs. These templates therefore act as a scaffold for DLM assembly, and so in some way this parallels the formation of secondary fibers on the primary fiber scaffold in vertebrates.

In contrast to the DLMs, all the other adult thoracic muscles are thought to develop de novo.[2] For both the DVMs and DFMs, cells that prefigure these muscles have been identified through marker gene morphology and expression.[135,136] Evidence is accumulating that these cells, and indeed others corresponding to adult abdominal muscles, are founder cells.[137,138] The pattern of adult myotubes is prefigured by a pattern of myoblasts at appropriate locations that express the *dumbfounded (duf) lacZ* transgene.[137] *duf* was characterized in the embryo as a founder cell specific gene involved in myoblast fusion.[139] Analysis of the DVMs shows that the number of "founders" corresponds to the number of fibers in a muscle. In the embryo, a key feature of founders is that they can form small muscles in the absence of fusion.[118] This is also true of these adult founders. When fusion is compromised they develop into mononucleate myosin-expressing fibers.[137] In summary, myotube formation in the adult appears to be initiated by single founders identifiable by *duf* expression, just as in the embryo, and the number of fibers per muscle is defined by an appropriate number of founders.

Comparison of Muscle Patterning in Adult *Drosophila* and Vertebrates

It is apparent that vertebrate primary fibers share some characteristics with *Drosophila* founder cells. This emphasizes the importance of understanding the mechanisms that direct primary fiber formation. There might be a seeding event, which could be through founder-type myoblasts or from an environmental cue. Evidence for the latter comes from the vertebrate limb.[140] When muscle cells differentiate here they immediately form a precisely oriented array that prefigures the future muscle pattern. This appears to be in response to a prepattern of the Wnt signaling effector, the TCF transcription factor, in lateral plate-derived cells in the developing limb.

In the *Drosophila* adult, it is not established how founder cells are selected from the population of muscle precursor cells, but as in vertebrates, it is suggested that the founder cell pattern may be specified by external cues.[2,135-138] A second issue is how the number of fibers per muscle is defined. In *Drosophila*, this appears to be through the number of founders corresponding to the number of fibers,[137] but how they are grouped to contribute to a single muscle is not yet known. The situation in vertebrates must be, at least temporally, different, as all muscles are formed in waves with no single period defining the number of fibers in adult muscle.[16]

How are specific attributes conferred on the muscles? In the *Drosophila* adult, there is evidence that at least some aspects of identity are specified in groups of myoblasts.[141,142] The myoblasts associated with the wing imaginal disc that will make the adult flight muscles are divided into two populations. Those that express the Cut and Apterous transcription factors form the DFMs, and those that express the Vestigial transcription factor form IFMs. Of note, both *vestigial* and *apterous* are founder cell identity genes in the embryo. However, this emerging picture contrasts with the embryo where muscle identity is specified by the founder cell, although even in the embryo there are indications that the fusion-competent myoblast population might be heterogeneous in its gene expression.[143]

In vertebrates, there is also some evidence for genetic differences in the myoblasts. In the developing mouse limb, one specific subset of muscles is affected in *Lbx1* mutants.[144-146] *Lbx1* is a homologue of *Drosophila ladybird*, a muscle identity gene in the embryo.[124] In the mouse, the myoblast characteristic affected appears to be migration. A different subset of limb muscles is affected in *Mox2* mutants, and in this case the effect is not on migration, but on another, not yet defined, aspect of muscle development.[147]

The specific muscle characteristic most worked on in vertebrates is whether fibers are fast or slow. The broad pattern of slow and fast fibers is defined during development,[148,149] although it can be strongly influenced in the adult through nerve activity.[10,150] Analysis in zebrafish has revealed that slow fibers are defined through the action of Hh signaling from axial midline tissue inducing the Blimp1 transcriptional repressor. Slow myogenesis can be driven by Blimp1 and ablated by Blimp1 down-regulation.[151] In mice, the PPARδ transcription factor can increase slow fiber number.[152]

Myoblast Fusion

A characteristic feature of almost all vertebrate skeletal muscles and the analogous somatic muscles in *Drosophila* is that they are multinucleate syncytia formed by myoblast fusion. There is a difficulty in studying the process in vivo during vertebrate development as fusion is asynchronous and takes place over a protracted period of weeks or longer. Following the finding that myoblast fusion occurs and can be manipulated in vitro, most vertebrate work has therefore been in cell culture. In contrast, in *Drosophila* much progress has resulted from an in vivo molecular genetic approach to analyzing muscle development. These studies have been driven by the powerful molecular genetics available, coupled to the fact that myoblast fusion to make the larval muscles occurs in a defined period of a few hours of embryonic development.

A consequence of the different experimental approaches is that the study of myoblast fusion in vertebrates and *Drosophila* has largely remained separate, and the extent of similarities and differences between them is not yet fully apparent. However, because of the underlying conservation of many aspects of muscle development, some significant similarities are anticipated. Already, it is clear that the basic cell biology of myoblast fusion is similar.[153] It starts with cell attraction, followed by adhesion, alignment and finally membrane breakdown and fusion itself.

In vertebrate cell culture, molecules that influence these events have been identified.[154] Examples include, cadherins and Cell Adhesion Molecules (CAMs) that are implicated in the recognition between newly differentiated myoblasts and fibers, and metalloproteinases called meltrins that are implicated in fusion itself.[155-158] However, the microanatomy of these cultures is very different to that in normal muscle development and so it is essential to assess the in vivo role. However, and with some exceptions (for example, see below), the in vivo role of specific molecules in vertebrate myoblast fusion is not yet established. In some cases where it has been explored, e.g., N-cadherin and N-CAM,[159,160] no phenotype is apparent and one explanation is genetic redundancy.

This is where studies of larval muscle development in *Drosophila* have had a significant impact. Genetic and expression based screens coupled to molecular genetic analysis have identified many molecules that play key roles in myoblast fusion in vivo.[5,154,161-163] This work is summarized in the following section. In *Drosophila*, each larval muscle develops through a specific founder cell fusing with fusion-competent myoblasts (Fig. 3). This requires specific molecules to mediate the initial attraction between founder cell and fusion-competent myoblast, and then others to forge links to the cytoskeletal reorganisation that underlies the cell shape and membrane changes of fusion. There are at least two distinguishable stages in myoblast fusions to produce a syncytial muscle. First, muscle precursors with 2-3 nuclei form, and second these precursors enlarge by further fusions.

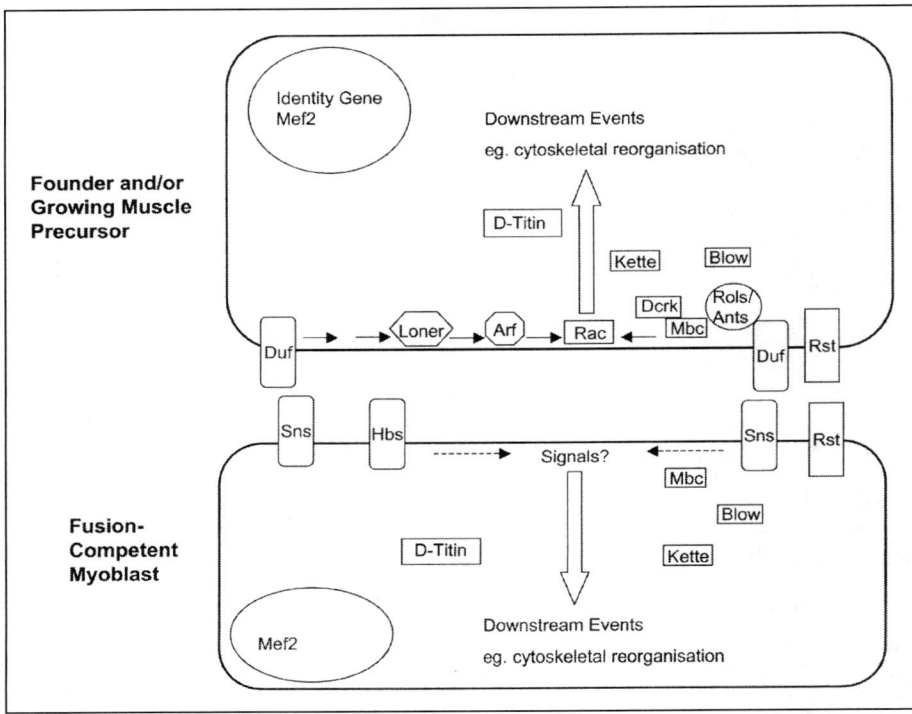

Figure 3. Molecular pathways of *Drosophila* myoblast fusion. There are two characterised molecular pathways from the cell surface molecule Duf to the Rac small GTPases that link with the cytoskeleton on the founder cell side of myoblast fusion. The first is via the adaptor protein Rols/Ants, the second is via the guanine nucleotide exchange factor Loner. Much less is known about events in the fusion-competent myoblasts. In this simplified summary, demonstrated direct protein interactions are indicated by the representations of the proteins touching their neighbour. Established pathways are indicated by solid arrows. Putative pathways are indicated by broken arrows. There is a molecular asymmetry that may underlie the observed asymmetry of *Drosophila* myoblast fusion. For example, Duf, Rols/Ants, and Loner are specific to the founder cell side. It is not yet known whether the illustrated chains of events have parallels in vertebrates. However, all the molecules shown, except Blow, have relatives in vertebrates. There is also evidence that vertebrate myoblast fusion is asymmetric, as in *Drosophila*. See the text for details and references.

Molecular Pathways of *Drosophila* Myoblast Fusion

One chain of molecules that links the cell surface to the cytoskeleton has been established on the founder cell side of the process (Fig. 3).[5,154,161-163] It starts with Duf, a founder cell-specific transmembrane protein belonging to the Immunoglobulin superfamily, which can function as a myoblast attractant and shares features with adhesion molecules. Duf binds to Rolling pebbles/Antisocial (Rols/Ants), a founder cell-specific intracellular adaptor protein, which in turn binds Myoblast city (Mbc), an SH3 domain-containing cytoplasmic protein. Mbc interacts with both D-crk, an SH2 and SH3 adaptor protein, and Rac small GTPases that influence both the rearrangement and function of the cytoskeleton. Included in the targets may be D-Titin. There is a second route from Duf to Rac via a guanine-nucleotide exchange factor called Loner and the ARF6 GTPase. The relationship of this second route to the first signaling pathway is not yet understood. For example, do the two routes operate downstream of different cell surface interactions or at different stages of the fusion process?

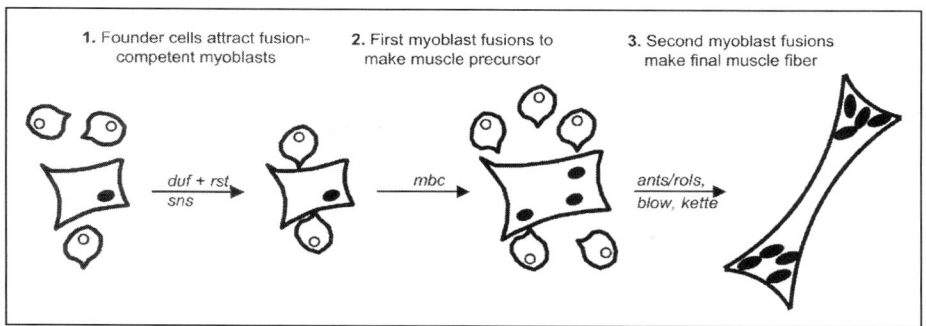

Figure 4. Sequential steps of myoblast fusion in *Drosophila*. 1) Fusion-competent myoblasts are attracted to a founder cell through cell surface molecules like Duf, Rst and SNS. 2) The first phase of fusion to produce bi- or tri-nucleated muscle precursors requires *mbc*. 3) Further fusions occur in a second phase that requires *ants/rols*, *blow* and *kette* to form the final syncytial muscle fiber. Vertebrate myoblast fusion follows a similar series of events of attraction and adhesion, followed by fusion itself. Moreover, analysis of the role of *NFAT2C* in mouse myoblast fusion indicates that vertebrates may also have two comparable sequential steps of fusion.

This body of work has uncovered some of the molecules behind the cell biology of fusion. In addition, it already provides insights into two important characteristics of fusion: its two-step nature and its asymmetry. First the two steps (Fig. 4) differ molecularly. Thus, *rols/ants*, *kette* and *blown fuse* (*blow*) are not required for (all) initial fusions, but are required for the subsequent enlargement of the muscle precursors.[164-166] Second, fusion is asymmetric. Founder cells fuse with fusion-competent myoblasts, but neither myoblast type fuses with itself.[123,167] Some of the molecules, both membrane and intracellular, identified as players in *Drosophila* myoblast fusion, are expressed asymmetrically (Fig. 3). This may lie behind the observed asymmetry of fusion. For example, at the cell surface, Duf is expressed in founder cells, but not in fusion-competent myoblasts. In contrast, two related molecules, Sticks and stones (SNS) and Hibris (Hbs), are expressed in fusion-competent myoblasts, but not in founder cells. Intracellularly, both Rols/Ants and Loner are expressed in founder cells and not fusion-competent myoblasts.

The extent to which the molecular chain of myoblast fusion in *Drosophila* is recapitulated in vertebrates is not yet determined. However, all the molecules, except Blow, in the above scheme (Fig. 3) have relatives in vertebrates.[163] Although it is not yet known whether they function similarly in muscle development, there are some relevant findings. For example, in vertebrate cell culture, there are indications that the roles of Mbc and ARF6 are conserved. Thus, DOCK 180, the vertebrate homologue of Mbc, affects cell morphology,[168] and a dominant negative *Drosophila* ARF6 inhibits myoblast fusion.[169] One mouse orthologue of Rols/Ants is expressed transiently in developing muscle.[170] In contrast, this is not the case with mouse SC-1 and Nephrin, the closest mammalian relatives to *Drosophila* Duf and SNS in the Immunoglobulin superfamily. Neither SC-1 nor Nephrin has been reported in the developing mesoderm. However, CDO and BOC, two other mammalian Immunoglobulin superfamily members that complex with cadherins are expressed during muscle development.[171] Moreover, they stimulate myoblast fusion in vitro. Like SNS, Hbs and N-CAM they contain both Immunoglobulin-like and Fibronectin type III-like domains, which suggests a conserved role for this type of cell surface protein in myoblast fusion in *Drosophila* and vertebrates.

In light of these similarities, it will be of great interest to assess the role of myoblast fusion genes identified in *Drosophila* in vertebrate development. Also the assessment in vivo of vertebrate genes so far only analyzed in cell culture is clearly important too. Cell culture has revealed much about myoblast fusion and will continue to be important for analyzing the cell biology of the process, and an incisive approach is the combination of developmental genetics and cell culture. Here the phenotype of mouse mutants is assessed alongside the behavior in cell culture of mutant myoblasts isolated from the mice, and this has made for some interesting comparisons with *Drosophila*.

The Genetics Plus Cell Culture Approach

The first example of this was myogenin, one of the MRFs. Myogenin is a key regulator of muscle differentiation. One aspect of this is the ability to fuse, as revealed by the *myogenin* "knock-out" phenotype of a large number of unfused myoblasts. However, the nature of the defect is unclear, as myoblasts isolated from these mice will fuse in vitro.[57] One can compare its position to *Drosophila mef2*, which functions after specification to drive differentiation, including fusion. Although myoblasts are specified in *mef2* mutants, they do not fuse. Both *mef2* in *Drosophila* and myogenin in mice sit near the top of a hierarchy of muscle differentiation of which fusion is a part, and provide links between the transcriptional network governing muscle differentiation and the process of making a functional muscle itself.

A second example is NFAT2C, a calcium sensitive transcription factor.[172] The adult mouse phenotype suggests myofiber formation in embryogenesis is normal, but subsequent growth is altered. There are no embryonic development studies yet, but culture of *NFAT2C* mutant cells shows that they form smaller myotubes, indicating a muscle intrinsic role for *NFAT2C* in regulating myotube size. Mutant cells can differentiate and fuse to form the initial multinucleate cell containing 2-4 nuclei, but are defective in recruiting myoblasts or myotubes for subsequent growth. These results echo the *Drosophila* findings with *rols/ants* and indicate that there is a two-step process in vertebrate myoblast fusion too, in that fusion of muscle cells with myotubes/myofibers is distinct from the initial fusion of myoblasts to form a multinucleate cell. One player in this is Interleukin-4 (IL-4), which lies downstream of *NFAT2C* and acts as a secreted myoblast recruitment factor.[173] IL-4 is specifically required for myoblast to myotube fusion, not for the distinct myoblast/myoblast fusion. A similar differential role for an attractant/recruitment factor in the two steps of fusion has not yet been attributed to any *Drosophila* molecule. Whether aspects of IL-4 function are similar to the *Drosophila* myoblast attractants Duf and Roughest (Rst) remains to be explored.

Mammalian fusion, like that in *Drosophila*, is also asymmetric and specific. For example, secondary myoblasts do not generally fuse with each other, instead they fuse primarily with the forming secondary myofiber.[15,174] Two candidate molecules for a role in this are the potential recognition molecules, vLA-4 and VCAM1.[175] They are asymmetrically expressed, with the integrin vLA-4 on secondary myotubes, and its receptor VCAM-1 on secondary myoblasts. This can be compared with the asymmetric distribution of cell surface molecules, and their intracellular links, between founder cells and fusion-competent myoblasts in *Drosophila* embryonic development.

A final example of the combination of developmental genetics and cell culture is integrin β1. The mouse Cre-lox system was used to inactivate integrin β1 specifically in skeletal muscle.[176] Unfused myoblasts and small syncytia accumulate during embryogenesis. A cell culture study of the mutant cells shows that integrin β1 is necessary for a step subsequent to myoblast adhesion in myoblast fusion, and also for sarcomere assembly. In contrast, in *Drosophila* there is no indication of a defect in fusion in integrin mutants, although integrin is required to assemble organized sarcomeres, and for attachment to tendon cells.[177]

There is much interest in myoblast fusion, both because of various potential clinical applications and because it is a fundamental characteristic of many muscles. However, not all muscle is syncytial. A prominent example from one of the stalwarts of developmental biology is that *Xenopus* muscle is not fused until late in development.[20] The slow muscle of zebrafish is also mononucleate.[178]

Drosophila *Adult Muscle Precursors and Vertebrate Satellite Cells*

In both vertebrates and *Drosophila* there are cells that do not immediately follow the differentiation pathway I have described, but instead remain single and undifferentiated (Fig. 1). In vertebrates, the "satellite cells", which lie under the basal lamina that surrounds myofibers, are a major population of this type of cell and appear at late fetal and postnatal stages. Satellite cells mediate the post-natal growth of muscle and are the primary means by which the bulk of adult muscle is formed.[179] They also have an essential role in both muscle hypertrophy and in muscle regeneration in damage and disease, and are activated by exercise or trauma to up-regulate *MyoD* or *Myf5* expression and reenter the cell cycle. They proliferate and differentiate, and yet the population is maintained, probably through self-renewal.[179]

Many, but not all, aspects of satellite cell differentiation, including the MRF expression program, recapitulate differentiation during embryogenesis.[180,181] However, although their developmental origin is not certain, it appears distinct from the embryonic myogenic lineage.[179] The separation of the two pathways of development is illustrated by *Pax7* mutant mice in which there are no satellite cells, but embryonic muscle development is relatively normal.[182]

Partly driven by the possibility of cell-based therapies for degenerative muscle disease, other studies have uncovered different cells that can contribute to muscle regeneration. First, in adult muscle there is a stem cell population distinct from satellite cells and often known as Side Population (SP) cells. In response to Wnt signaling and the activation of *Pax7*, they can replenish the satellite cell population during muscle regeneration.[183] Second, nonmuscle, bone marrow cells can give rise to myogenic cells and repopulate damaged muscle.[184]

In *Drosophila*, the adult muscle precursors (AMPs), which arise during embryonic development, have similarities with vertebrate satellite cells. Like satellite cells they are quiescent, undifferentiated cells that are triggered to proliferate and eventually differentiate to make the muscle of the adult fly.[2] In both *Drosophila* and vertebrates the bulk of adult muscle is produced from these cells either through the formation of new fibers or through fusion with existing muscle fibers. In contrast to vertebrates, the developmental origin of the AMPs is established in *Drosophila*, at least in the abdomen. Abdominal AMPs arise from an asymmetric division of a muscle progenitor cell that produces a founder cell and an AMP.[4,123,185]

Inhibition of Muscle Development

The molecular analysis of muscle differentiation has moved a long way from a simple model in which the key events were expression of pivotal positive regulators, the MRF family and Mef2, in the right time and place. It is apparent that there are multiple mechanisms to fine tune muscle differentiation. Many of these involve inhibition of differentiation. They are responsible for both spatial and temporal restriction of myogenesis. First, they ensure that the muscle development pathway is restricted to the appropriate group of cells, and second they ensure that some cells fated to become muscle do not differentiate immediately. An example of the latter is the maintenance of cells, e.g., vertebrate satellite cells and *Drosophila* AMPs, that are required for making or repairing muscle at a later time. There must be mechanisms to maintain these single cells in an undifferentiated state until appropriately triggered. Similarly, vertebrate muscle development occurs in waves.[10,14-16] The first wave produces the primary muscle fibers, and the second produces the secondary muscle fibers and uses cells that have avoided earlier differentiation. Another example of escape from premature differentiation is the development of limb muscle. In the somite when some muscle differentiation is underway, other as yet undifferentiated cells have to migrate from the somite into the limb, proliferate and then make muscle there.[7,13]

These examples indicate that mechanisms to restrain muscle differentiation are required. In vertebrates, in some of the situations one would anticipate that Myf5 and MyoD are likely targets. These two MRFs are expressed while myoblasts are still proliferating and before their target genes are activated, in both cell culture and during development.[70,186,187] Similarly, one anticipates mechanisms to down-regulate *Drosophila* Mef2, which is expressed significantly before muscle differentiation commences.[100,111] Much of the early work to implicate specific molecules, e.g., FGF and TGFβ, in this crucial inhibitory function was undertaken in cell culture.[186] This revealed that there are multiple levels at which inhibition of muscle differentiation can occur, including the inhibition of both the expression of myogenic genes and the activity of the encoded proteins. However, there are many gaps in knowledge of the in vivo importance and mechanism of action of specific molecules in muscle differentiation inhibition. Here I have selected examples where there is some in vivo information.

Molecules that Inhibit Muscle Development

Twist proteins are bHLH molecules. In cultured cells, mouse Twist inhibits muscle differentiation, and can inhibit the function of both MyoD and Mef2 proteins.[188,189] Its in vivo role in muscle development is not established, but included in the complex phenotype of *twist* mutant mouse embryos there is a somite defect.[190] Moreover, its expression pattern is suggestive. Mouse *twist* is expressed throughout the somite and then is excluded from the myotome at the start of myogenesis when MyoD and Myf5 are up-regulated, persisting only in the dermomyotome and sclerotome.[191] A target for Shh in the sclerotome, where myogenesis must be suppressed, may be another inhibitor of MRF protein activity, I-mf.[192] Together with Twist, and perhaps other proteins, it may restrict the population of cells that will go on to make muscle.

In *Drosophila, twist* appears to play a similar inhibitory role in the development of the adult Indirect Flight Muscles (IFMs). First, its expression declines in IFM progenitors prior to their fusion. Second, persistent Twist expression arrests IFM development, indicating that a decline in Twist is a requirement for differentiation of these adult muscles.[193] This inhibitory effect appears to contrast with the positive muscle differentiation role for *twist* in embryonic development described earlier. However, this can be explained by consideration of the Twist dimerization partner. Thus, in the embryo whereas Twist homodimers promote myogenesis, a heterodimer of Twist and Daughterless, the homologue of vertebrate E-proteins, can inhibit it.[80] The parsimonious model is that the mesoderm domains that make somatic muscle and which express high levels of Twist favor Twist homodimer formation, whereas the domains that make visceral muscle and fat body and which express low levels of Twist favor Twist/Daughterless heterodimer formation.[80]

Vertebrate Id proteins and *Drosophila* Extra macrochaetae (Emc) are related HLH proteins that lack a basic DNA-binding domain. Id can inhibit MyoD function through binding E-proteins, the MyoD heterodimerization partners, both in vitro and in cell culture.[84,194] However, its role in vertebrate muscle development is not established. Nevertheless, after early widespread Id expression, Id and MRFs are expressed mutually exclusively in mouse development and Id is lost on myoblast differentiation in culture.[195] A more general role for Id is indicated by its suppression of embryonic stem cell differentiation.[196] In *Drosophila, emc* does have a documented role in muscle development, although the mechanism is not established. *emc* mutants have an extreme disruption of the somatic muscle pattern with muscle losses and detachment.[197]

In contrast to the proposed mechanism of action of Id, Twist and I-mf, which target the protein, ZEB/zfh1 is a conserved transcriptional regulator that might down-regulate muscle gene expression through binding to promoters/enhancers. Vertebrate ZEB is a zinc finger/homeodomain transcriptional repressor that binds to E-boxes and blocks myotube formation in culture.[198] *Drosophila* Zfh1 is also a transcriptional repressor.[199] In development its expression declines prior to muscle differentiation and loss-of-function mutants have aberrant muscles.

The Notch signaling pathway is a widely used route to influence differentiation.[200] In vertebrate muscle, activation of the Notch pathway keeps cultured cells undifferentiated.[187] In development, it may prevent premature differentiation as activation of the Notch pathway down-regulates MyoD and inhibits muscle differentiation in both chick somite and limb bud.[70,201] Similarly, in adult *Drosophila* muscle development, persistent activated Notch expression causes a failure of IFM differentiation.[193] There are likely to be multiple targets for Notch in its effects on muscle development. In the IFMs one link may be Twist, which inhibits muscle development at this stage. Persistent Notch signaling causes continued Twist expression, and reduced Notch signaling reduces *twist* expression.[193]

Analysis of the effects of Notch on embryonic *Drosophila* muscle development has revealed a complex situation with effects at different stages of the process.[90,202-206] One aspect that has some parallels with vertebrates is that during the subdivision of the mesoderm Notch down-regulates *twist* expression. In the embryo, it is the domains expressing high Twist levels that go on to make muscle. Notch represses *twist* both directly through its nuclear effector Suppressor of Hairless (Su(H)) and indirectly through activating *emc*.[90]

There is only limited information on the mechanisms that hold *Drosophila* AMPs and vertebrate satellite cells in an undifferentiated state, but some of the same players are implicated. For example, Notch is activated in satellite cells as they progress from quiescence to active proliferation, and attenuation of Notch signaling leads to MRF expression and commitment to muscle fate.[207] In an assessment of gene expression changes in an in vivo muscle regeneration system, both *twist* and *Id* were induced at early time-points.[181] One striking characteristic of *Drosophila* AMPs is that they continue to express Twist when quiescent or proliferating, but it declines when they differentiate.[2,131,208]

An emerging angle likely to be critical for muscle differentiation is the role of chromatin. For example, a specific linker histone, via an interaction with the homeodomain protein Msx1, can repress MyoD expression and inhibit muscle differentiation.[209] More generally, there is the role of Histone Acetylases and Deacetylases, which can regulate muscle differentiation through a variety of interactions with MRFs and Mef2 proteins.[210] In vivo assessments of their role in muscle development are awaited with interest.

In summary, there is much to learn about the crucial aspect of inhibition of muscle differentiation. It is already apparent that inhibition occurs in many ways and although similar players are used in *Drosophila* and vertebrates, exactly how they are used may differ as the example of Twist illustrates. Moreover, even in one animal what holds for one muscle group might not hold for another. Thus, although in *Drosophila* adult muscle development Notch inhibits IFM differentiation, the nearby DFMs appear unaffected.[193]

Heart Development in *Drosophila* and Vertebrates

The considerable recent advances in the molecular genetics of heart development in both *Drosophila* and various vertebrates have been extensively reviewed.[211-215] In this comparison of *Drosophila* and vertebrates I will highlight just some aspects of particular interest. The heart (or dorsal vessel) of *Drosophila* functions analogously to the vertebrate heart. Both pump in a posterior to anterior direction, although *Drosophila* is an open circulation without blood vessels, in contrast to the closed circulation of vertebrates. In both the *Drosophila* and vertebrate heart there are two major cell types. In *Drosophila,* they are the cardioblasts, which are the contractile muscle cells with a similar ultrastructure to mammalian cardiomyocytes,[211] and the pericardial cells, which form a layer outside the muscle cells. In contrast, in vertebrates the second cell type, the endothelial cells, are interior to the muscle. Although the structure of the *Drosophila* heart at first appears very different to that of vertebrate hearts, which as a group differ substantially themselves, there are considerable similarities in how the heart in all these species develops (Fig. 5).[211]

Morphogenetic Movements of Heart Development

In both *Drosophila* and vertebrates, e.g., mice, heart precursors are generated bilaterally in the mesoderm under the influence of inductive signaling from adjacent germ layers (Fig. 5).[211] These precursors then move towards the midline where they form a linear, contractile tube. In vertebrates, it is only subsequently that this tubular heart loops and develops into the multi-chambered and physiologically complex organ with which we are familiar. The heart develops dorsally in *Drosophila*, but ventrally in vertebrates. This is considered to be an illustration of the inversion of the D/V axis between arthropods and vertebrates, which was proposed in 1822,[216] and which has received support from more recent studies.[217]

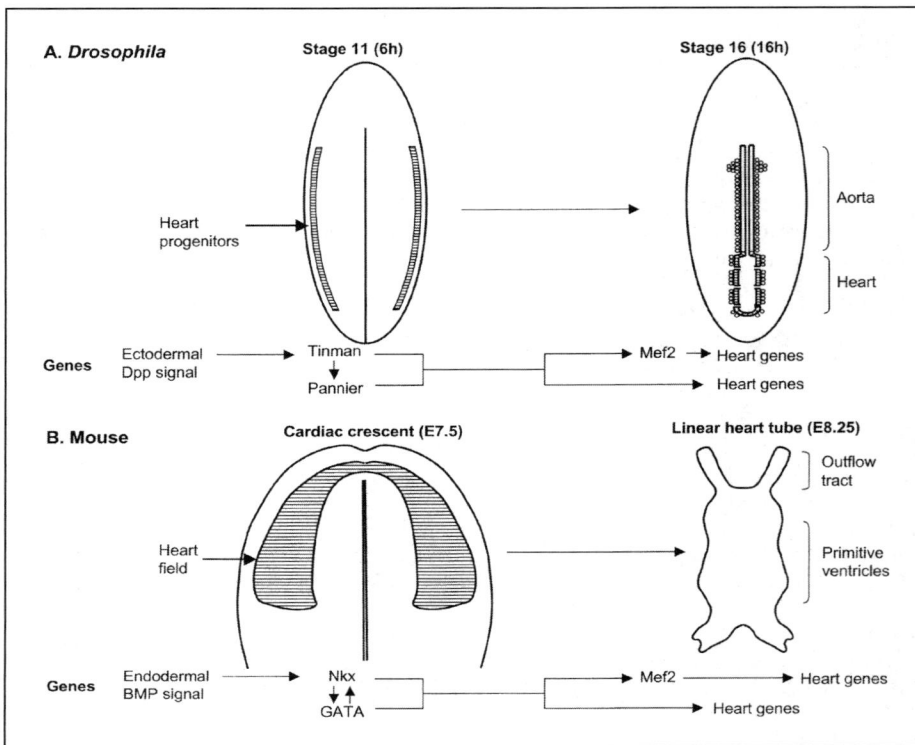

Figure 5. A comparison of the morphogenesis and genetic control of heart development in *Drosophila* and vertebrates. In both *Drosophila* (A) and mouse (B) the heart develops from bilateral precursors that migrate towards the midline to form a linear tube with an anterior aorta or outflow tract. The underlying similarity of some of the signals early in development and the transcription factors that function subsequently is indicated. In *Drosophila*, stage 11 of embryogenesis corresponds to the extended germ band stage at about 6 hours of development at 25°C. Stage 16 is towards the end of embryogenesis at about 16 hours. The two cell-types are shown. Pericardial cells (O) are external to the cardioblasts. In mice, the cardiac crescent at about 7.5 days of embryonic development develops into the linear heart tube by about 8.25 days. Inspired by reference 211.

The similarity between *Drosophila* and vertebrates extends to the molecular genetic basis of heart development (Fig. 5). This was first analyzed in *Drosophila*, where studies identified a number of genes with key roles that proved to be similar in *Drosophila* and vertebrates. They have been assembled into a conserved pathway controlling heart development centered on signaling molecules, for example the BMP family, and several transcription factors, notably the Tinman and GATA families.

Signals in Heart Development

The heart develops from a population of mesodermal cells whose commitment to heart fate depends on signaling from adjacent tissues. In *Drosophila*, a key signal is Dpp, a member of the BMP superfamily with a major role in D/V patterning in early development and a subsequent role in the subdivision of the mesoderm that refines cell fate. *dpp* mutants do not form heart progenitor cells.[211] The Dpp signal is from the overlying ectoderm. This contrasts with vertebrates where it is signals from the adjacent endoderm that have a key role.[213] However, evidence is accumulating that one of these endodermal signals is also a BMP. BMPs are expressed in the right time and place and can ectopically induce heart genes.[214] Loss-of-function analyses are, however, less clear-cut. BMP antagonists suppress early heart development in chick,[218] and heart mesoderm is not formed in *BMP2* mutants in zebrafish,[219] nor after expression of dominant negative BMP receptors in *Xenopus*.[220] However, interpretation of the phenotypes is complicated by the signals acting at multiple points in development and the possibility that the phenotypes are a consequence of more general effects.[214,221] In mice, *BMP2* null mutants have abnormal heart development, but, contrary to the proposed BMP role in early heart development, produce cardiac mesoderm.[222] Again, the interpretation of this phenotype of a single gene "knock-out" is compromised by other related genes with overlapping expression patterns and functions.[214]

Other signals have important roles. One example is the Wnt family. In *Drosophila*, *wg* is required at multiple times for heart development.[223,224] Its effect, at least for the early phase, is mediated by the canonical *wingless* signaling pathway.[225] Wnt signaling is also important in vertebrates. However, here it is a noncanonical Wnt signaling pathway that promotes cardiogenesis, whereas canonical Wnt signaling inhibits it.[211,213,214,226]

Transcription Factors in Heart Specification

A conserved target for Dpp/BMP signaling in both *Drosophila* and vertebrates is the homeobox containing transcription factor Tinman/Nkx2-5. The *tinman (tin)* gene was discovered in *Drosophila*. It is expressed in heart progenitors, and mutants show it is required for the specification of all heart cells.[211,227,228] However, Tin is not sufficient to promote cardiogenesis, in contrast to MRFs and Twist in skeletal muscle. Dpp directly activates *tin* expression via its transcription factor effectors, the Smads, which bind directly to a *tin* enhancer.[229] Here they synergize with Tin itself, bound to adjacent sites. This recalls the auto-regulation in the transcriptional control of skeletal muscle development. A second gene essential for *Drosophila* heart development is *pannier*, which encodes a GATA family transcription factor. Pannier together with Tin can induce expanded heart gene expression.[230,231]

Transcription factors related to Tin and Pannier also have roles in vertebrate heart formation, illustrating the underlying similarity of the transcriptional regulatory circuits of heart development in *Drosophila* and vertebrates (Fig. 5). In vertebrates, the *Nkx2* family are the *tin*-related genes. Analysis is again complicated by the number of genes and their overlapping expression patterns. Moreover, different combinations of *Nkx* genes are expressed in the developing heart in different species.[232] However, *Nkx2-5* stands out as the gene expressed in all vertebrate hearts. The mouse "knock-out" of *Nkx2-5* is embryonic lethal. However, in contrast to *Drosophila tin*, *Nkx2-5* is not required for heart specification, instead mutants have severe early disruption of heart tube morphogenesis.[232-234] Is there redundancy obscuring an earlier *Nkx2-5* role? It is not clear yet, although in mouse, the only other *Nkx* gene expressed in the myocardium is *Nkx2-6*, and *Nkx2-6* mutants have no heart phenotype. Nevertheless, an earlier role for *tin* homologues in vertebrate development is suggested, but not proven, in *Xenopus*, where dominant negative or repressor Nkx constructs block heart formation.[235,236] Another, later role is indicated by the association of defects in heart valve and septal development with *Nkx2.5* mutations in humans.[237,238]

Despite the conserved role of *tin* and *Nkx2-5* in cardiogenesis, the mouse Nkx2-5 protein does not have all the same functions as Tin. When tested for its ability to rescue the *tin* mutant phenotype in *Drosophila*, although it can rescue some aspects, it cannot rescue the heart phenotype.[239,240] In contrast, zebrafish *Nkx2-3* can.[239] *Nkx2-5*, like *tin*, is a direct target for transcriptional activation by Smad proteins.[241] Despite this similarity, the independence of this *Nkx2-5* enhancer from Nkx2-5 binding indicates a difference in the transcriptional control of *Nkx2-5* and *tin* expression during heart development in vertebrates and *Drosophila*. In summary, Tin has many parallels with its vertebrate counterparts, but there are clearly differences too.

GATA transcription factors are implicated in vertebrate heart development. Three GATA genes, GATA 4, 5 and 6, are expressed in the developing heart,[214] and they can activate a range of heart genes in cultured cells.[221,242] Mouse "knock-outs" of each of GATA4, 5 and 6 have been analyzed to make an in vivo assessment of their role in heart development.[214] However, interpretation is again complicated by the overlapping expression and possible function of different family members, and also by their roles in other events and tissues, for example in endoderm differentiation. Apart from extrapolating from the documented role of *Drosophila pannier*, there are two striking reasons for pursuing the analysis of the precise role of GATA factors in vertebrate heart development. Firstly, overexpression of GATA4 in *Xenopus* and of GATA5 in zebrafish can induce ectopic, beating tissue.[221,243] Secondly, GATA4 mutations are responsible for a class of congenital heart defects in humans.[244]

In both *Drosophila* and vertebrates heart development, there is evidence for a mutually reinforcing regulatory network centered on Tin and GATA transcription factors. This has some parallels with the relationship between MRFs and MEF2 in skeletal muscle. For example, in *Drosophila*, *pannier* is regulated by Tin.[245] In mice, the GATA6 promoter contains functionally important Nkx2-5 binding sites, and the *Nkx2-5* promoter contains GATA sites involved in early heart field expression.[214]

Heart Differentiation

In *Drosophila*, the Mef2 transcription factor is required for proper differentiation of the heart, but not for its specification and basic morphogenesis.[96,97] Many, but not all, genes encoding components of the contractile apparatus are not expressed in *mef2* mutants.[212] *Drosophila mef2* is a target of both Tin and Pannier and so lies downstream of these regulators in the genetic pathway controlling heart development.[230,246,247]

Mef2 genes are also important in vertebrate heart development, although the picture of its role is less complete. In mouse *Mef2c* mutants, there is a failure in normal cardiogenesis and differentiation. For example, looping does not occur and some cardiac muscle genes are down-regulated.[248] In part of the developing heart, one GATA factor target is *Mef2c*.[249]

It is now apparent that cardiac gene transcription is regulated by a number of transcription factors acting in various combinations.[212,250] This includes members of the Nkx2, GATA and Mef2 families, together with SRF, Myocardin and the T-box factor, Tbx5. Many of these transcription factors form complexes through protein/protein interactions.[212,250] Prominent amongst these is Myocardin, which binds to SRF to regulate gene transcription and is implicated in both *Xenopus* heart development and in cardiomyocyte differentiation.[251,252]

Patterning

Some progress has been made into the control of development of different regions of the heart. In *Drosophila*, the most obvious subdivision of its linear heart tube is into a posterior "heart proper" and a more anterior aorta (Fig. 5). This heart proper is specified by the Hox gene, *AbdA*.[253-255] The aorta region itself is also subdivided along the A/P axis, and Hox genes function in this too.[256] In vertebrates at the beating heart stage, there is polarity along the A/P axis. The regions that will form the different subdivisions of the final heart structure, e.g., right ventricle, atria, are found at defined positions along this axis.[211] There are suggestive, but not compelling, indications of Hox gene involvement in A/P subdivisions of the developing vertebrate heart.[253,254,257]

One gene with a clear region-specific role is that encoding the dHAND bHLH transcription factor. In mouse "knock-outs" for this gene, development of the right ventricle fails.[258] This phenotype reflects the specific expression of *dHAND* in the developing ventricle. A related factor, eHAND, is expressed in the developing left ventricle. *Drosophila* has a single *Hand* gene, which is expressed throughout the heart, but no functional information is available yet.[259]

Visceral Muscle Development

In general, much less is known about vertebrate smooth muscle than about skeletal and heart muscle. Nevertheless, smooth muscle is very important clinically and detailed knowledge

Table 1. A comparison of visceral muscle development in Drosophila and vertebrates

| | Localization of Primordium | Specification and Early Differentiation | | Terminal Differentiation |
		Signals	Transcription Factors	Transcription Factors
Drosophila	Inner (splanchnic) layer of mesoderm	Dpp Hh Jeb	Tin Bap Bin	Mef2
Vertebrates	Inner (splanchnic) layer of lateral plate mesoderm	BMP Hh	Bapx1/Nkx3.2? Foxf1	SRF Myocardin Mef2??

of its development and differentiation is likely to have significant applications in many diseases. It is the muscle of the blood vessel wall and the digestive tract, and also of the respiratory and urogenital systems. Here I will compare vertebrate digestive tract muscle with the visceral muscle surrounding the *Drosophila* midgut (Table 1). *Drosophila* hindgut visceral muscle is described elsewhere.[260,261]

Gut muscle has analogous developmental origins in vertebrates and *Drosophila*. In vertebrates, precursors are lateral plate mesoderm cells from the splanchnic, or inner, layer,[7] and in *Drosophila*, visceral muscle precursors also derive from the splanchnic layer of the mesoderm.[3] The differentiated muscle has similarities too, and in both *Drosophila* and vertebrates there is a lattice of circular and longitudinal musculature.[3,262] There are also differences. *Drosophila* midgut visceral muscle is syncytial,[263,264] whereas vertebrate visceral muscle is generally regarded as unicellular. Also, in *Drosophila*, both larval and adult visceral muscle is striated,[2,17] although other aspects are like vertebrate smooth muscle.[265]

In *Drosophila*, the circular visceral muscle develops from clusters of cells, one per hemisegment along each side of the embryo. These clusters derive from part of the *eve* domain of each hemi-segment defined by the intersecting A/P and D/V cues described earlier and express the homeodomain transcription factor Bagpipe (Bap).[23] Hh and Dpp signals from the ectoderm, together with Bap are sufficient for the development of the circular muscle primordia.[23] Although few details are established, similar signals are implicated in vertebrates. For example, there is Hh family signaling between the epithelium and muscle in the intestinal wall, and both *Shh* and *Indian hedgehog* mutants have reduced smooth muscle.[266] Furthermore, BMP signaling lateralizes developing mesoderm and lateral plate may require high BMP4.[267]

The Role of Bagpipe and FoxF Family Transcription Factors

Bagpipe (*bap*) is required for *Drosophila* visceral mesoderm development,[228] and is part of a transcriptional hierarchy. It is a target of Tin whose expression is in turn dependent on Twist. *bap* expression is not maintained during visceral muscle development and so differentiation itself must occur through downstream genes, e.g., *vimar* and *β3 tubulin*.[268,269] A key downstream target is *biniou* (*bin*), which is specifically expressed in all visceral muscle and encodes a protein that belongs to the FoxF subfamily of Forkhead transcription factors. In its absence, differentiation of visceral muscle fails, and ectopic Bin expression in the somatic mesoderm ectopically activates visceral mesoderm genes.[265]

Two vertebrate Bin orthologues, FoxF1 and Fox F2, are expressed in the splanchnic mesoderm and the derived intestinal smooth muscle (see ref. 265). Consistent with this, *foxF1* has a role in lateral plate differentiation. In mouse *foxF1* mutants, the separation into somatic and splanchnic layers is often absent or incomplete,[270] while in *Xenopus foxF1* "knock downs" the visceral mesoderm does not differentiate normally.[271] Another point of similarity between vertebrates and *Drosophila* is the relationship between BMP family signals and the *bin* orthologues. First, in *Drosophila*, Dpp regulates *bin* expression,[261] and in vertebrates there appear to be many links between BMP4 signaling and *foxF1*.[270,271] Second, a specific enhancer of *dpp* is a direct target of Bin,[265] and normal BMP4 expression in the vertebrate lateral plate requires *foxF1*.[270] Moreover, in the mouse Bin is coexpressed with the *bap* orthologue *bapx1* in the splanchnic mesoderm, although *bapx1* is expressed elsewhere too.[272] These and other findings have led to the suggestion that the "splanchnic mesoderm layers in *Drosophila* and vertebrates are homologous structures whose development into gut muscle and other visceral organs is critically dependent on FoxF genes".[265] There are some differences of course, such as in *Xenopus* where *bap* is not expressed until after *foxF1*.[271]

There have been significant advances in understanding the gene expression pathways that govern determination and differentiation in skeletal and cardiac muscle. However, much less is known about gene expression in smooth muscle differentiation. Notwithstanding this, targets of Bap and Bin and their orthologues are clearly going to be important. Moreover, in vertebrates, many, but not all, smooth muscle genes have functionally important CArG boxes in their promoter/enhancer regions.[273] These bind the widely expressed SRF transcription factor together with the smooth and cardiac muscle restricted coactivator, Myocardin.[251,274] It is not known whether this regulation has parallels in *Drosophila*.

Mef2, which is structurally related to SRF, is important in differentiation. In *Drosophila*, and as with somatic muscle and the heart, *mef2* is required for the differentiation, but not the specification, of visceral muscle.[91] In vertebrates, I am not aware of any functional analysis of *mef2* on digestive tract muscle development, although in mice, *mef2c* mutants have defects in smooth muscle differentiation in the vasculature,[95] and in humans, *mef2a* mutations are linked to coronary artery disease.[275]

Visceral Muscle Founder Cells

As with *Drosophila* larval somatic muscle, in *Drosophila* visceral muscle development there are two types of cell, founder cells and fusion-competent myoblasts, that specifically express *duf* and *sns* respectively. There are two classes of visceral founders: those for the circular musculature develop from clusters in each trunk hemisegment and then form a continuous ordered file of cells along the A/P axis of the embryo; those for the longitudinal muscles develop at the posterior end of the germ band and subsequently migrate. There appears to be a common pool of fusion-competent myoblasts for both classes of muscle in the trunk.[264,276] It has been argued that the founder cells really are true founders, not simply pioneers.[264] Thus, in the absence of fusion they make mononucleate circular or longitudinal fibers, as appropriate, and fusion-competent myoblasts remain undifferentiated. Moreover, localized gene expression is initiated in subsets of the founders and spreads throughout the muscles arising from them by fusion.[264] The finding of founder cells in another example of the development of syncytial muscles makes it clear that founder cells are not unique to *Drosophila* somatic muscle.

There is a novel signal implicated in the development of visceral founder cells. It is called Jelly belly (Jeb). It contains a LDL receptor motif, and is secreted by the nearby somatic mesoderm.[277] In *jeb* mutants, visceral muscle is not produced because all precursors become fusion-competent myoblasts and founder cells do not develop.[278,279] The Jeb receptor is Alk, the *Drosophila* version of the tyrosine kinase receptor encoded by the human proto-oncogene *Anaplastic Lymphoma Kinase*.[278-280] It is unknown whether there are vertebrate Jeb-like proteins that bind to Alk to affect muscle development, although there are other known ligands for vertebrate Alk.[281,282]

Analysis of *jeb* highlights the closeness of the relationship between visceral and somatic muscle and raises the question of how distinct the myoblast populations of somatic and visceral muscle are. Thus, in *jeb* mutants, with visceral founders not specified, the visceral fusion-competent myoblasts become incorporated into the somatic musculature.[278-280] There is a close relationship in vertebrates too. Thus, transdifferentiation of smooth muscle cells into skeletal muscle is reported, for example in the mouse oesophagus,[283,284] and there are cell lines with characteristics of both smooth and skeletal muscle.[285] A counter example from *Drosophila* is in *eve* mutants where there is no midgut visceral mesoderm of the trunk. Longitudinal muscle founder cells still migrate and yet remain mononucleate despite the nearby somatic fusion-competent myoblasts.[264]

Concluding Remarks

In comparing diverse aspects of muscle development in *Drosophila* and in vertebrates it is apparent that there are many similarities. These include the developmental strategies, some of the cell biology, and the underlying transcriptional networks. Of course there are differences too, and I have highlighted some examples. Nevertheless, the overriding conclusion is that muscle development in *Drosophila* and in vertebrates shares fundamental similarities. This suggests that findings in one organism will continue to advance understanding in others. This can happen in many ways. For example, the orthologues of key genes identified in screens in *Drosophila* can be analyzed in vertebrates to provide new entry points into studies in these species. The direction of information transfer is certainly not just from *Drosophila* to vertebrates. Genes identified in vertebrates may be difficult and/or expensive to study in vertebrate systems and so a helpful strategy is to analyze orthologues in *Drosophila*. Mef2, which was first described in mouse, is an example of this. One major outcome of comparative analyses of different aspects of muscle development in different species is to gain insights into how different animals develop their distinct morphologies and functional attributes. Another important outcome is the opening of new approaches to the treatment of muscle-linked diseases.

A prominent example of how studies of muscle development in both *Drosophila* and vertebrates have advanced clinical understanding is the heart.[286] Mutations in some of the key transcription factors involved in heart development, Nkx2-5, TBX5 and GATA4, are linked to human congenital heart disease.[215] The Nkx2-5 story illustrates the synergism between *Drosophila* research and clinical investigations.[286] This greatly accelerated the normally lengthy procedure of discovering the genetic basis of a disease, in this case, atrial septal defects and conduction abnormalities. Conventional linkage analysis identified a chromosomal region associated with these heart abnormalities. Of the many genes in the region, *Nkx2-5* quickly became the focus of attention, as it was highly related to *tin*, which was already known to have a pivotal role in *Drosophila* heart development. In another aspect of heart disease, hypertrophy of the adult heart, it has also become apparent that some transcription factors that function in heart development, e.g., GATA4 and Mef2, play a central role here too.[215,286]

One possible route for the future repair of damaged and diseased skeletal, cardiac or smooth muscle tissue is the use of stem cell based therapies.[184,287] Central to this will be a detailed knowledge of muscle cell differentiation both to monitor and to control the differentiation of stem cells. Approaches might involve introducing cells by transplantation or systemic injection, or signaling to endogenous stem cell populations. In skeletal muscle, for example, the latter approach might exploit the natural, but limited, repair capacity of adult muscle via its satellite cells, although much remains to be understood about this regenerative ability, including how close the parallels are with embryonic development. A final specific aspect of skeletal muscle differentiation of great clinical interest is myoblast fusion. This will be critical to the repair of damaged or diseased muscle tissue, and may also be a novel route for stable expression of molecules to treat various nonmuscle conditions.[288,289]

Whether your interest is in development, in comparative zoology, in the basic biology of muscle or the clinical treatment of muscle disease, there is much still to be uncovered from future studies of muscle development in *Drosophila* and the variety of vertebrate species.

References

1. Rubin GM. The draft sequences. Comparing species. Nature 2001; 409(6822):820-821.
2. Bate M. The mesoderm and its derivatives. In: Bate M, Martinez-Arias A, eds. The Development of Drosophila melanogaster. New York: Cold Spring Harbor Press, 1993:1013-1090.
3. Poulson DF. Histogensis, organogenesis and differentiation in the embryo of Drosophila melanogaster. In: Demerec M, ed. Biology of Drosophila. Facsimile ed. New York: Cold Spring Harbor Press, 1950:168-274.
4. Ruiz-Gomez M. Muscle patterning and specification in Drosophila. Int J Dev Biol 1998; 42(3):283-290.
5. Taylor MV. Muscle differentiation: How two cells become one. Curr Biol 2002; 12(6):R224-228.
6. Carmena A, Baylies M. The development of the Drosophila larval somatic musculature. In: Sink H, ed. Muscle Development in Drosophila. Georgetown: Landes Bioscience, 2006:79-91.
7. Christ B, Ordahl CP. Early stages of chick somite development. Anat Embryol (Berl) 1995; 191(5):381-396.
8. Marcelle C, Lesbros C, Linker C. Somite patterning: A few more pieces of the puzzle. Results Probl Cell Differ 2002; 38:81-108.
9. Redkar A, Montgomery M, Litvin J. Fate map of early avian cardiac progenitor cells. Development 2001; 128(12):2269-2279.
10. Wigmore PM, Evans DJ. Molecular and cellular mechanisms involved in the generation of fiber diversity during myogenesis. Int Rev Cytol 2002; 216:175-232.
11. Pourquie O. The segmentation clock: Converting embryonic time into spatial pattern. Science 2003; 301(5631):328-330.
12. Brent AE, Tabin CJ. Developmental regulation of somite derivatives: Muscle, cartilage and tendon. Curr Opin Genet Dev 2002; 12(5):548-557.
13. Buckingham M, Bajard L, Chang T et al. The formation of skeletal muscle: From somite to limb. J Anat 2003; 202(1):59-68.
14. Kelly AM, Zacks SI. The histogenesis of rat intercostal muscle. J Cell Biol 1969; 42(1):135-153.
15. Harris AJ, Duxson MJ, Fitzsimons RB et al. Myonuclear birthdates distinguish the origins of primary and secondary myotubes in embryonic mammalian skeletal muscles. Developments 1989; 107(4):771-784.
16. Patel K, Christ B, Stockdale FE. Control of muscle size during embryonic, fetal, and adult life. Results Probl Cell Differ 2002; 38:163-186.
17. Miller A. The internal anatomy and histology of the imago of Drosophila melanogaster. In: Demerec M, ed. Biology of Drosophila. Facsimile ed. New York: Cold Spring Harbor Press, 1950:420-534.
18. Lassar AB, Munsterberg AE. The role of positive and negative signals in somite patterning. Curr Opin Neurobiol 1996; 6(1):57-63.
19. Tajbakhsh S, Cossu G. Establishing myogenic identity during somitogenesis. Curr Opin Genet Dev 1997; 7(5):634-641.
20. Chanoine C, Hardy S. Xenopus muscle development: From primary to secondary myogenesis. Dev Dyn 2003; 226(1):12-23.
21. Beer J, Technau G, Campos Ortega J. Lineage analysis of transplanted individual cells in embryos of Drosophila melanogaster. IV. Commitment and proliferative capabilities of individual mesodermal cells. Roux's Arch Dev Biol 1987; 196:222-230.
22. Borkowski OM, Brown NH, Bate M. Anterior-posterior subdivision and the diversification of the mesoderm in Drosophila. Development 1995; 121(12):4183-4193.
23. Azpiazu N, Lawrence PA, Vincent JP et al. Segmentation and specification of the Drosophila mesoderm. Genes Dev 1996; 10(24):3183-3194.
24. Riechmann V, Irion U, Wilson R et al. Control of cell fates and segmentation in the Drosophila mesoderm. Development 1997; 124(15):2915-2922.
25. Pourquie O, Fan CM, Coltey M et al. Lateral and axial signals involved in avian somite patterning: A role for BMP4. Cell 1996; 84(3):461-471.
26. Marcelle C, Stark MR, Bronner-Fraser M. Coordinate actions of BMPs, Wnts, Shh and noggin mediate patterning of the dorsal somite. Development 1997; 124(20):3955-3963.
27. Dietrich S, Schubert FR, Healy C et al. Specification of the hypaxial musculature. Development 1998; 125(12):2235-2249.
28. Bock E. Wechselbeziehungen zwischen den Keimblättern bei der Organbildung von Chrysopa perla L. Die Entwicklung des Ektoderms in mesodermdefekten Keimteilen. Wilhelm Roux' Arch Entwicklungsmech Org 1941; 141:159-247.
29. Baker R, Schubiger G. Ectoderm induces muscle-specific gene expression in Drosophila embryos. Development 1995; 121(5):1387-1398.

30. Maggert K, Levine M, Frasch M. The somatic-visceral subdivision of the embryonic mesoderm is initiated by dorsal gradient thresholds in Drosophila. Development 1995; 121(7):2107-2116.
31. Gisselbrecht S, Skeath JB, Doe CQ et al. Heartless encodes a fibroblast growth factor receptor (DFR1/DFGF-R2) involved in the directional migration of early mesodermal cells in the Drosophila embryo. Genes Dev 1996; 10(23):3003-3017.
32. Beiman M, Shilo BZ, Volk T. Heartless, a Drosophila FGF receptor homolog, is essential for cell migration and establishment of several mesodermal lineages. Genes Dev 1996; 10(23):2993-3002.
33. Fan CM, Lee CS, Tessier-Lavigne M. A role for WNT proteins in induction of dermomyotome. Dev Biol 1997; 191(1):160-165.
34. Munsterberg AE, Lassar AB. Combinatorial signals from the neural tube, floor plate and notochord induce myogenic bHLH gene expression in the somite. Development 1995; 121(3):651-660.
35. Fan CM, Tessier-Lavigne M. Patterning of mammalian somites by surface ectoderm and notochord: Evidence for sclerotome induction by a hedgehog homolog. Cell 1994; 79(7):1175-1186.
36. Johnson RL, Laufer E, Riddle RD et al. Ectopic expression of Sonic hedgehog alters dorsal-ventral patterning of somites. Cell 1994; 79(7):1165-1173.
37. Linker C, Lesbros C, Stark MR et al. Intrinsic signals regulate the initial steps of myogenesis in vertebrates. Development 2003; 130(20):4797-4807.
38. Ikeya M, Takada S. Wnt signaling from the dorsal neural tube is required for the formation of the medial dermomyotome. Development 1998; 125(24):4969-4976.
39. Borycki AG, Brunk B, Tajbakhsh S et al. Sonic hedgehog controls epaxial muscle determination through Myf5 activation. Development 1999; 126(18):4053-4063.
40. Staehling-Hampton K, Hoffmann FM, Baylies MK et al. Dpp induces mesodermal gene expression in Drosophila. Nature 1994; 372(6508):783-786.
41. Frasch M. Induction of visceral and cardiac mesoderm by ectodermal Dpp in the early Drosophila embryo. Nature 1995; 374(6521):464-467.
42. Lee HH, Frasch M. Wingless effects mesoderm patterning and ectoderm segmentation events via induction of its downstream target sloppy paired. Development 2000; 127(24):5497-5508.
43. Baylies MK, Martinez Arias A, Bate M. Wingless is required for the formation of a subset of muscle founder cells during Drosophila embryogenesis. Development 1995; 121(11):3829-3837.
44. Ranganayakulu G, Schulz RA, Olson EN. Wingless signaling induces nautilus expression in the ventral mesoderm of the Drosophila embryo. Dev Biol 1996; 176(1):143-148.
45. Lawrence PA, Bodmer R, Vincent JP. Segmental patterning of heart precursors in Drosophila. Development 1995; 121(12):4303-4308.
46. Greig S, Akam M. Homeotic genes autonomously specify one aspect of pattern in the Drosophila mesoderm. Nature 1993; 362(6421):630-632.
47. Michelson AM. Muscle pattern diversification in Drosophila is determined by the autonomous function of homeotic genes in the embryonic mesoderm. Development 1994; 120(4):755-768.
48. Capovilla M, Kambris Z, Botas J. Direct regulation of the muscle-identity gene apterous by a Hox protein in the somatic mesoderm. Development 2001; 128(8):1221-1230.
49. Alvares LE, Schubert FR, Thorpe C et al. Intrinsic, Hox-dependent cues determine the fate of skeletal muscle precursors. Dev Cell 2003; 5(3):379-390.
50. Davis RL, Weintraub H, Lassar AB. Expression of a single transfected cDNA converts fibroblasts to myoblasts. Cell 1987; 51(6):987-1000.
51. Weintraub H, Davis R, Tapscott S et al. The myoD gene family: Nodal point during specification of the muscle cell lineage. Science 1991; 251(4995):761-766.
52. Braun T, Rudnicki MA, Arnold HH et al. Targeted inactivation of the muscle regulatory gene Myf-5 results in abnormal rib development and perinatal death. Cell 1992; 71(3):369-382.
53. Tajbakhsh S, Rocancourt D, Cossu G et al. Redefining the genetic hierarchies controlling skeletal myogenesis: Pax-3 and Myf-5 act upstream of MyoD. Cell 1997; 89(1):127-138.
54. Kaul A, Koster M, Neuhaus H et al. Myf-5 revisited: Loss of early myotome formation does not lead to a rib phenotype in homozygous Myf-5 mutant mice. Cell 2000; 102(1):17-19.
55. Rudnicki MA, Braun T, Hinuma S et al. Inactivation of MyoD in mice leads to up-regulation of the myogenic HLH gene Myf-5 and results in apparently normal muscle development. Cell 1992; 71(3):383-390.
56. Rudnicki MA, Schnegelsberg PN, Stead RH et al. MyoD or Myf-5 is required for the formation of skeletal muscle. Cell 1993; 75(7):1351-1359.
57. Nabeshima Y, Hanaoka K, Hayasaka M et al. Myogenin gene disruption results in perinatal lethality because of severe muscle defect. Nature 1993; 364(6437):532-535.
58. Hasty P, Bradley A, Morris JH et al. Muscle deficiency and neonatal death in mice with a targeted mutation in the myogenin gene. Nature 1993; 364(6437):501-506.

59. Kassar-Duchossoy L, Gayraud-Morel B, Gomes D et al. Mrf4 determines skeletal muscle identity in Myf5: Myod double-mutant mice. Nature 2004; 431(7007):466-471.
60. Hadchouel J, Tajbakhsh S, Primig M et al. Modular long-range regulation of Myf5 reveals unexpected heterogeneity between skeletal muscles in the mouse embryo. Development 2000; 127(20):4455-4467.
61. Summerbell D, Ashby PR, Coutelle O et al. The expression of Myf5 in the developing mouse embryo is controlled by discrete and dispersed enhancers specific for particular populations of skeletal muscle precursors. Development 2000; 127(17):3745-3757.
62. Buchberger A, Nomokonova N, Arnold HH. Myf5 expression in somites and limb buds of mouse embryos is controlled by two distinct distal enhancer activities. Development 2003; 130(14):3297-3307.
63. Dechesne CA, Wei Q, Eldridge J et al. E-box- and MEF-2-independent muscle-specific expression, positive autoregulation, and cross-activation of the chicken MyoD (CMD1) promoter reveal an indirect regulatory pathway. Mol Cell Biol 1994; 14(8):5474-5486.
64. Asakura A, Lyons GE, Tapscott SJ. The regulation of MyoD gene expression: Conserved elements mediate expression in embryonic axial muscle. Dev Biol 1995; 171(2):386-398.
65. Kucharczuk KL, Love CM, Dougherty NM et al. Fine-scale transgenic mapping of the MyoD core enhancer: MyoD is regulated by distinct but overlapping mechanisms in myotomal and nonmyotomal muscle lineages. Development 1999; 126(9):1957-1965.
66. Maroto M, Reshef R, Munsterberg AE et al. Ectopic Pax-3 activates MyoD and Myf-5 expression in embryonic mesoderm and neural tissue. Cell 1997; 89(1):139-148.
67. Thayer MJ, Tapscott SJ, Davis RL et al. Positive autoregulation of the myogenic determination gene MyoD1. Cell 1989; 58(2):241-248.
68. Lun Y, Sawadogo M, Perry M. Autoactivation of Xenopus MyoD transcription and its inhibition by USF. Cell Growth Differ 1997; 8(3):275-282.
69. Tajbakhsh S, Borello U, Vivarelli E et al. Differential activation of Myf5 and MyoD by different Wnts in explants of mouse paraxial mesoderm and the later activation of myogenesis in the absence of Myf5. Development 1998; 125(21):4155-4162.
70. Hirsinger E, Malapert P, Dubrulle J et al. Notch signalling acts in postmitotic avian myogenic cells to control MyoD activation. Development 2001; 128(1):107-116.
71. Fisher ME, Isaacs HV, Pownall ME. eFGF is required for activation of XmyoD expression in the myogenic cell lineage of Xenopus laevis. Development 2002; 129(6):1307-1315.
72. Bergstrom DA, Penn BH, Strand A et al. Promoter-specific regulation of MyoD binding and signal transduction cooperate to pattern gene expression. Mol Cell 2002; 9(3):587-600.
73. Michelson AM, Abmayr SM, Bate M et al. Expression of a MyoD family member prefigures muscle pattern in Drosophila embryos. Genes Dev 1990; 4(12A):2086-2097.
74. Paterson BM, Walldorf U, Eldridge J et al. The Drosophila homologue of vertebrate myogenic-determination genes encodes a transiently expressed nuclear protein marking primary myogenic cells. Proc Natl Acad Sci USA 1991; 88(9):3782-3786.
75. Zhang JM, Chen L, Krause M et al. Evolutionary conservation of MyoD function and differential utilization of E proteins. Dev Biol 1999; 208(2):465-472.
76. Wei Q, Marchler G, Edington K et al. RNA interference demonstrates a role for nautilus in the myogenic conversion of Schneider cells by daughterless. Dev Biol 2000; 228(2):239-255.
77. Keller CA, Erickson MS, Abmayr SM. Misexpression of nautilus induces myogenesis in cardioblasts and alters the pattern of somatic muscle fibers. Dev Biol 1997; 181(2):197-212.
78. Balagopalan L, Keller CA, Abmayr SM. Loss-of-function mutations reveal that the Drosophila nautilus gene is not essential for embryonic myogenesis or viability. Dev Biol 2001; 231(2):374-382.
79. Baylies MK, Bate M. Twist: A myogenic switch in Drosophila. Science 1996; 272(5267):1481-1484.
80. Castanon I, Von Stetina S, Kass J et al. Dimerization partners determine the activity of the Twist bHLH protein during Drosophila mesoderm development. Development 2001; 128(16):3145-3159.
81. Murre C, McCaw PS, Vaessin H et al. Interactions between heterologous helix-loop-helix proteins generate complexes that bind specifically to a common DNA sequence. Cell 1989; 58(3):537-544.
82. Lassar AB, Davis RL, Wright WE et al. Functional activity of myogenic HLH proteins requires hetero-oligomerization with E12/E47-like proteins in vivo. Cell 1991; 66(2):305-315.
83. Neuhold LA, Wold B. HLH forced dimers: Tethering MyoD to E47 generates a dominant positive myogenic factor insulated from negative regulation by Id. Cell 1993; 74(6):1033-1042.
84. Benezra R, Davis RL, Lockshon D et al. The protein Id: A negative regulator of helix-loop-helix DNA binding proteins. Cell 1990; 61(1):49-59.
85. Mitsui K, Shirakata M, Paterson BM. Phosphorylation inhibits the DNA-binding activity of MyoD homodimers but not MyoD-E12 heterodimers. J Biol Chem 1993; 268(32):24415-24420.
86. Li FQ, Coonrod A, Horwitz M. Preferential MyoD homodimer formation demonstrated by a general method of dominant negative mutation employing fusion with a lysosomal protease. J Cell Biol 1996; 135(4):1043-1057.

87. Hopwood ND, Gurdon JB. Activation of muscle genes without myogenesis by ectopic expression of MyoD in frog embryo cells. Nature 1990; 347(6289):197-200.
88. Rashbass J, Taylor MV, Gurdon JB. The DNA-binding protein E12 cooperates with XMyoD in the activation of muscle-specific gene expression in Xenopus embryos. EMBO J 1992; 11(8):2981-2990.
89. Faerman A, Pearson-White S, Emerson C et al. Ectopic expression of MyoD1 in mice causes prenatal lethalities. Dev Dyn 1993; 196(3):165-173.
90. Tapanes-Castillo A, Baylies MK. Notch signaling patterns Drosophila mesodermal segments by regulating the bHLH transcription factor twist. Development 2004; 131(10):2359-2372.
91. Taylor MV. Muscle development. Making Drosophila muscle. Curr Biol 1995; 5(7):740-742.
92. Black BL, Olson EN. Transcriptional control of muscle development by myocyte enhancer factor-2 (MEF2) proteins. Annu Rev Cell Dev Biol 1998; 14:167-196.
93. Molkentin JD, Black BL, Martin JF et al. Cooperative activation of muscle gene expression by MEF2 and myogenic bHLH proteins. Cell 1995; 83(7):1125-1136.
94. Edmondson DG, Lyons GE, Martin JF et al. Mef2 gene expression marks the cardiac and skeletal muscle lineages during mouse embryogenesis. Development 1994; 120(5):1251-1263.
95. Lin Q, Lu J, Yanagisawa H et al. Requirement of the MADS-box transcription factor MEF2C for vascular development. Development 1998; 125(22):4565-4574.
96. Bour BA, O'Brien MA, Lockwood WL et al. Drosophila MEF2, a transcription factor that is essential for myogenesis. Genes Dev 1995; 9(6):730-741.
97. Lilly B, Zhao B, Ranganayakulu G et al. Requirement of MADS domain transcription factor D-MEF2 for muscle formation in Drosophila. Science 1995; 267(5198):688-693.
98. Ranganayakulu G, Zhao B, Dokidis A et al. A series of mutations in the D-MEF2 transcription factor reveal multiple functions in larval and adult myogenesis in Drosophila. Dev Biol 1995; 171(1):169-181.
99. Ornatsky OI, Andreucci JJ, McDermott JC. A dominant-negative form of transcription factor MEF2 inhibits myogenesis. J Biol Chem 1997; 272(52):33271-33278.
100. Taylor MV, Beatty KE, Hunter HK et al. Drosophila MEF2 is regulated by twist and is expressed in both the primordia and differentiated cells of the embryonic somatic, visceral and heart musculature. Mech Dev 1995; 50(1):29-41.
101. Cripps RM, Black BL, Zhao B et al. The myogenic regulatory gene Mef2 is a direct target for transcriptional activation by Twist during Drosophila myogenesis. Genes Dev 1998; 12(3):422-434.
102. Wang DZ, Valdez MR, McAnally J et al. The Mef2c gene is a direct transcriptional target of myogenic bHLH and MEF2 proteins during skeletal muscle development. Development 2001; 128(22):4623-4633.
103. Dodou E, Xu SM, Black BL. Mef2c is activated directly by myogenic basic helix-loop-helix proteins during skeletal muscle development in vivo. Mech Dev 2003; 120(9):1021-1032.
104. Cripps RM, Lovato TL, Olson EN. Positive autoregulation of the Myocyte enhancer factor-2 myogenic control gene during somatic muscle development in Drosophila. Dev Biol 2004; 267(2):536-547.
105. Cheng TC, Wallace MC, Merlie JP et al. Separable regulatory elements governing myogenin transcription in mouse embryogenesis. Science 1993; 261(5118):215-218.
106. Yee SP, Rigby PW. The regulation of myogenin gene expression during the embryonic development of the mouse. Genes Dev 1993; 7(7A):1277-1289.
107. Spitz F, Demignon J, Porteu A et al. Expression of myogenin during embryogenesis is controlled by Six/sine oculis homeoproteins through a conserved MEF3 binding site. Proc Natl Acad Sci USA 1998; 95(24):14220-14225.
108. Lin MH, Nguyen HT, Dybala C et al. Myocyte-specific enhancer factor 2 acts cooperatively with a muscle activator region to regulate Drosophila tropomyosin gene muscle expression. Proc Natl Acad Sci USA 1996; 93(10):4623-4628.
109. Damm C, Wolk A, Buttgereit D et al. Independent regulatory elements in the upstream region of the Drosophila beta 3 tubulin gene (beta Tub60D) guide expression in the dorsal vessel and the somatic muscles. Dev Biol 1998; 199(1):138-149.
110. Kelly KK, Meadows SM, Cripps RM. Drosophila MEF2 is a direct regulator of Actin57B transcription in cardiac, skeletal, and visceral muscle lineages. Mech Dev 2002; 110(1-2):39-50.
111. Taylor MV. A novel Drosophila, mef2-regulated muscle gene isolated in a subtractive hybridization-based molecular screen using small amounts of zygotic mutant RNA. Dev Biol 2000; 220(1):37-52.
112. Vivian JL, Gan L, Olson EN et al. A hypomorphic myogenin allele reveals distinct myogenin expression levels required for viability, skeletal muscle development, and sternum formation. Dev Biol 1999; 208(1):44-55.
113. Gunthorpe D, Beatty KE, Taylor MV. Different levels, but not different isoforms, of the Drosophila transcription factor DMEF2 affect distinct aspects of muscle differentiation. Dev Biol 1999; 215(1):130-145.

114. Bate M. The embryonic development of larval muscles in Drosophila. Development 1990; 110(3):791-804.
115. Kardon G. Muscle and tendon morphogenesis in the avian hind limb. Development 1998; 125(20):4019-4032.
116. Leiss D, Hinz U, Gasch A et al. Beta 3 tubulin expression characterizes the differentiating mesodermal germ layer during Drosophila embryogenesis. Development 1988; 104(4):525-531.
117. Dohrmann C, Azpiazu N, Frasch M. A new Drosophila homeo box gene is expressed in mesodermal precursor cells of distinct muscles during embryogenesis. Genes Dev 1990; 4(12A):2098-2111.
118. Rushton E, Drysdale R, Abmayr SM et al. Mutations in a novel gene, myoblast city, provide evidence in support of the founder cell hypothesis for Drosophila muscle development. Development 1995; 121(7):1979-1988.
119. Ho RK, Ball EE, Goodman CS. Muscle pioneers: Large mesodermal cells that erect a scaffold for developing muscles and motoneurones in grasshopper embryos. Nature 1983; 301(5895):66-69.
120. Ball EE, Ho RK, Goodman CS. Muscle development in the grasshopper embryo. I. Muscles, nerves, and apodemes in the metathoracic leg. Dev Biol 1985; 111(2):383-398.
121. Prokop A, Landgraf M, Rushton E et al. Presynaptic development at the Drosophila neuromuscular junction: Assembly and localization of presynaptic active zones. Neuron 1996; 17(4):617-626.
122. Bourgouin C, Lundgren SE, Thomas JB. Apterous is a Drosophila LIM domain gene required for the development of a subset of embryonic muscles. Neuron 1992; 9(3):549-561.
123. Ruiz-Gomez M, Romani S, Hartmann C et al. Specific muscle identities are regulated by Kruppel during Drosophila embryogenesis. Development 1997; 124(17):3407-3414.
124. Jagla T, Bellard F, Lutz Y et al. Ladybird determines cell fate decisions during diversification of Drosophila somatic muscles. Development 1998; 125(18):3699-3708.
125. Knirr S, Azpiazu N, Frasch M. The role of the NK-homeobox gene slouch (S59) in somatic muscle patterning. Development 1999; 126(20):4525-4535.
126. Halfon MS, Carmena A, Gisselbrecht S et al. Ras pathway specificity is determined by the integration of multiple signal-activated and tissue-restricted transcription factors. Cell 2000; 103(1):63-74.
127. Knirr S, Frasch M. Molecular integration of inductive and mesoderm-intrinsic inputs governs even-skipped enhancer activity in a subset of pericardial and dorsal muscle progenitors. Dev Biol 2001; 238(1):13-26.
128. Ross JJ, Duxson MJ, Harris AJ. Formation of primary and secondary myotubes in rat lumbrical muscles. Development 1987; 100(3):383-394.
129. Duxson MJ, Usson Y. Cellular insertion of primary and secondary myotubes in embryonic rat muscles. Development 1989; 107(2):243-251.
130. Bodenstein D. The postembryonic development of Drosophila. In: Demerec M, ed. Biology of Drosophila. Facsimile ed. New York: Cold Spring Harbor Press, 1950:275-367.
131. Fernandes J, Bate M, Vijayraghavan K. Development of the indirect flight muscles of Drosophila. Development 1991; 113(1):67-77.
132. Peckham M, Molloy JE, Sparrow JC et al. Physiological properties of the dorsal longitudinal flight muscle and the tergal depressor of the trochanter muscle of Drosophila melanogaster. J Muscle Res Cell Motil 1990; 11(3):203-215.
133. Silva R, Sparrow JC, Geeves MA. Isolation and kinetic characterisation of myosin and myosin S1 from the Drosophila indirect flight muscles. J Muscle Res Cell Motil 2003; 24(8):489-498.
134. Farrell ER, Fernandes J, Keshishian H. Muscle organizers in Drosophila: The role of persistent larval fibers in adult flight muscle development. Dev Biol 1996; 176(2):220-229.
135. Rivlin PK, Schneiderman AM, Booker R. Imaginal pioneers prefigure the formation of adult thoracic muscles in Drosophila melanogaster. Dev Biol 2000; 222(2):450-459.
136. Kozopas KM, Nusse R. Direct flight muscles in Drosophila develop from cells with characteristics of founders and depend on DWnt-2 for their correct patterning. Dev Biol 2002; 243(2):312-325.
137. Dutta D, Anant S, Ruiz-Gomez M et al. Founder myoblasts and fibre number during adult myogenesis in Drosophila. Development 2004; 131(15):3761-3772.
138. Dutta D, VijayRaghavan K. Metamorphosis and the formation of the adult musculature. In: Sink H, ed. Muscle development in Drosophila. Georgetown: Landes Bioscience, 2006:125-142.
139. Ruiz-Gomez M, Coutts N, Price A et al. Drosophila dumbfounded: A myoblast attractant essential for fusion. Cell 2000; 102(2):189-198.
140. Kardon G, Harfe BD, Tabin CJ. A Tcf4-positive mesodermal population provides a prepattern for vertebrate limb muscle patterning. Dev Cell 2003; 5(6):937-944.
141. Ghazi A, Anant S, VijayRaghavan K. Apterous mediates development of direct flight muscles autonomously and indirect flight muscles through epidermal cues. Development 2000; 127(24):5309-5318.
142. Sudarsan V, Anant S, Guptan P et al. Myoblast diversification and ectodermal signaling in Drosophila. Dev Cell 2001; 1(6):829-839.

143. Artero R, Furlong EE, Beckett K et al. Notch and Ras signaling pathway effector genes expressed in fusion competent and founder cells during Drosophila myogenesis. Development 2003; 130(25):6257-6272.

144. Schafer K, Braun T. Early specification of limb muscle precursor cells by the homeobox gene Lbx1h. Nat Genet 1999; 23(2):213-216.

145. Gross MK, Moran-Rivard L, Velasquez T et al. Lbx1 is required for muscle precursor migration along a lateral pathway into the limb. Development 2000; 127(2):413-424.

146. Brohmann H, Jagla K, Birchmeier C. The role of Lbx1 in migration of muscle precursor cells. Development 2000; 127(2):437-445.

147. Mankoo BS, Collins NS, Ashby P et al. Mox2 is a component of the genetic hierarchy controlling limb muscle development. Nature 1999; 400(6739):69-73.

148. Stockdale FE. Myogenic cell lineages. Dev Biol 1992; 154(2):284-298.

149. Nikovits Jr W, Cann GM, Huang R et al. Patterning of fast and slow fibers within embryonic muscles is established independently of signals from the surrounding mesenchyme. Development 2001; 128(13):2537-2544.

150. Schiaffino S, Serrano A. Calcineurin signaling and neural control of skeletal muscle fiber type and size. Trends Pharmacol Sci 2002; 23(12):569-575.

151. Baxendale S, Davison C, Muxworthy C et al. The B-cell maturation factor Blimp-1 specifies vertebrate slow-twitch muscle fiber identity in response to Hedgehog signaling. Nat Genet 2004; 36(1):88-93.

152. Wang YX, Zhang CL, Yu RT et al. Regulation of muscle fiber type and running endurance by PPARdelta. PLoS Biol 2004; 2(10):e294.

153. Doberstein SK, Fetter RD, Mehta AY et al. Genetic analysis of myoblast fusion: Blown fuse is required for progression beyond the prefusion complex. J Cell Biol 1997; 136(6):1249-1261.

154. Abmayr SM, Balagopalan L, Galletta BJ et al. Cell and molecular biology of myoblast fusion. Int Rev Cytol 2003; 225:33-89.

155. Knudsen KA, Myers L, McElwee SA. A role for the Ca2(+)-dependent adhesion molecule, N-cadherin, in myoblast interaction during myogenesis. Exp Cell Res 1990; 188(2):175-184.

156. Dickson G, Peck D, Moore SE et al. Enhanced myogenesis in NCAM-transfected mouse myoblasts. Nature 1990; 344(6264):348-351.

157. Mege RM, Goudou D, Diaz C et al. N-cadherin and N-CAM in myoblast fusion: Compared localisation and effect of blockade by peptides and antibodies. J Cell Sci 1992; 103(Pt 4):897-906.

158. Yagami-Hiromasa T, Sato T, Kurisaki T et al. A metalloprotease-disintegrin participating in myoblast fusion. Nature 1995; 377(6550):652-656.

159. Charlton CA, Mohler WA, Radice GL et al. Fusion competence of myoblasts rendered genetically null for N-cadherin in culture. J Cell Biol 1997; 138(2):331-336.

160. Charlton CA, Mohler WA, Blau HM. Neural cell adhesion molecule (NCAM) and myoblast fusion. Dev Biol 2000; 221(1):112-119.

161. Paululat A, Holz A, Renkawitz-Pohl R. Essential genes for myoblast fusion in Drosophila embryogenesis. Mech Dev 1999; 83(1-2):17-26.

162. Dworak HA, Sink H. Myoblast fusion in Drosophila. Bioessays 2002; 24(7):591-601.

163. Chen EH, Olson EN. Towards a molecular pathway for myoblast fusion in Drosophila. Trends Cell Biol 2004; 14(8):452-460.

164. Rau A, Buttgereit D, Holz A et al. Rolling pebbles (rols) is required in Drosophila muscle precursors for recruitment of myoblasts for fusion. Development 2001; 128(24):5061-5073.

165. Menon SD, Chia W. Drosophila rolling pebbles: A multidomain protein required for myoblast fusion that recruits D-Titin in response to the myoblast attractant Dumbfounded. Dev Cell 2001; 1(5):691-703.

166. Schroter RH, Lier S, Holz A et al. Kette and blown fuse interact genetically during the second fusion step of myogenesis in Drosophila. Development 2004; 131(18):4501-4509.

167. Landgraf M, Baylies M, Bate M. Muscle founder cells regulate defasciculation and targeting of motor axons in the Drosophila embryo. Curr Biol 1999; 9(11):589-592.

168. Hasegawa H, Kiyokawa E, Tanaka S et al. DOCK180, a major CRK-binding protein, alters cell morphology upon translocation to the cell membrane. Mol Cell Biol 1996; 16(4):1770-1776.

169. Chen EH, Pryce BA, Tzeng JA et al. Control of myoblast fusion by a guanine nucleotide exchange factor, loner, and its effector ARF6. Cell 2003; 114(6):751-762.

170. Chen EH, Olson EN. Antisocial, an intracellular adaptor protein, is required for myoblast fusion in Drosophila. Dev Cell 2001; 1(5):705-715.

171. Kang JS, Feinleib JL, Knox S et al. Promyogenic members of the Ig and cadherin families associate to positively regulate differentiation. Proc Natl Acad Sci USA 2003; 100(7):3989-3994.

172. Horsley V, Friday BB, Matteson S et al. Regulation of the growth of multinucleated muscle cells by an NFATC2-dependent pathway. J Cell Biol 2001; 153(2):329-338.
173. Horsley V, Jansen KM, Mills ST et al. IL-4 acts as a myoblast recruitment factor during mammalian muscle growth. Cell 2003; 113(4):483-494.
174. Duxson MJ, Usson Y, Harris AJ. The origin of secondary myotubes in mammalian skeletal muscles: Ultrastructural studies. Development 1989; 107(4):743-750.
175. Rosen GD, Sanes JR, LaChance R et al. Roles for the integrin VLA-4 and its counter receptor VCAM-1 in myogenesis. Cell 1992; 69(7):1107-1119.
176. Schwander M, Leu M, Stumm M et al. Beta1 integrins regulate myoblast fusion and sarcomere assembly. Dev Cell 2003; 4(5):673-685.
177. Bokel C, Brown NH. Integrins in development: Moving on, responding to, and sticking to the extracellular matrix. Dev Cell 2002; 3(3):311-321.
178. Roy S, Wolff C, Ingham PW. The u-boot mutation identifies a Hedgehog-regulated myogenic switch for fiber-type diversification in the zebrafish embryo. Genes Dev 2001; 15(12):1563-1576.
179. Seale P, Rudnicki MA. A new look at the origin, function, and "stem-cell" status of muscle satellite cells. Dev Biol 2000; 218(2):115-124.
180. Cornelison DD, Wold BJ. Single-cell analysis of regulatory gene expression in quiescent and activated mouse skeletal muscle satellite cells. Dev Biol 1997; 191(2):270-283.
181. Zhao P, Hoffman EP. Embryonic myogenesis pathways in muscle regeneration. Dev Dyn 2004; 229(2):380-392.
182. Seale P, Sabourin LA, Girgis-Gabardo A et al. Pax7 is required for the specification of myogenic satellite cells. Cell 2000; 102(6):777-786.
183. Seale P, Ishibashi J, Scime A et al. Pax7 is necessary and sufficient for the myogenic specification of CD45(+): Sca1(+) stem cells from injured muscle. PLoS Biol 2004; 2(5):E130.
184. Cossu G, Mavilio F. Myogenic stem cells for the therapy of primary myopathies: Wishful thinking or therapeutic perspective? J Clin Invest 2000; 105(12):1669-1674.
185. Carmena A, Bate M, Jimenez F. Lethal of scute, a proneural gene, participates in the specification of muscle progenitors during Drosophila embryogenesis. Genes Dev 1995; 9(19):2373-2383.
186. Olson EN. Interplay between proliferation and differentiation within the myogenic lineage. Dev Biol 1992; 154(2):261-272.
187. Kopan R, Nye JS, Weintraub H. The intracellular domain of mouse Notch: A constitutively activated repressor of myogenesis directed at the basic helix-loop-helix region of MyoD. Development 1994; 120(9):2385-2396.
188. Hebrok M, Wertz K, Fuchtbauer EM. M-twist is an inhibitor of muscle differentiation. Dev Biol 1994; 165(2):537-544.
189. Spicer DB, Rhee J, Cheung WL et al. Inhibition of myogenic bHLH and MEF2 transcription factors by the bHLH protein Twist. Science 1996; 272(5267):1476-1480.
190. Chen ZF, Behringer RR. Twist is required in head mesenchyme for cranial neural tube morphogenesis. Genes Dev 1995; 9(6):686-699.
191. Fuchtbauer EM. Expression of M-twist during postimplantation development of the mouse. Dev Dyn 1995; 204(3):316-322.
192. Chen CM, Kraut N, Groudine M et al. I-mf, a novel myogenic repressor, interacts with members of the MyoD family. Cell 1996; 86(5):731-741.
193. Anant S, Roy S, VijayRaghavan K. Twist and Notch negatively regulate adult muscle differentiation in Drosophila. Development 1998; 125(8):1361-1369.
194. Jen Y, Weintraub H, Benezra R. Overexpression of Id protein inhibits the muscle differentiation program: In vivo association of Id with E2A proteins. Genes Dev 1992; 6(8):1466-1479.
195. Wang Y, Benezra R, Sassoon DA. Id expression during mouse development: A role in morphogenesis. Dev Dyn 1992; 194(3):222-230.
196. Ying QL, Nichols J, Chambers I et al. BMP induction of Id proteins suppresses differentiation and sustains embryonic stem cell self-renewal in collaboration with STAT3. Cell 2003; 115(3):281-292.
197. Cubas P, Modolell J, Ruiz-Gomez M. The helix-loop-helix extramacrochaetae protein is required for proper specification of many cell types in the Drosophila embryo. Development 1994; 120(9):2555-2566.
198. Postigo AA, Dean DC. ZEB, a vertebrate homolog of Drosophila Zfh-1, is a negative regulator of muscle differentiation. EMBO J 1997; 16(13):3935-3943.
199. Postigo AA, Ward E, Skeath JB et al. Zfh-1, the Drosophila homologue of ZEB, is a transcriptional repressor that regulates somatic myogenesis. Mol Cell Biol 1999; 19(10):7255-7263.
200. Artavanis-Tsakonas S, Rand MD, Lake RJ. Notch signaling: Cell fate control and signal integration in development. Science 1999; 284(5415):770-776.

201. Delfini M, Hirsinger E, Pourquie O et al. Delta 1-activated notch inhibits muscle differentiation without affecting Myf5 and Pax3 expression in chick limb myogenesis. Development 2000; 127(23):5213-5224.
202. Corbin V, Michelson AM, Abmayr SM et al. A role for the Drosophila neurogenic genes in mesoderm differentiation. Cell 1991; 67(2):311-323.
203. Bate M, Rushton E, Frasch M. A dual requirement for neurogenic genes in Drosophila myogenesis. Dev Suppl 1993; 149-161.
204. Rusconi JC, Corbin V. Evidence for a novel Notch pathway required for muscle precursor selection in Drosophila. Mech Dev 1998; 79(1-2):39-50.
205. Fuerstenberg S, Giniger E. Multiple roles for notch in Drosophila myogenesis. Dev Biol 1998; 201(1):66-77.
206. Brennan K, Baylies M, Arias AM. Repression by Notch is required before Wingless signalling during muscle progenitor cell development in Drosophila. Curr Biol 1999; 9(13):707-710.
207. Conboy IM, Rando TA. The regulation of Notch signaling controls satellite cell activation and cell fate determination in postnatal myogenesis. Dev Cell 2002; 3(3):397-409.
208. Bate M, Rushton E, Currie DA. Cells with persistent twist expression are the embryonic precursors of adult muscles in Drosophila. Development 1991; 113(1):79-89.
209. Lee H, Habas R, Abate-Shen C. MSX1 cooperates with histone H1b for inhibition of transcription and myogenesis. Science 2004; 304(5677):1675-1678.
210. McKinsey TA, Zhang CL, Olson EN. Control of muscle development by dueling HATs and HDACs. Curr Opin Genet Dev 2001; 11(5):497-504.
211. Zaffran S, Frasch M. Early signals in cardiac development. Circ Res 2002; 91(6):457-469.
212. Cripps RM, Olson EN. Control of cardiac development by an evolutionarily conserved transcriptional network. Dev Biol 2002; 246(1):14-28.
213. Solloway MJ, Harvey RP. Molecular pathways in myocardial development: A stem cell perspective. Cardiovasc Res 2003; 58(2):264-277.
214. Brand T. Heart development: Molecular insights into cardiac specification and early morphogenesis. Dev Biol 2003; 258(1):1-19.
215. Olson EN. A decade of discoveries in cardiac biology. Nat Med 2004; 10(5):467-474.
216. Saint-Hilaire EG. Mem, du Mus Hist Nat 1822; 9:89-119.
217. De Robertis EM, Sasai Y. A common plan for dorsoventral patterning in Bilateria. Nature 1996; 380(6569):37-40.
218. Schlange T, Andree B, Arnold HH et al. BMP2 is required for early heart development during a distinct time period. Mech Dev 2000; 91(1-2):259-270.
219. Kishimoto Y, Lee KH, Zon L et al. The molecular nature of zebrafish swirl: BMP2 function is essential during early dorsoventral patterning. Development 1997; 124(22):4457-4466.
220. Shi Y, Katsev S, Cai C et al. BMP signaling is required for heart formation in vertebrates. Dev Biol 2000; 224(2):226-237.
221. Latinkic BV, Kotecha S, Mohun TJ. Induction of cardiomyocytes by GATA4 in Xenopus ectodermal explants. Development 2003; 130(16):3865-3876.
222. Zhang H, Bradley A. Mice deficient for BMP2 are nonviable and have defects in amnion/chorion and cardiac development. Development 1996; 122(10):2977-2986.
223. Wu X, Golden K, Bodmer R. Heart development in Drosophila requires the segment polarity gene wingless. Dev Biol 1995; 169(2):619-628.
224. Jagla K, Frasch M, Jagla T et al. Ladybird, a new component of the cardiogenic pathway in Drosophila required for diversification of heart precursors. Development 1997; 124(18):3471-3479.
225. Park M, Wu X, Golden K et al. The wingless signaling pathway is directly involved in Drosophila heart development. Dev Biol 1996; 177(1):104-116.
226. Pandur P, Lasche M, Eisenberg LM et al. Wnt-11 activation of a noncanonical Wnt signalling pathway is required for cardiogenesis. Nature 2002; 418(6898):636-641.
227. Bodmer R. The gene tinman is required for specification of the heart and visceral muscles in Drosophila. Development 1993; 118(3):719-729.
228. Azpiazu N, Frasch M. Tinman and bagpipe: Two homeo box genes that determine cell fates in the dorsal mesoderm of Drosophila. Genes Dev 1993; 7(7B):1325-1340.
229. Xu X, Yin Z, Hudson JB et al. Smad proteins act in combination with synergistic and antagonistic regulators to target Dpp responses to the Drosophila mesoderm. Genes Dev 1998; 12(15):2354-2370.
230. Gajewski K, Fossett N, Molkentin JD et al. The zinc finger proteins Pannier and GATA4 function as cardiogenic factors in Drosophila. Development 1999; 126(24):5679-5688.
231. Klinedinst SL, Bodmer R. Gata factor Pannier is required to establish competence for heart progenitor formation. Development 2003; 130(13):3027-3038.

232. Evans SM. Vertebrate tinman homologues and cardiac differentiation. Semin Cell Dev Biol 1999; 10(1):73-83.
233. Biben C, Harvey RP. Homeodomain factors Nkx2-5 controls left/right asymmetric expression of bHLH gene eHand during murine heart development. Genes Dev 1997; 11(11):1357-1369.
234. Tanaka M, Chen Z, Bartunkova S et al. The cardiac homeobox gene Csx/Nkx2.5 lies genetically upstream of multiple genes essential for heart development. Development 1999; 126(6):1269-1280.
235. Fu Y, Yan W, Mohun TJ et al. Vertebrate tinman homologues XNkx2-3 and XNkx2-5 are required for heart formation in a functionally redundant manner. Development 1998; 125(22):4439-4449.
236. Grow MW, Krieg PA. Tinman function is essential for vertebrate heart development: Elimination of cardiac differentiation by dominant inhibitory mutants of the tinman-related genes, XNkx2-3 and XNkx2-5. Dev Biol 1998; 204(1):187-196.
237. Benson DW, Silberbach GM, Kavanaugh-McHugh A et al. Mutations in the cardiac transcription factor NKX2.5 affect diverse cardiac developmental pathways. J Clin Invest 1999; 104(11):1567-1573.
238. Schott JJ, Benson DW, Basson CT et al. Congenital heart disease caused by mutations in the transcription factor NKX2-5. Science 1998; 281(5373):108-111.
239. Park M, Lewis C, Turbay D et al. Differential rescue of visceral and cardiac defects in Drosophila by vertebrate tinman-related genes. Proc Natl Acad Sci USA 1998; 95(16):9366-9371.
240. Ranganayakulu G, Elliott DA, Harvey RP et al. Divergent roles for NK-2 class homeobox genes in cardiogenesis in flies and mice. Development 1998; 125(16):3037-3048.
241. Lien CL, McAnally J, Richardson JA et al. Cardiac-specific activity of an Nkx2-5 enhancer requires an evolutionarily conserved Smad binding site. Dev Biol 2002; 244(2):257-266.
242. Evans T. Regulation of cardiac gene expression by GATA-4/5/6. Trends in Cardiovascular Medicine 1997; 7(3):75-83.
243. Reiter JF, Alexander J, Rodaway A et al. Gata5 is required for the development of the heart and endoderm in zebrafish. Genes Dev 1999; 13(22):2983-2995.
244. Garg V, Kathiriya IS, Barnes R et al. GATA4 mutations cause human congenital heart defects and reveal an interaction with TBX5. Nature 2003; 424(6947):443-447.
245. Gajewski K, Zhang Q, Choi CY et al. Pannier is a transcriptional target and partner of Tinman during Drosophila cardiogenesis. Dev Biol 2001; 233(2):425-436.
246. Gajewski K, Kim Y, Lee YM et al. D-mef2 is a target for Tinman activation during Drosophila heart development. EMBO J 1997; 16(3):515-522.
247. Gajewski K, Kim Y, Choi CY et al. Combinatorial control of Drosophila mef2 gene expression in cardiac and somatic muscle cell lineages. Dev Genes Evol 1998; 208(7):382-392.
248. Lin Q, Schwarz J, Bucana C et al. Control of mouse cardiac morphogenesis and myogenesis by transcription factor MEF2C. Science 1997; 276(5317):1404-1407.
249. Dodou E, Verzi MP, Anderson JP et al. Mef2c is a direct transcriptional target of ISL1 and GATA factors in the anterior heart field during mouse embryonic development. Development 2004; 131(16):3931-3942.
250. Bruneau BG. Transcriptional regulation of vertebrate cardiac morphogenesis. Circ Res 2002; 90(5):509-519.
251. Wang D, Chang PS, Wang Z et al. Activation of cardiac gene expression by myocardin, a transcriptional cofactor for serum response factor. Cell 2001; 105(7):851-862.
252. Ueyama T, Kasahara H, Ishiwata T et al. Myocardin expression is regulated by Nkx2.5, and its function is required for cardiomyogenesis. Mol Cell Biol 2003; 23(24):9222-9232.
253. Lo PC, Skeath JB, Gajewski K et al. Homeotic genes autonomously specify the anteroposterior subdivision of the Drosophila dorsal vessel into aorta and heart. Dev Biol 2002; 251(2):307-319.
254. Lovato TL, Nguyen TP, Molina MR et al. The Hox gene abdominal-A specifies heart cell fate in the Drosophila dorsal vessel. Development 2002; 129(21):5019-5027.
255. Ponzielli R, Astier M, Chartier A et al. Heart tube patterning in Drosophila requires integration of axial and segmental information provided by the Bithorax Complex genes and hedgehog signaling. Development 2002; 129(19):4509-4521.
256. Perrin L, Monier B, Ponzielli R et al. Drosophila cardiac tube organogenesis requires multiple phases of Hox activity. Dev Biol 2004; 272(2):419-431.
257. Lo PC, Frasch M. Establishing A-P polarity in the embryonic heart tube: A conserved function of Hox genes in Drosophila and vertebrates? Trends Cardiovasc Med 2003; 13(5):182-187.
258. Srivastava D. HAND proteins: Molecular mediators of cardiac development and congenital heart disease. Trends Cardiovasc Med 1999; 9(1-2):11-18.
259. Kolsch V, Paululat A. The highly conserved cardiogenic bHLH factor Hand is specifically expressed in circular visceral muscle progenitor cells and in all cell types of the dorsal vessel during Drosophila embryogenesis. Dev Genes Evol 2002; 212(10):473-485.

260. San Martin B, Bate M. Hindgut visceral mesoderm requires an ectodermal template for normal development in Drosophila. Development 2001; 128(2):233-242.
261. Lee H-H, Zaffran S, Frasch M. Development of the larval visceral musculature. In: Sink H, ed. Muscle Development in Drosophila. Georgetown: Landes Bioscience, 2006:62-78.
262. Masumoto K, Nada O, Suita S et al. The formation of the chick ileal muscle layers as revealed by alpha-smooth muscle actin immunohistochemistry. Anat Embryol (Berl) 2000; 201(2):121-129.
263. Klapper R, Heuser S, Strasser T et al. A new approach reveals syncytia within the visceral musculature of Drosophila melanogaster. Development 2001; 128(13):2517-2524.
264. Martin BS, Ruiz-Gomez M, Landgraf M et al. A distinct set of founders and fusion-competent myoblasts make visceral muscles in the Drosophila embryo. Development 2001; 128(17):3331-3338.
265. Zaffran S, Kuchler A, Lee HH et al. Biniou (FoxF), a central component in a regulatory network controlling visceral mesoderm development and midgut morphogenesis in Drosophila. Genes Dev 2001; 15(21):2900-2915.
266. Ramalho-Santos M, Melton DA, McMahon AP. Hedgehog signals regulate multiple aspects of gastrointestinal development. Development 2000; 127(12):2763-2772.
267. Tonegawa A, Funayama N, Ueno N et al. Mesodermal subdivision along the mediolateral axis in chicken controlled by different concentrations of BMP-4. Development 1997; 124(10):1975-1984.
268. Lo PC, Frasch M. Bagpipe-Dependent expression of vimar, a novel Armadillo-repeats gene, in Drosophila visceral mesoderm. Mech Dev 1998; 72(1-2):65-75.
269. Zaffran S, Frasch M. The beta 3 tubulin gene is a direct target of bagpipe and biniou in the visceral mesoderm of Drosophila. Mech Dev 2002; 114(1-2):85-93.
270. Mahlapuu M, Ormestad M, Enerback S et al. The forkhead transcription factor Foxf1 is required for differentiation of extra-embryonic and lateral plate mesoderm. Development 2001; 128(2):155-166.
271. Tseng HT, Shah R, Jamrich M. Function and regulation of FoxF1 during Xenopus gut development. Development 2004; 131(15):3637-3647.
272. Tribioli C, Frasch M, Lufkin T. Bapx1: An evolutionary conserved homologue of the Drosophila bagpipe homeobox gene is expressed in splanchnic mesoderm and the embryonic skeleton. Mech Dev 1997; 65(1-2):145-162.
273. Miano JM. Serum response factor: Toggling between disparate programs of gene expression. J Mol Cell Cardiol 2003; 35(6):577-593.
274. Wang Z, Wang DZ, Hockemeyer D et al. Myocardin and ternary complex factors compete for SRF to control smooth muscle gene expression. Nature 2004; 428(6979):185-189.
275. Wang L, Fan C, Topol SE et al. Mutation of MEF2A in an inherited disorder with features of coronary artery disease. Science 2003; 302(5650):1578-1581.
276. Klapper R, Stute C, Schomaker O et al. The formation of syncytia within the visceral musculature of the Drosophila midgut is dependent on duf, sns and mbc. Mech Dev 2002; 110(1-2):85-96.
277. Weiss JB, Suyama KL, Lee HH et al. Jelly belly: A Drosophila LDL receptor repeat-containing signal required for mesoderm migration and differentiation. Cell 2001; 107(3):387-398.
278. Lee HH, Norris A, Weiss JB et al. Jelly belly protein activates the receptor tyrosine kinase Alk to specify visceral muscle pioneers. Nature 2003; 425(6957):507-512.
279. Englund C, Loren CE, Grabbe C et al. Jeb signals through the Alk receptor tyrosine kinase to drive visceral muscle fusion. Nature 2003; 425(6957):512-516.
280. Stute C, Schimmelpfeng K, Renkawitz-Pohl R et al. Myoblast determination in the somatic and visceral mesoderm depends on Notch signalling as well as on milliways(mili(Alk)) as receptor for Jeb signalling. Development 2004; 131(4):743-754.
281. Stoica GE, Kuo A, Aigner A et al. Identification of anaplastic lymphoma kinase as a receptor for the growth factor pleiotrophin. J Biol Chem 2001; 276(20):16772-16779.
282. Stoica GE, Kuo A, Powers C et al. Midkine binds to anaplastic lymphoma kinase (ALK) and acts as a growth factor for different cell types. J Biol Chem 2002; 277(39):35990-35998.
283. Patapoutian A, Wold BJ, Wagner RA. Evidence for developmentally programmed transdifferentiation in mouse esophageal muscle. Science 1995; 270(5243):1818-1821.
284. Stratton CJ, Bayguinov Y, Sanders KM et al. Ultrastructural analysis of the transdifferentiation of smooth muscle to skeletal muscle in the murine esophagus. Cell Tissue Res 2000; 301(2):283-298.
285. Miano JM, Thomas S, Disteche CM. Expression and chromosomal mapping of the mouse smooth muscle calponin gene. Mamm Genome 2001; 12(3):187-191.
286. Epstein JA. Developmental cardiology comes of age. Circ Res 2000; 87(10):833-834.
287. Rudnicki MA. Marrow to muscle, fission versus fusion. Nat Med 2003; 9(12):1461-1462.
288. Partridge TA, Davies KE. Myoblast-based gene therapies. Br Med Bull 1995; 51(1):123-137.
289. Blau HM. A twist of fate. Nature 2002; 419(6906):437.
290. Baylies MK, Bate M, Ruiz Gomez M. Myogenesis: A view from Drosophila. Cell 1998; 93(6):921-927.

Index